Victor Manuel García Barrios

Personalisation Systems

Victor Manuel García Barrios

Personalisation Systems

Multi-purpose User Modelling

VDM Verlag Dr. Müller

Imprint

Bibliographic information by the German National Library: The German National Library lists this publication at the German National Bibliography; detailed bibliographic information is available on the Internet at http://dnb.d-nb.de.

Any brand names and product names mentioned in this book are subject to trademark, brand or patent protection and are trademarks or registered trademarks of their respective holders. The use of brand names, product names, common names, trade names, product descriptions etc. even without a particular marking in this works is in no way to be construed to mean that such names may be regarded as unrestricted in respect of trademark and brand protection legislation and could thus be used by anyone.

Cover image: www.purestockx.com

Publisher:
VDM Verlag Dr. Müller Aktiengesellschaft & Co. KG , Dudweiler Landstr. 125 a, 66123 Saarbrücken, Germany,
Phone +49 681 9100-698, Fax +49 681 9100-988,
Email: info@vdm-verlag.de

Zugl.: Graz, TU, Diss., 2007

Produced in USA and UK by:
Lightning Source Inc., La Vergne, Tennessee, USA
Lightning Source UK Ltd., Milton Keynes, UK
BookSurge LLC, 5341 Dorchester Road, Suite 16, North Charleston, SC 29418, USA

ISBN: 978-3-8364-6623-3

A mis padres y hermanos.

Contents

1 Introduction

> *"If we knew what it was we were doing,*
> *it would not be called research, would it?"*
>
> (Albert Einstein)

This chapter provides an overview on the main aspects of the research presented in this book. Firstly, the general and the specific scopes for the research work are introduced, whereby the targets of investigation and application are outlined. Secondly, the stipulated main research goals and a summary of the achieved results are depicted. This chapter concludes describing the chosen research methodology and the structure of this book.

1.1   Research Scope

The need of adaptive e-learning
Long existing educational paradigms stipulate that distinct learners follow distinct strategies during learning and show distinct preferences in the consumption of learning materials. Evaluation results indicate that individual learning styles are identifiable and that adaptation towards personal styles and traits (i.e. personalisation) increases the learning performance of some learners [Rasmussen and Davidson-Shivers 1998]. Both, scientific and industrial communities working on e-learning have identified the aforementioned issue and address this new topic under the label Adaptive E-Learning.

Thus, one of the major benefits of using adaptive e-learning systems is seen in the efficient delivery of those materials that best suit to personal traits of learners. And this is not a trivial task; moreover, it is not the only task. This book provides a comprehensive view on the general topics of Adaptive E-Learning. On the one hand, the concepts of adaptation and e-learning are clarified, and on the other side, the application and objectives of personalisation methods in adaptive e-learning are extensively analysed.

Benefits of adaptive e-learning
Adaptive e-learning research and development work aim at solving problems of traditional learning and traditional e-learning by utilising techniques that enable an efficient learner-tailored (i.e. personalised) knowledge transfer process. For example, Web-based explorative e-learning shows problems if the corpus of learning materials is too open, i.e. cognitive overload or the lost in hyperspace problems may arise. In these particular cases, an adaptive e-learning system aims at overcoming these difficulties of comprehension and disorientation reducing and optimising the material repertory.

Further, the continuous achievements in communication technology have influenced traditional learning, as teachers have to reconceptualise their notion of didactical goals. For example, the location-distributed and asynchronous character of e-learning leads to asymmetric learning [Jain et al. 2002], i.e. distinct persons learning at own pace reflect distinct learning performances at a certain time, a fact that in turn could be hard to solve by the teacher. In this case, adaptive e-learning may assist a teacher in this context through a continuous observation of individual learning progress.

In traditional classrooms, learners perform also differently over time, but the continuous presence of the teacher allows an immediate intervention and thus, symmetry can be recovered. This book inspects advantages and drawbacks of applying adaptive e-learning, e.g. a review on adaptive systems and user modelling systems provides insights into the potential of personalisation technologies.

The point is: know the learner To enable optimally personalised learning, adaptive e-learning systems ought to be aware of 'the rules of the game', i.e. it should be able to make ad-hoc decisions based on didactically stipulated rules. Within this context, teachers and computer scientists have been working since decades on the investigation of most effective methods for the distinct didactical tasks at hand. For example: the best instructional alternative may be selected according to the *individual intellectual abilities of a learner*; the navigation path through learning materials can be personalised according to the *individual expertise level* of the learner; parts of learning pages may be removed for a specific learner according to the already mastered course topics (i.e. *individual learning performance*).

On the one hand, these adaptations imply that some component in the system must be capable of deciding which next step is the most suitable for the task at hand. On the other hand, and this is the reason for having highlighted some words of the previous examples *in italic*, it implies that the system ought to *know each individual learner* as well as possible. This latter aspect is covered by the research field of Learner Modelling (also known as Student Modelling) and represents one main focus of this book. Hence, the technological outcome is a system that models individual information about learner traits.

The point is: know the user In fact, personalisation in the field of adaptive e-learning is not the sole target of this book. Generally speaking, any adaptation-pertinent interactive system needs a component that supports User Modelling. In order to *know the user*, an adaptive system needs an internal representation for each interacting person, i.e. an individual user model. The information about single users can be gained e.g. explicitly by asking the user or indirectly by inferring assumptions through the observation of user behaviour.

In this context, also some examples: the contents of Web pages can be adapted in order to better match the *preferred media type* of a user; the presentation of user interface elements may be personalised in accordance with *individual layout preferences*; the provisioning of help to find information can be personalised according to assumptions about the *individual interests* of the user. Therefore, this book also investigates the techniques utilised within the field of User Modelling. The developed system integrates distinct user modelling techniques for distinct purposes.

User modelling vs. the rest of duties In connection with the observations stated so far, an additional relevant aspect arises as essential part of the research scope of this book. As previously indicated, the outcome of an adaptive e-learning system is measured in terms of its usefulness at the front-end. If the system undertakes an adaptation (i.e. it personalises) and individual learners perform well, then the system is meant to be 'fit', i.e. the system adapts successfully. This fitness of an overall adaptive system depends mainly on the efficient and accurate collaboration work between the adaptive component and the modelling components.

To give an example, the knowledge about individual users is a duty of the user modelling component, whereas the knowledge about the overall personalisation goals of the system is a duty of the adaptive component (usually called adaptive or personalisation engine). Hence, this book aims also at identifying which are the functional interdependencies and boundaries among user modelling components and the other internal parts of an adaptive system.

Towards a multi-purpose solution The Web has been the Big Thing since the beginning of the 1990's. Though, the new century and its adaptive hypermedia architectures should lead to Adaptive Web-based Systems as the Next Big Thing [deBra et al. 2004]. Indeed, looking at the present and in order to give one example, personalised Web-based (search) services like Google[1] or Yahoo[2] dominate the market of adaptive Web-based systems at present. Almost all users of adaptive systems are aware of the increasing amount of well-tailored information they may access for their particular needs. And moreover, they are aware of the fact that in the majority of cases, they have

---

[1] http://www.google.com/ig, http://mail.google.com
[2] http://my.yahoo.com, http://groups.yahoo.com

to pay the price or hazard the consequences of delivering personal data to ensure those services.

According to several surveys, as shown e.g. in [ChoiceStream 2004], most of the users are willing to do so. Specifically, the utilisation of personalisation methods can be identified in various application fields. No doubt, rather than a trend, there is a need for personalisation-pertinent systems in distinct situations of modern life, thus personalisation systems must be flexible and reusable to reach a high degree of application and domain-independence. Furthermore, the need of multi-purpose capabilities in modelling systems becomes clear when taking a partial view on the broad applicability spectrum: adaptive user interfaces [Langley 1999] [Thevenin and Coutaz 1999], context-aware systems [Assad et al. 2006], context management systems [Zimmermann et al. 2005], mobile computing [Holmén 2000] [Lankhorst et al. 2002] [Kinshuk and Goh 2003], multi-model adaptive systems [Conlan et al. 2003], personalised collection management in museums [Zimermann et a. 2003] [Filippini-Fantoni et al. 2005], personalisation and adaptation in digital libraries [Hicks and Tochtermann 2001] [Krottmaier 2004] [Kolbitsch and Maurer 2005], personalisation in e-commerce applications [Ardissono et al. 2002] [Perugini and Ramakrishnan 2003], personalised information and multi-media retrieval [Grcar et al. 2001] [Li et al. 2001], personalised multi-modal semantic retrieval systems [García-Barrios 2006d] [García-Barrios and Gütl 2006], personalised news delivery [Billsus and Pazzani 2000] [Shepherd et al. 2002], ubiquitous systems [Carmichael et al. 2005] [Heckmann 2005], and so forth.

The modelling system proposed in this book supports multi-purposefulness in terms of the following two goals: (a) a multi-purpose modelling system should be capable of modelling not only users, but also other physical and abstract entities the adaptive system must be aware of (e.g. groups of persons, sensors, interacting devices, respectively, information or knowledge structures, activity workflows, roles), and (b) the utilisation scope of multi-purpose modelling systems should not be restricted to a certain application type or a single domain.

Identifying technological challenges

Designing and developing user modelling components for adaptive systems are not trivial tasks. They imply various challenges, such as how to identify relevant individual traits, how to model the distinct types of traits, how and why to distinguish between facts and assumptions about a user, how to integrate distinct modelling techniques in single systems, how and when to protect private information, how to give users a human-readable view on the data collected about them, how to identify complex user goals from low-level interactions.

This book aims at considering current software development paradigms on distributed systems that enable the highest transparency at distinct levels of its architecture, reusability and flexibility of modules as well as extensibility of functionalities and overall performance at the level of communication. In concrete, the development of the multi-purpose modelling system presented in this book is based on the service-oriented software programming approach.

Application area 1: user modelling for adaptive e-learning

The AdeLE research project (*Adaptive e-Learning with Eye-tracking*) aims at developing an adaptive e-learning system, which uses fine-grained user profiling techniques for the analysis of learner behaviour and learner states in real-time. One of the main goals of the AdeLE system is to be capable of personalising the presentation of learning material in real-time according to individual learning styles and to inferences gained from individual gaze movements.

This book explains how eye-tracking technology is used for real-time user modelling as well as presents the implementation and results of gaze-behaviour in the developed user modelling system. Further, also within the scope of AdeLE, the multi-purpose capability of its modelling system is utilised to implement a rather distinct tool, as depicted in the ensuing paragraph.

Application area 2: knowledge exploration at own pace

Regarding the applicability of e-learning, it is highly relevant to note that *"computers cannot improve the knowledge acquisition (i.e. learning) per se.*

Learning is a basic cognitive process, which has to be carried out by the learners themselves. This means that learning is an active process from a learner's perspective" [Ebner et al. 2006]. Thus, not only conveying knowledge but also involving learners in practical tasks (i.e. learning by doing) is the key for successful e-learning [Ebner et al. 2006]. One possibility to achieve this goal is by means of Exploratory Learning [diSessa et al. 1995]. Among others, this theory is related to adaptive e-learning regarding the following premises: learning can and should be done at own pace; learners may approach a learning task in very diverse ways; and, learning does not have to be forced. This theory can be applied by giving learners the freedom to discover new knowledge in open (but topic-relevant) information spaces.

Based on this idea, the first prototype of a Web-based tool, called Dynamic Background Library [Dietinger et al. 1998], was developed at Graz University of Technology in 2001 [García-Barrios et al. 2002]. The tool offers to learners a set of topic-relevant hyperlinks that are attached to their used learning materials and might be followed to explore additional materials residing outside the mandatory learning repository. These hyperlinks are defined in advance by teachers, but are not connected to specific Web resources, rather they point at smart search engines. Thus, each hyperlink is a specialised search query that can lead learners to relevant, topical and contextualised resources. On the one hand, the search results ought to be consumed by learners, but on the other hand, they represent a guided motivation and starting point to leave the mandatory repository and discover new knowledge at own pace [García-Barrios et al. 2002]. This book shows a proposed modelling system solution as enhancement for this tool.

Application area 3: personalised cross-media retrieval The Mistral research project aims at smart semi-automatic solutions for semantic annotation and enrichment of multi-modal data from meeting recordings and meeting-related documents. In order to efficiently and contextually manage the knowledge addressed and generated in meetings, sophisticated mechanisms are needed for the semantic processing, enrichment and integration of multi-modal data.

Therefore, Mistral's overall system prototype consists of units for data management, uni-modal stream processing, multi-modal merging of extracted semantics, semantic enrichment of concepts, and semantic applications. Mistral's semantic applications unit defines a set of prototypes at the front–end of the system. This book focuses on the applicability of the modelling system within this unit. The modelling system resulting from the research of this book enhances the overall Mistral system with personalisation features for cross-media information retrieval.

All aspects presented so far embrace the general as well as specific research scope of this book. The ensuing part of this introduction gives an overview on the stipulated main research goals and achieved results.

1.2   Research Goals & Results

The research presented in this book concentrates on the issue of designing and developing a user modelling system that meets the requirements of distinct domain and application areas. Thereby, in accordance to the aspects introduced so far in this chapter, special focus is set on the following main research goals. The corresponding achieved results are briefly mentioned after each goal.

General research goals

How to reach a high level of abstraction in the design and implementation of a modelling system that ensures multi-purposefulness.

A comprehensive investigation of terminological and functional distinctions was conducted. Based on that, main research and technological challenges have been stipulated. The proposed solution is applied within three distinct research projects.

How to identify and establish a clear separation of duties between a modelling component and the other parts of an adaptive system.

The identified most relevant tasks and main constituent elements of a user modelling component draw the outline of internal and external duties. The strict modularity of the proposed architecture enables a clear separation of duties.

How to find a flexible and reusable solution for a modelling system that ensures its domain- and application-independence.

The service-oriented solution approach and the proposed architectural framework enabled a flexible and reusable decomposition.

Specific research questions

How is it possible to support gaze-tracking in real-time in order to enhance the learning performance of users of the AdeLE system.

In spite of the high complexity of gaze-tracking technology, a simple and innovative technique (called Behaviour Tracking) has been developed and implemented.

How can knowledge of users be well organised for or in a user modelling system.

A solution based on the field of Concept Modelling has been developed and integrated into the modelling system providing this required additional functionality.

How can a modelling system contribute to improve knowledge acquisition in an adaptive system and which implications does it have on the adaptive behaviour of the overall system.

Based on a concept-oriented approach, a so-called Concept-based Context (CO2) modelling system has been designed and implemented. The CO2 system is not an additional solution; rather it represents a sibling of the implemented multi-purpose modelling system.

How can a modelling system support company members in their working tasks.

A conceptual design for an overall solution has been elaborated as well as integrative experiments have been successfully conducted within the Semantic Applications Unit of the Mistral system.

Summary of Contributions

In addition to the previously summarised main research results, let me also mention the following technical and scientific contributions of this research work.

The technical outcome of the research work is the **"yo?"** multi-purpose profiling and modelling system. This system has been designed, imple-

mented, evaluated and improved in several development phases within the last four years. I have designed (and redesigned) the overall system mainly according to the requirements of the AdeLE and Mistral research projects. The implementation of system versions as well as their evaluations and improvements comprised the practical part of my dissertation work.

Beside my own programming contributions, the overall development work represented the basis for several student works at the Institute for Information Systems and Computer Media (Graz University of Technology). So, I have supervised two MSc. theses ([Fröschl 2005], [Safran 2006]) and more than twenty BSc. theses, all of them within the Bachelor and Master programmes for the studies 'Telematics', 'Software Development - Economy' as well as 'Software Development and Knowledge Management'. At this point, let me emphasise that I am indebted to my students for their contributions to the implementation and enhancements of the system.

Further, as main author or co-author, several parts of the chapters in this book have been published in (twenty-eight) peer-reviewed scientific journals and conference proceedings. The most relevant scientific contributions include publications on the following topics:

- adaptive e-learning and personalisation systems [García-Barrios et al. 2004b] [García-Barrios et al. 2005] [Gütl et al. 2005b] [Pivec et al. 2005] [García-Barrios 2006b] [García-Barrios 2006c] [Mödritscher et al. 2006a],

- knowledge transfer, discovery and retrieval [Pivec et al. 2004] [Gütl and García-Barrios 2005c] [Mödritscher et al. 2006c],

- concept-based explorative learning [García-Barrios et al. 2004a] [Gütl and García-Barrios 2005b] [Mödritscher et al. 2005] [García-Barrios 2006a] [Safran et al. 2006],

- multi-purpose modelling systems [Gütl et al. 2004] [Gütl and García-Barrios 2005a] [García-Barrios 2006d] [García-Barrios 2006e] [García-Barrios and Gütl 2006], and

- personalised cross-media applications [Gütl and García-Barrios 2005d] [Gütl and García-Barrios 2006].

1.3 Methodology & Structure

This section describes the chosen methodology to meet the aforementioned research goals and to accordingly present the results in a concise manner along this book. Thus, a constant focus is retained on the field of adaptive e-learning, but in parallel, the general scope of multi-purpose user modelling gains priority at some parts of the book (e.g. at the core of technological analyses and presentations).

The overall methodology and structure of this work is shown in Figure 1.1. The left column in the figure represents the main chapters of the book, which in turn reflect the four basic conceptual parts of investigation (conceptual framework, technological investi-gation, solution framework and proof of concept). The middle column in the figure (in relation with the left column) shows the main specific topics treated in each conceptual part. The right column shows the main general research scopes as described in the previous paragraph.

Hence, this book is organised as follows:

- The Conceptual Framework of the book is represented by chapter 2. This part embraces the theoretical investigation regarding the topics of general Adaptive Systems, E-Learning, and Adaptive E-Learning. The aim is to approach to the subject of adaptive e-learning from its terminological, historical and theoretical roots concerning the general notions of Adaptation and E-Learning.

- Based on the conceptual framework, a more detailed Technological Investigation is conducted in chapter 3 (Personalisation and User Modelling). After a brief presentation of the distinctions between personalisation and adaptation, the focus is set on the application of adaptive e-learning. Hereby, the basic principles of modern adaptive e-learning systems are presented as well as distinct well-known models are reviewed in order to identify the main role and functions of user modelling components in such systems. Thereafter, the book concentrates on the analysis of the main

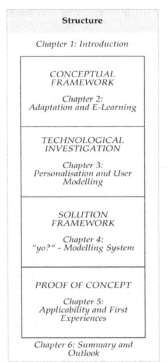
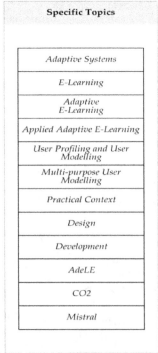
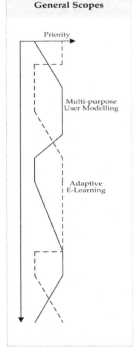

Figure 1.1: Methodology and Structure of the Book.

technological issues regarding the topics of User Profiling and User Modelling, i.e. underlying principles and techniques. This part of the book concludes with a review on Multi-purpose User Modelling System, a topic usually coined as Generic User Modelling Systems. The aim of this chapter is mainly to collect and infer most challenging technological aspects for the design and development of multi-purpose user modelling systems.

- Against the theoretical and technological background collected in the first two main parts, chapter 4 presents the proposed Solution Framework, which is the basis for the development of the service-oriented Multipurpose Profiling and Modelling System called **"yo?"**. Because the practical field of the system is given by the research projects AdeLE, CO2 and Mistral, these projects are briefly presented at the beginning of this part. The corresponding specific topic is called Practical Context in the middle column of Figure 1.1 and comprises two sub-

chapters, one that narrows the focus to the scope of the projects, and a second one opening the scope again to the topic of general adaptive systems. This last step is necessary to stipulate the theoretical adaptive environment that is covered by the multipurposefulness of the **"yo?"** system. Given that, detailed insights are given into the most relevant design and development issues of the system.

- Chapter 5 of this book (Applicability and First Experiences) represents the practical Proof of Concept of the system. The results as well as the role of the **"yo?"** system are described by means of the research projects at hand.

The last chapter comprises the following issues: some general conclusions are drawn, the corroboration of research goals is discussed, related work is analysed, and finally, open issues and an outlook for the road ahead is presented.

2 Adaptation and E-Learning

The object of teaching a child

is to enable him to get along without his teacher.

(Elbert Hubbard)

The topic of adaptation-pertinent systems was treated intensively by researchers before the era of computers, also in the field of teaching and learning. Though, Adaptive E-Learning (AEL) is a relatively young term. Further, since the beginning of computer science in the middle of the past century, many software systems and theoretical solution approaches were developed to enable a user-targeted technology-based instruction. Thus, three main questions arise: Is AEL just a buzzword? What is new in AEL? What means 'Adaptive' and 'E-Learning'? In order to find an accurate answer to these questions, a concrete common understanding of the terminology behind this topic is necessary. There is no doubt that a contextual linkage exists between AEL and much longer existing research disciplines that are either related with technology-based instruction or with adaptation-pertinent systems. This chapter investigates if there are relevant issues missing at present that can be found in the past, and if any, which implications they have. The aim is to present and clarify the general and specific notions around the topics of 'adaptation' and 'e-learning', and consequently of 'adaptive e-learning'. The results of this investigation are: (a) a comprehensive terminological basis for common understanding in the aforementioned topics, (b) a collection of most relevant aspects regarding user-adapting systems, and (c) an overview on the evolution of AEL systems.

2.1 Adaptation

Adaptive systems are in the present a common and trendy concept within the field of Computer Science, especially in the sub-field of Information Systems. Though, many distinct definitions are floating around this fascinating term. This section aims at clarifying its denotations and connotations.

2.1.1 Definition and Concepts

In general, *to adapt* can be interpreted as to 'make suitable by modifying the current state'. But this is only one of the innumerable explanations around the concept of adaptation. Nevertheless, very specific interpretations of terms, which are related with adaptation, can be identified within particular contexts or disciplines (like Biology, Climatology and others). From the linguistic point of view, a set of morphologically related terms can be reduced to a common stem[3].

The stem *adapt* makes several word combinations possible, which according to their type (verbs, noun, adjective, ..) may conceal different semantics (meanings) and consequently, different pragmatics (intentions) while utilising them. As it is not the objective of this section to list all available definitions related to the stem *adapt*, only those morphological variants relevant to the context of this book are given in the following. According to [WordNet 2007], the verb *to adapt* means

"to conform oneself to new or different conditions"

and

"to make fit for, or change to suit a new purpose".

---

[3] In linguistics, an affix is a word element (prefix or suffix), that can only occur attached to a stem. Thus, a stem is the main semantic part of a word to which affixes can be added. [Oxford 2007]

On the one hand, these definitions imply the notion of changing something to meet some specific requirements or purposes. On the other hand, it embraces the act of becoming used to a changing environment. Further, these notions of the verb lead to the need of explaining the adjectives *adaptive* and *adaptable*, because they represent the most used terms within the context of adaptive systems in computer sciences. Very briefly, and taking into account the definitions in several dictionaries, the first term means 'capable of adapting', and the second 'capable of being adapted' (see e.g. [Britannica 2007], [Merrian-Webster 2007], [Oxford 2007] or [WordNet 2007]).

In accordance to the above depicted terms, the nouns *adaptation*, *adaptivity* and *adaptability* are defined as follows[4]. In [Merrian-Webster 2007], adaptation is defined as

"the act or process of adapting or fitting"

and

"the state of being adapted or fitted".

Therefore, adaptation refers to both the process of adapting and the condition of being adapted. Next, following the definitions given so far, adaptivity and adaptability refer to the quality and capacity of being adaptive, respectively adaptable. A distinction is made in this book between *adaptor* and *adapter* as well. Although both terms are used as synonyms in everyday language, *adaptor* is used to describe that component within an adaptive system, which recognises, starts and fulfils the adaptation, i.e. the 'decision-maker' and 'executor' of adaptations. From a technological point of view, an adaptor is usually referred to as *adaptive engine*. Finally, consider that an adaptable entity must not be deformed or shaped through adaptation; it might be also complemented, completed or replaced with 'something' else. This additional thing is called *adapter*. The following example clarifies the use of the terms by means of a real-life scenario. Consider the author of this book travelling home, i.e. from Austria to Guatemala, and taking his laptop with him. The laptop cannot be connected to some electric socket in Guatemala, because the plug of the cable of this European product will not fit. Thus, an adaptation of the plug is needed. Indeed, the plug is adaptable, because the traveller (adaptor) knows how to adapt it and has an artefact (adapter) to make the plug suitable. The traveller gets adaptive when fulfilling the adaptation, and if he takes the correct adapter, the adaptation will be successful as well.

Before moving towards the technological context of adaptive systems, the next section gives an overview on some disciplines outside the context of computer sciences, which contribute with relevant findings and terminological issues that help to identify the most relevant general characteristics of adaptive systems.

2.1.2 General Characteristics of Adaptive Systems

As mentioned in the introductory part of this chapter, adaptive systems were not born within the computerised era. In this section, going back to the roots of adaptive systems and leaving the context of computer sciences aims at finding out relevant issues that have not been treated or taken into account by researchers in the last decades. Thus, the findings of this section open new perspectives on the thematic and do not represent a confrontation between 'old good research' and 'bad new research'. To generalise the notion of adaptive systems, the term *adaptation-pertinent systems* is used.

Well-established research fields, as for example Climatology, Biology, Evolutionary Research, Cybernetics, Control Systems, Systems Theory, or even Information Theory and Complex Adaptive Systems, allow a deeper insight into the thematic of adaptation-pertinent systems. This insight makes it possible to discover and understand the processes and the components generally involved in adaptational procedures. The following disciplines have been chosen, because of their universal recognition:

- Biology

- Climatology

- Cybernetics

---

[4] At this point, for the sake of simplicity, many synonyms (e.g. adaptedness, adaptableness) and related verbs (e.g. to accommodate, to adjust, to conform or to acclimate) are not taken into account.

- Evolutionary Research
- Information Theory
- Systems Theory

Biology

In [Frank 1996], an article about The Design of Natural and Artificial Adaptive Systems, an adaptive system is defined as

"a population of entities fulfilling the three requirements of natural selection: they (a) are variable, (b) have continuity, i.e. are capable of heritability, and (c) may show differences in their success".

In other words, successful entities integrate the adapted traits into their genetic code in order to transfer it to their descendants. In this sense, population entities can be computers processes or anything else capable of satisfying those three conditions. The relevant keywords behind this definition are defined by the terms *population of entities* (indicating the participation of many components in the system), *variability* (describing the ability to change) and *success differences* (indicating that different implications of distinct reactions are possible).

Another definition underlines the setting of adaptation in an evolutionary context and allows determining *acclimatisation* as a synonymic term [Haubelt et al. 2004]:

"Adaptation to changes in environmental conditions is the driving force behind the evolutionary development of all species from a common predecessor".

The concept of adaptation is characterised by the property of a response to an extra-cellular *stimulus* to return to its pre-stimulus value even for the case that the stimulating signal reports a continuing presence [Haubelt et al. 2004]. This leads to the very important finding that the system must not always react to stimuli, particularly if a successful reaction to a stimulus took place previously, further identical stimuli can be ignored. The key aspect derived from this context is *sensitivity* and will be explained later within this section.

According to [Chang and Gütl 2007], the notion of an ecosystem, which is an assemblage of living beings and its associated physical environment in a specific place, was coined firstly by A.G. Tansley in 1935. In a (biological) ecosystem, individuals form groups spontaneously and may perform, adopt or adapt behaviours that contribute to or perturb the *success of the system*.

Climatology

In the context of climatology, adaptation refers to adjustments in natural or human systems (i.e. ecological, social or economic systems) in response to current or expected climatic stimuli and their effects or impacts (see [Smit et al. 1999]). Many other definitions have been proposed and can be found in 'climate-change' literature, as in [Burton 1992], [Smith et al. 1996] or [Stakhiv 1993].

In [Watson et al. 1996] the key term adaptability is introduced and defined as

"the degree to which adjustments are possible in practices, processes, or structures of systems to projected or actual changes of climate".

This observation leads to the statement that adaptation has *measurable* parameters, because adaptability describes a *dimensional* feature. In addition, [Smit et al. 1999] state that adaptation can be e.g. *spontaneous, anticipatory, autonomous, planned* or *reactive*. This means that adaptations occur in (many) different forms, i.e. a variety of *adaptation types* may be determined according to specific attributes.

Finally, also from [Smit et al. 1999], relevant specific aspects regarding adaptations can be gained; they describe 'key questions' regarding the treatment of adaptive systems and are exposed in the following citation:

"[...] a rigorous description of any adaptation would specify the system of interest (who or what adapts?), the climate-related stimulus (adaptation to what?), and the processes and forms involved (how does adaptation occur?). The task of developing or facilitating adaptation options or measures as part of a response strategy [...] involves the additional step of evaluation to judge the merit of potential adaptations (how good is the adaptation?)".

Cybernetics

The field of cybernetics is a wide and complex one. In order to give a simple and short description of the aims of this discipline and according to the definitions found in [Heylighen 2004], it can be stated that within cybernetics, theories tend to rest on four basic pillars: (a) *variety*, (b) *circularity*, (c) *process* and (d) *observation*. Furthermore, almost all theories are directly related to notions regarding processes and changes, like the notion of information, the difference between two states of uncertainty, or theories of adaptation, evolution and growth processes. Relevant in the context of this book is that cybernetics explains such processes in terms of the structure and organisation of the system manifesting them, e.g. the circular causality of *feedback* loops is taken into account for processes of *regulation* and a system's effort to maintain an *equilibrium* or to reach a specific *goal* [Heylighen 2004].

According to [Melnikov 1978], a form of behaviour is adaptive, if it maintains the essential variables of the adapted entity within physiological limits. Further,

"adaptation does not occur instantaneously, in that it is connected with internal re-structuring".

Thus, adaptation is time-dependent and implies the property of *dynamics* in the system itself. In addition, [Melnikov 1978] calls *reception zone* the receptive area of a system (with external spatial and temporal boundaries) that reacts to external causes. This concept indicates that in adaptive systems some kind of *sensing components* are included in order to perceive stimuli from its external environment.

Evolutionary Research

In the field of evolutionary research the concept of adaptation may be examined from two different points of view: Evolutionary History and Evolutionary Biology[5]. Considering the field of evolutionary history and as stated in [Reeve and Sherman 1993], Gould and Vrba specified an enhanced model for defining adaptations and

exaptations: (a) adaptations are features built by natural selection for their current role, and (b) exaptations are features originally built for something other than their current role.

Also in this context, Greene, Coddington and others, define adaptations as *traits* with current utility, which are derived in their phylogenetic group [Reeve and Sherman 1993]. Thus, they strive for developing answers by looking at traits from the *taxonomic level*. In the teleonomic camp - represented e.g. by Williams and Thornhill - the emphasis is put on the relationships between form and function. Here, no differentiation is done between adaptations and exaptations. Further, Sober defines adaptation as a

"trait, which is spread through a population as a result of natural selection for the performance of a specific task that increases fitness".

From the point of view of evolutionary biology, and again according to [Reeve and Sherman 1993], adaptation is defined as

"a phenotypic variant that results in the highest fitness among a specified set of variants in a given environment".

This definition embraces the following three components:

- a set of phenotypes, i.e. a group of specific, alternative, demonstrable and observable traits with fitness measures that must be compared to see if the expressed trait is an adaptation,

- a fitness measure, which depends on the reproductive success of traits (short-term or long-term measures), and

- the environmental context, represented by the environmental situation in which the phenotypes are being evaluated.

For this book, the relevant keywords from the point of view of evolutionary biology are given firstly by the time-dependent attributes *short-term* and *long-term measures*: a short-term adaptation implies an easier isolation of impact types on variants, a long-term adaptation (or long-term fitness) represents the ultimate goal for the successful evolution of a population. The second interesting issue is given by the keyword *phenotype*, which can be interpreted as 'how an organism

---

[5] All statements and findings in this section regarding Evolutionary Research are based on [Reeve and Sherman 1993].

looks like as a consequence of its genetic traits and its changing environment'. Thus, generally speaking, a phenotype is described by its typical characteristics, its environment and its interactions with the environment. This observation indicates that it is possible and necessary for the adaptive system to *model* the environment to which is to adapt by means of an internal representation of its traits using the given interaction capabilities.

Finally, it is important at this point to show the differences between evolutionary adaptation and acclimatisation. According to [Britannica 2007], acclimatisation refers to

"any of numerous gradual, long-term responses of an individual organism to changes in its environment".

Furthermore, these responses are *habitual* and *reversible*, so that under certain conditions the system may revert to an earlier state. These criteria differentiate acclimatisation from e.g. growth and development (which is not reversible) and from evolutionary adaptation (which occurs in a population of individuals and over many generations).

Information Theory

The concept of adaptation developed in [Mullen 2003] has the 'advantage' of being exactly expressible in terms of mathematics and is applicable to all system models where the Shannon Information between system components is well-defined. Mullen's definition states that adaptation between system components implies a *change in the quality of the shared information*.

From the general point of view on adaptive systems, the key thought behind Mullen's definition relies on the fact that adaptation can be 'shown' by the system (i.e. it is *demonstrable*) and in addition, if expressed as shared information, its quality is *measurable*. Furthermore, this 'mutual information' embraces and requires both, *communication* and *interaction*.

Systems Theory

As shown in [Haubelt et al. 2004], the discipline of systems theory delivers a number of

ways to interpret adaptation. [Ashby 1960] defines adaptation as

"the ability of a system to compensate for changes in its environment".

Further, [Åström and Hägglund 1995] state firstly, that adaptation is equivalent to *adaptive control*, where the controller parameters inside the system are constantly adjusted to accommodate changes in the dynamics, parameters or disturbances of processes. And secondly, that after responding to a persisting stimulus, the output of the system returns to its pre-stimulus state.

From these issues, [Haubelt et al. 2004] conclude two observations: (a) in Ashby's definition, the changes to which the system adapts take place outside the system, and (b) in Åström's interpretation, the changes are internal to the system, and a controlling component adapts to them, meaning that the main purpose is to keep controlling the system (i.e. the stimulus is an internal change and it initiates a reaction that works against the external disturbance). Consequently, the ultimate goal of these processes of adaptation is to guarantee that the system keeps *stable* and *controllable*. Thus, as [Ashby 1960] stated, by means of adaptive mechanisms the

"essential variables of the system should stay within acceptable (i.e. stable / non-dangerous / uncritical) limits".

General characteristics of adaptive systems

The following list gives an overview on the most relevant findings gained so far within the context of this investigation and comprises the general characteristics for adaptive systems:

- *Sensitivity* is the degree to be affected by or responsive to some environmental stimuli, i.e. the impact potential of the system.

- *Susceptibility* is the extent to which a system is open, liable, or sensitive to environmental stimuli. Thus, similar to sensitivity, but with some connotations toward damage.

- *Robustness*, as in a general context, is the strength or degree to which a system is not given to influence.

- The potential or capability to adapt something according to environmental stimuli or their effects or impacts, is called *adaptive capacity*.

- *Adaptability* is the property of being adaptable. It may be also seen as the sum of adaptable entities. It is worth mentioning at this point that if something adapts itself, then it is said to be adaptive and adaptable.

- The *responsiveness* of an adaptive system is the magnitude or degree to react to stimuli. It is a broader term than adaptability, because in the context of general adaptation responses do not need to be 'successful'.

- *Stability* describes to which extent a system is not easily moved or modified from a stable state.

- *Feedback*: The circular causality of feedback loops is taken into account for regulation processes, i.e. for ensuring the success of system's efforts to maintain equilibrium or to reach a goal. Further, integral feedback control is not only sufficient but also necessary for robust adaptation.

- *Fitness* is a measurable degree that depends on the reproductive success of adaptations. Equivalently to fitness, sometimes the term efficiency is used as a synonym.

- *Predictability* comprises the a-priori work in order to pre-compute future behavioural steps, e.g. in order to optimise some results, enhance the fitness or foster the autonomy of the system.

2.1.3 Adaptive Systems within Computer Science

As shown in the previous section, many findings can be gained from the investigation of the general concept of adaptation from the point of view of disciplines that are not tightly related with Computer Science.

In this section, a chronological review over relevant milestones in the evolution of computer-based adaptive systems is given. This methodology was chosen, because it is an illustrative way to demonstrate how research areas related with Computer Science have contributed to the technological character of Adaptive E-Learning.

The 1940's and 1950's

Although the fields of Cybernetics and Systems Theory were already mentioned in the previous section, it is important to start here also with these fields, but from a technological point of view. The first reason for this methodology is that these two disciplines have a large history and are often seen as the 'root-disciplines' for modern and well-established fields concerning computer sciences, such as Artificial Intelligence, Control Systems or Evolutionary Algorithms. And the second reason is that they have contributed directly to the context of adaptive systems with definitions, methods and innovation impulses.

As stated in [Francois 1999], although the term 'cybernetics' was not known at the beginning of the 1940's, both root-disciplines were represented in that decade by a melting pot of research groups, where ideas boiled and the vocabularies of engineering and physiology were used interchangeably. In accordance with [Francois 1999], the term was born in 1948, as Norbert Wiener wrote his 'Cybernetics or Control and Communication in the Animal and the Machine'. Thus, for the 1950's, the fundaments of a common language were created and terms like *learning, regulation, adaptation, self-organisation, perception, memory*, and so forth, begin to have a technical connotation (see [deRosnay 1978] for more details). Further, in 1949, Shannon and Weaver published their work 'Mathematical Theory of Communication' (which is frequently referred to as Theory of Information) where they explain and specify the concept of *communication*. They also introduce other concepts, such as *source, message, transmitter, signal, channel* and *receptor* (please refer to [Mullen 2003] for details).

Consequently, there is no doubt that cybernetics had the highest influence to the origin of adaptive systems, but until the late 1950's no concrete definition can be found. For example, [Ashby 1956] described and defined many concepts, which have direct relation with the context of adaptive systems, such as *stability, regulation, interaction, communication, feedback* or *control*.

Nonetheless, according to [Isermann et al. 1992], although the concept of 'adaptive control' (which represented the key issue in Control Theory during the 1960's) was first introduced in the late 1950's as the

> *"ability of a controller to adjust its parameters to the process statics and dynamics"*,

the concept 'automatic adjustment' of controller parameters was first discussed in the late 1940's. Many sub-areas of Control Theory contributed with issues for the design of adaptive control systems, like *state space, feedback* and *stability theory*.

The 1960's!

In his 'Outline for a Logical Theory of Adaptive Systems', John Holland stipulates a complete theory on adaptation-capable automata (see [Holland 1962]). His work is considered as one of the most important in the field of Genetic Algorithms, but finds also repercussion in other fields, such as Complex Adaptive Systems, Control Systems, Evolutionary Computing or Theory of Automata. In short, Holland's study of adaptation is based on *generation procedures* and *generated populations* of *interaction programmes*. In general terms, the study shows how a system can generate procedures that can help itself to efficiently adjust to its environment.

Also at the beginning of the 1960's, first serious discussions have been conducted around the field of Artificial Intelligence, whereby first attempts were done to write a general overview on the main issues regarding this field (see e.g. [Travis 1962] and [Tonge 1966]). Concepts like *machine learning, pattern recognition* or *self-organisation* are mentioned in those papers. At present, adaptive systems are often interpreted as self-organising systems, because of its possible independent and self-adaptive behaviour (see e.g. [Lendaris 1964] for details).

The origins of Information Retrieval date back to the time before the computer era [Gütl 2002]. Nonetheless, within the scope of computer-based systems, [Cooper 1964] is one of the first authors that classified *Information Retrieval System*s either as Document Retrieval Systems or Fact Retrieval Systems and analyses basic requirements on such systems. Thus, in the context of this book such systems might be coined as the first *content-based user-adapting systems*; the following two citations

from [Cooper 1964] give a basement for this assumption: first,

> *"the purpose of a Document Retrieval system is to refer the user to books, articles, or other documents which might turn out to be of interest to him, on the basis of a request which indicates his topic of interest. The system does not supply directly any specific information in the subject area of interest, but only information about what to read in order to find such information [...]"*

and second,

> *"A Fact Retrieval system, by contrast, yields specific facts in the subject area of interest in direct response to the user's specific questions".*

According to [Slagle 1965], the LISP program DEDUCOM had the purpose of showing how an 'intelligent' machine could solve questioning-answering problems by *deduction*. Thus, answering to users' questions by means of adaptation. Slagle states further that he eventually wanted to

> *"get a computer to 'learn from its experience' at least as well as a human does. [...] In particular, the computer should learn how to learn better".*

Other adaptation-based issues in the field of Dialoguing Systems can be found in [Simmons 1965] and [Green and Raphael 1968]. Also in this context, as pioneers in the field of Man-Machine-Interaction (nowadays called Human-Computer-Interaction, HCI), [Corbin and Frank 1966] address the importance of *use-oriented terminal devices* and identify many requirements on software tools for computers using such devices. Interesting issues are found here, as for example *conversational procedures, responsive online multi-user environment, problem oriented operations*, etc.

Progress in the field of HCI moved towards the experimentation with different user interaction devices and their implications on different users. For example, as stated in [Duchowski 2000], early diagrammatic depictions of recorded eye movements, as undertaken by Noton and Stark between 1967 and 1971, represent one of the roots of the field of Visual Attention. Since then, the field of Visual Attention has been contributing to advanced interface design as well as to modern user-adapting systems. Noton and Stark performed their own eye-movement measurements over images and coined the observed patterns *scan-paths*, which can be used to calculate some cognitive traits of users, and in turn to personalise the presentation of digital content according to these traits (e.g. as done in the adaptive e-learning system AdeLE, which is presented in chapter 5).

The 1970's

Lyle Smith's 'A Survey of Interactive Graphical Systems for Mathematics' gives a comprehensive review that shows the state of the art in Human-Computer-Interaction at that time [Smith 1970]. Several problems, as the need to *shape the information of processing procedures* are addressed and evaluated towards a number of interactive systems of those days. Many of the systems presented in [Smith 1970] (and the list is long), reflect adaptational features by means of adapting to two different target environments: on the one hand, the systems adapt to the context in where they are developed by changing their functionality, and on the other hand, they try to fit their output for the interaction devices for different types of user (experts and casual users).

Further, in 'The Use of Interactive Graphics To Solve Numerical Problems' Lyle Smith presents PEG (an online data-fitting program) and states two properties of interactive systems, which are primordial to the usability of the system: effectiveness and simplicity of use. Regarding the latter property, Smith stated that

"as more generality and flexibility are put in a system, more choices (actions) must be made and more complex information must be given by the user to the system to 'tailor' its power to the particular problem at hand".

The methodology solving this problem is also given by the author:

"anticipating the desires of a user and presenting him with a corresponding list (menu) of options

can be called 'interaction by anticipation' [...]",

a known type of adaptation mentioned in the previous section. [Smith 1970a]

Moving towards the didactical side of educational sciences, an interesting issue related to adaptive systems is the concept of *observation* (meaning e.g. the observation of the environment

or of the adaptable structures). According to [Francois 1999], Von Glasersfeld developed since 1976 his Constructivism Theory as a general reflection on the conditions of learning and knowing, and states:

"Objectivity is the delusion that observations could be made without an observer". [Francois 1999]

In addition, [Francois 1999] explains that Von Glasersfeld's aim is

"to discover how we perceive and construct reality, to retrace the ways we follow to construct concepts and to elaborate abstractions, and to better understand the relation of the self with others and with the environment in general",

and thus, as already mentioned within the context of evolutionary research in section 2.1.1, the need to internally model the external environment (to which is to adapt) by means of observation and interaction is shown again. A few years later, within the field of Computer Aided Instruction (CAI), which is an 'early' notion for e-learning systems, the requirement of adapting the system to user's needs is fundamental. [Hoffman 1978] talks about *individualised instruction* within the context of CAI.

But also aspects regarding the architectural design of adaptive systems can be identified in the 1970's. [Brandwajn et al. 1979] present AR-CADE, a research project with the aim of studying the feasibility and effectiveness of an adaptive computer system, focusing on the topic of *adaptive computer architecture*. One of the most interesting issues in [Brandwajn et al. 1979] is given by the *separation of slow and fast computational components* within the functional architecture in order to enhance the performance of the adaptational procedures and to reduce the complexity of the system. Interesting are the features in the system's component, which models its operation. The system measures constantly specific parameters of that model and is thus, able to dynamically optimise its performance based on given criteria to modify other components.

The 1980's

In 1981, a relationship between adaptive systems and dynamic systems can be identified.

[Rouse 1981] analysed models of HCI within the context of 'control of dynamic systems' and states:

> "A system is dynamic to the extent that its future outputs depend on its past outputs, as well as on inputs generated by humans, computers, and the environment. Because the outputs of dynamic systems depend on more than just the systems' inputs, they continue to evolve regardless of whether or not any control is exercised."

In their work 'A Self-Regulating Adaptive System', [Trevellyan and Browne 1986] investigate the viability of providing *adaptive user interfaces*. Such self-regulating adaptive systems differentiate among users' traits in order to *force* the system to change its user interface according to different *objectives*. Thus, [Trevellyan and Browne 1986] demonstrate that

> "adaptive systems can be built so that they regulate their own behaviour by assessing whether their adaptations are being successful in meeting these objectives".

Two years later, [Halasz 1988] introduced important aspects regarding Hypermedia Systems, and presented several reflections that had direct influence in the field of what he called *Next Generation Hypermedia Systems* (NGHS). Halasz' 'Seven Issues' are summarised as follows:

- Search and Query in a Hypermedia Network,
- Composites (augmenting the basic node and link model),
- Virtual Structures for Dealing with Changing Information,
- Computation in Hypermedia Networks,
- Versioning,
- Support for Collaborative Work and
- Extensibility and Tailorability.

The last issue in the list shows the need of user-adapting mechanisms, as in e.g. *Adaptive Hypermedia Systems* (AHS), which belong to NGHS. Further, also within the scope of user tailorability of software systems, [Oppermann 1994] stated in his work 'Adaptively supported Adaptability' that since 1981

> "adaptivity in the form of an adaptive system was based on the assumption that the system is able to adapt itself to the wishes and tasks of the user by an evaluation of user behaviour, thus breaking down a communication barrier between man and machine".

This is a key observation applicable for user-adapting systems, indicating first notions towards the topic of *personalisation systems*. It is worth mentioning at this point, that a well-known (and probably the most cited) denotation for distinguishing between adaptable and adaptive systems also comes from [Oppermann 1994], and it indicates that

> "A system is called 'adaptable' if it provides the user with tools that make it possible to change the system characteristics. […] This kind of individualization gives the control over the adaptation to the user. It is up to the user to initiate the adaptation and to employ it. […]
>
> A system is called 'adaptive' if it is able to change its own characteristics automatically according to the user's needs. The self–adapting system is the most common conception of adaptivity. Modification of interface presentation or system behaviour depends on the way the user interacts with the system. The system initiates and performs changes appropriate to the user, his tasks, and specific demands".

Oppermann derives his denotation from the context of HCI (see e.g. [Feeney and Hood 1977] or [Edmonds 1981]) and emphasises the relevance of adaptive systems at the beginning of the 1990s referring to [Browne et al. 1990]; there it says that the dream of adaptivity is not only based on

> "everyone should be computer literate",

but also on

> "computers should be user literate".

The 1990's and 2k+

Although [Holland 1992] mentions a number of adaptive techniques from the point of view of Complex Adaptive Systems, they all involve *one autonomous process* that represents the basic element on which other system components *self-organise* themselves towards solutions with higher fitness by means of following an *adaptive plan*. Further, [Holland 1992] identifies four main components in an adaptive system:

- the environment of the system to which the system adapts,

- the structures in the system (i.e. a set of components which adapt themselves),

- the adaptive plan (i.e. the managing plan that controls the modifications of the system structures), and

- a measurement component for the fitness or performance of each structure (which gives feedback information on how well the new structure has solved the problem).

With respect to terminological differences, [Oppermann 1994] states that adaptable and adaptive systems (as defined also there) should not be seen as 'alternatives'. The arguments are that

"adaptable features are not a solution, because they are not a common part of user interfaces, so users are not (yet) familiar with their existence, operation, and benefits",

and

"adaptive features are not a solution either, because they keep the users dependent on suggestions with respect to time and content of (tips for) changes. Users refuse to be at a mercy of the system".

As tailorability has become an essential issue for hypermedia-based software environments during the 1980's, the need of defining rules and models for such systems has arisen. Peter Brusilovsky can be seen as one of the precursors in the field of *Adaptive Hypermedia Systems* (AHS). To give an example, [Brusilovsky 1998] identified, among other things, (a) possible classifications for user-adapting hypermedia methods and techniques as well as (b) a differentiation of adaptation technologies from the point of view of adaptive hypermedia. These topics will be treated in detail along the next sections of this book.

Another research stream of high relevance within the scope of this book can be identified in the 1990's: one of the first notions on *Recom-*

mender Systems originate from [Resnick and Varian 1997], defining such systems as software environments, where people provide recommendations as inputs, and based on that, the system aggregates and/or directs them to appropriate recipients. But also in the late 1990's, Mobile or Wearable Computing and Wireless Multimedia Communication emerged as new forms of *adaptive nomadic information systems* (as called in [Specht and Oppermann 1999]).

Arriving at the present century and from the software engineering point of view, the newly emerging field of *Service-Oriented Programming* wins increasingly more relevance within the research community. Within the scope of *user-adapting systems*, [Conlan et al. 2003] stated that

"adaptive services base their adaptivity on user and context information, as well as on an encapsulation of the expertise that support the adaptation".

Further, current modelling techniques, adaptive dimensions and personalisation techniques, as e.g. used in AHS, may supply the basis for *next generation adaptive collaborative services*. The dynamic capture of users' requirements, the design of adaptive services and a meaningful combination of services lead to a number of challenges in different research areas, such as User or Context Modelling, Semantic Interoperability, Multi-modal Information Presentation, Service Composition and Self-management of Services. [Conlan et al. 2003]

An investigation on a new way of how computer systems can better meet users' requirements is presented in [Klann et al. 2003]. The basis of their observations is given by the notion of *situation-aware adaptivity*; on the one hand, adaptivity

"is automatically carried out by the system",

and on the other hand, adaptability is

"consciously carried out by the users" [Klann et al. 2003].

Thus, the advantages of the synergy of adaptivity and adaptability in software systems are essential to meet specific users' requirements; in other words, the shared adoption of static and dynamic configurability of systems is significant for the fulfilment of dynamically changing users' situations.

A last aspect must be considered regarding adaptive systems: which are their functions at present? So far it is clear that adaptive systems are used in many application domains in order to perform distinct tasks. [Weibelzahl 2002] adopted a list of functions of adaptive systems from [Jameson 2001] and listed the most relevant of them within the scope of user-adaptation. These functions can be summarised as follows:

- Personalised help (to find information, with recurrent tasks, or by virtual assistants)

- Tailored information content

- Recommendation of products

- Adaptation of user interfaces

- Support learning (e.g. adaptive annotations, adaptive link-hiding, adaptive curriculum sequencing, and the like).

- Conduction of goal-oriented dialogs.

- Support of collaboration (e.g. in distributed workspace environments).

Thus, when building a single system for various adaptive functions, another important aspect arises: inclusion of *multi-purposefulness* in the functionality of the modelling and/or the adaptive component in order to achieve different or combined goals of the system.

Through the findings presented so far, the general complexity of adaptation-pertinent systems has been shown. Also, a first approach towards a chronological roadmap of adaptive systems within the context of Computer Science was given. With the conformation of this conceptual background, the reminder of this chapter shifts the focus of investigation to the scope of adaptive e-learning.

2.2 Adaptive E-Learning

At present, the success of adaptive e-learning systems is seen in the efficient delivery of courseware by means of advanced personalisation techniques. Further, experts from different research disciplines are of the same opinion with respect to personalisation in e-learning environ-ments: *the tenet of modern teaching and learning paradigms is that different learning goals require different didactical approaches.*

Thus, from the technological point of view, one of the main challenges relies on the fact that the didactical expectations of teachers must be differentiated from and adapted to the learners' opportunities and objectives. This section covers the main aspects of *Adaptive E-Learning* from the point of view of technology-based learner-adapting systems. After clarifying the most relevant terminological issues and giving an overview on some didactical aspects, the topic is investigated in terms of a chronological analysis of adaptive e-learning systems by means of adaptive learning theories. Based on that, the relevance of multi-purpose solutions is depicted through the variety of goals and functionalities of the different components.

2.2.1 Terminological Issues

First of all, it must be clarified at this point that this section does not aim at presenting all possible terminological denotations and connotations of the concept of *E-Learning*. Though, it must be clear what it means within the context of this book.

The denotation of the term *E-Learning* reflects only one of both aspects of its real meaning, i.e. 'electronic learning', so it should be complemented by a proper second term, such as E-Instruction or E-Teaching. Hence, some integrative terms like *E-Education* or *E-Didactics* or *E-Pedagogy* would better fit to real connotations. Nevertheless, these integrative terms will be neither treated nor used in this book due to the existing 'universal convention' of understanding them as synonyms to E-Learning. Rather, the main focus is set on *didactical and technological issues regarding the transfer of knowledge*. Within this chapter, the interpretation and usage of the term *knowledge* follows restrictively the connotation and scope of conceptual thinking.

As stated in [Bransford et al. 2000], from the didactical point of view, teachers require a high expertise in their particular areas, and this need involves more than a set of general problem-solving skills. It requires also well-organised knowledge of *concepts* in the specific domain and in experience-based assessment methods. From a general point of view, it is known that people tend to simplify *knowledge assets* and to interrelate them into conceptual structures for their thoughts and notions, i.e. people arrange concepts in their minds (e.g. consider the use of concept maps for didactical purposes [Novak 2003]).

As for the general term of *E-Learning* and according to [García-Barrios et al. 2002], the notion of the *knowledge transfer process* within the context of technology-based teaching and learning environments can be also interpreted as a holistic phenomenon composed of two tightly related streams:

- the *teaching process*, concerning mainly the generation and delivery of knowledge, and

- the *learning process*, concerning knowledge acquisition.

On the one side, the teaching process, together with the aspect of assessing the success of the learning process, represent key didactical issues for effective adaptive e-learning systems. On the other side, the learning process embraces more than simply reading online lessons. Thus, e-learning is a large and complex field of research comprising a variety of learning and teaching paradigms, such as constructivist, serial, symmetric [Jain et al. 2002], cognitive, face-to-face [Pivec 2000], discovery, and managed e-learning [Lennon and Maurer 2003]. Therefore, before analysing the technological perspective on adaptive e-learning, at least the most relevant didactical aspects should be investigated by means of well-known paradigms.

2.2.2 Learning and Teaching Paradigms

Teaching itself is a very complex action and can not be fully realisable within e-learning systems (see e.g. [Spector and Ohrazda, 2003], [Mödritscher et al. 2006]). Thus, from a didactical point of view, the first questions that arise are: what to teach? How to teach? Teach for which purposes? How to measure the acquisition of learning?

An answer to the first question can be found taking into account that learning is associated with building competences. This idea is the basis of Thomas Durand's competences model ([Durand 1998a], [Durand 1998b]), which comprises the following three generic dimensions of what can be mediated to learners (see Figure 2.1):

- *knowledge*, as denoted in the previous section, enables the modelling of (parts of) the real world,

- *know-how* (or skills), relates to the capacity to act in a specific way according to predefined goals or processes, thus, including the application and use of acquired knowledge, and

- *attitudes*, comprises commitment and culture, but also social or affective behaviours.

According to this classification, a *competence* is defined by the combination of at least two of these classes [Durand, 1998a]. Thus, within the scope of e-learning, one of the main goals of teaching is to convey competencies, mostly through the delivery of digitalised data. Competence advancement enables people to (a) assimilate information and/or (b) to acquire or improve their skills and capabilities and/or (c) to share activities conformable to some mutual behaviours. However, a 'usual' interpretation for a competence consists of the combination of knowledge and skills, meaning respectively, *know what and why* and *know how*.

The second and third questions mentioned at the beginning of this section deal with the problem of how and why to teach in dependence of what to impart to a learner. In literature, for explaining this problem, most experts shift the point of view from the teaching to the learning

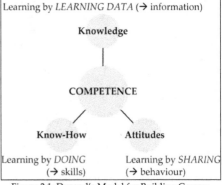

Figure 2.1: Durand's Model for Building Competencies (adapted from [Durand 1998b]).

process and utilise *learning styles*. As stated in [Mödritscher et al. 2006b], this is done, because in many cases teachers have to define their intended didactic objectives following some *taxonomy of learning objectives*, as for example Bloom's Taxonomy. In the 1950's and 1960's, Benjamin Bloom was the head of a group, which created a theory of learning goals. The resultant taxonomy comprises three *domains* that correspond to the previously depicted Durand's classification (as shown e.g. in [Bloom et al. 1956]):

- cognitive domain, regarding knowledge assets (see Table 2.1),

- psychomotor domain, regarding physical skills, and

- affective domain, regarding personal attitudes.

The idea behind the taxonomies defined in [Bloom et al. 1956] is based on different *levels* of verbs (also called 'actions' or 'action verbs'), which indicate that some specific cognitive activity happens in the brain of learners.

These levels can be interpreted as degrees of difficulty, i.e. the lowest one must be mastered before entering the next upper one. In other words, when teachers define learning objectives using this taxonomy, they stipulate what learners

must do in order to express learning, critical thinking or reasoning skills in accordance to the chosen level. To give an example, objectives for an introductory overview into a topic can be assigned to lower levels, whereas objectives for an expert course should be assigned to the upper levels. Though, teachers face often the problem that one action verb within Bloom's taxonomy can be assigned to multiple levels depending on the teaching topic, and so, a more intuitive definition of domains and levels might be needed.

A former student of Bloom, Lorin Anderson, led a new group during the 1990's for the purpose of enhancing and updating Bloom's taxonomy. Basically, most levels were changed from noun to verb forms and a few were renamed. For the levels in the Cognitive Domain (see first row in Table 2.1) the new names are Remembering, Understanding, Applying, Analysing, Evaluating, and Creating. In general, some changes were also made in structure and emphasis of the taxonomy. [Forehand 2005]

Thus, what to teach and for which objective has been answered so far. As the psychomotoric skills and attitude issues of learning are not within the scope of this book, the following observations focus on the 'knowledge domain'. The last question stipulated at the beginning of this section regards the problem of how to assess the

Knowledge	Comprehension	Application	Analysis	Synthesis	Evaluation
lower ←		levels of expression of critical thinking		→	higher
action verbs					
define	convert	demonstrate	calculate	arrange	appraise
identify	describe	discover	categorise	assemble	assess
label	distinguish	modify	compare	collect	criticise
list	exemplify	operate	discriminate	compile	defend
match	extend	prepare	experiment	compose	discriminate
name	infer	produce	infer	formulate	evaluate
recall	interpret	schedule	outline	manage	justify
record	locate	show	relate	organise	measure
recognise	report	solve	separate	plan	rate
repeat	translate	use	test	summarise	value
...	...	...	...	...	...

Table 2.1: Bloom's Taxonomy – Cognitive Domain Levels (and some Examples of Action Verbs)

acquisition of knowledge. Though, the problem is not only 'how to assess', 'when and what to assess' should be also considered.

As stated in [IDS 2002], there exists not only the necessity of assessing knowledge in order to grade learners, but also to measure the learning process. Further, [Scouller 1998] indicates that the different assessment methods have an influence on the learning behaviour. According to [Mödritscher et al. 2006b], within traditional instruction teachers assess the learning process by means of *limited-choice* or *open-ended* questions, e.g. multiple choices or respectively, sentence completion, short answers or essays. On the one hand, limited-choice questions are intended to reach Bloom's lowest level objectives, i.e. recalling data and information. And on the other hand, open-ended questions require higher cognitive skills in order to e.g. reformulate the acquired knowledge in own words. Thus, the solutions of open-ended questions are not always pre-determined and depend on the learned experiences. Summarising, teachers should choose the assessment method (and consequently the types of questions) so that the level of learning objectives suits to e.g. the overall performance of the class, the reliability of the grading technique, but also 'the prevention of cheating' and other criteria. [Mödritscher et al. 2006b]

So far, some observations have been stated concerning the 'way of teaching', but another important perspective on the topic is represented by so called Learning Theories. There exist many schools of thought on learning, and in the era of e-learning, no one of these schools is applied exclusively to design e.g. online learning materials. At present, 'e-teachers' use a combination of theories and further, as research progresses, new theories are evolving. Learning theories are chosen in order to e.g. motivate learners, facilitate deep processing, cater for individual differences, promote a company-specific expertise, or to encourage interaction within knowledge workers. According to literature, the most relevant Learning Theories are Behaviourism, Cognitivism and Constructivism.

Behaviourism, which was influenced by e.g. [Thorndike 1913], [Pavlov 1927], and [Skinner 1974], postulates that learning is a modification

in the observable behaviour of learners as a consequence of external stimuli in the environment. Nevertheless, many teachers claim that it is not the learning behaviour which indicates if or not somebody has learned something, but what is going on in the learner's mind. Thus, there was a shift away from behaviouristic to a cognitive perspective. Cognitivists claim that in learning the use of memory, motivation, thinking as well as reflection play an the most important role. Hence, learning is an internal process and its successful conclusion depends on the capacity of the learner to process knowledge, the amount of effort made during the process of learning as well as the existing conceptual structures in the head of the learners ([Ausubel 1960], [Craik and Lockhart 1972]). In the last decades, there has been a shift to Constructivism. According to e.g. [Cooper 1993] and [Wilson 1997], constructivists claim that a learner interprets acquired information and its surrounding world in dependence of her personal reality. Consequently, learners first observe, process and interpret information, and then personalise it into the own experienced world. Therefore, constructivism supports the idea that learners will learn at best when they contextualise what they have learned within the own real world, applying and personalising it immediately.

After this brief overview on the didactical aspects of education, the next section introduces into the concept of e-learning in order to conclude the fundaments needed to, finally, look inside the evolution of adaptive e-learning systems.

2.2.3 E-Learning and Education as a System

This section is based on the following hypothesis: *E-learning is technology-based distance education.* The intention is to complement the findings from the previous section with the essence of adaptive e-learning. General adaptational issues have been presented in section 2.1, didactical aspects in 2.2.1 and 2.2.2. The next section (2.2.4) focuses on the main aspects of adaptive systems for e-learning form a historical and conceptual point of view. Thus, taking into account the hypothesis stated above, this section bridges the gap from the field of e-learning as a 'mixture of distance education and learning machine' to the field of adaptive e-learning as 'virtual environments that automatically adapt teaching to different types of learners'.

Thus, the field of *Distance Learning* represents the starting point. [Keegan 1980] identified six key aspects of distance education:

- the physical separation of teacher and learner
- the implication of an educational organisation,
- the use of a media to link teacher and learner,
- a two-way exchange of communication,
- the treatment of learners as individuals rather than groups, and
- the perspective on education as an industrialised form.

The continuous improvements in communication technology, which increased the interest in the unlimited possibilities of individualised distance learning, have changed this traditional definition as teachers had to reconceptualise the idea of schooling and lifelong learning [Keegan 1980].

Thus, space-distributed and time-delayed (asynchronous) conferencing shows the possibility to interconnect learners for certain time periods. This paradigm showing the need of learning at own pace in an individualised manner is referred to as asymmetric learning in [Jain et al. 2002], because, in contrast to symmetric learning in traditional classes, learners perform differently over time.

Thus, [Holmberg 1989] refined the definition of distance learning by stating that it

"is a concept that covers the learning-teaching activities in the cognitive and/or psycho-motor and affective domains of an individual learner and a supporting organization. It is characterized by non-contiguous communication and can be carried out anywhere and at any time [...]".

The fundament of this definition can be identified also in [Keegan 1986], where three historical approaches to the stipulation of a theory for distance education can be identified:

- autonomy and independence reflect the primordial role of individualisation of learners (see [Moore 1973], [Wedemeyer 1977]);
- the industrialisation theory (see e.g. [Peters 1971]) attempts to view distance education as an industrialised form of instruction; and
- the approach to interpret the theories of interaction and communication in an integrative manner (as stated in e.g. [Sewart 1987] and [Daniel and Marquis 1979]).

According to [Keegan 1986], the focus at that time was set on the concepts of industrialised, open, and non-traditional learning. And these issues concentrated on media like radio and television. But by bringing these aspects to the present Internet-era, we can automatically get a definition for the concept of E-Learning. In agreement with [Dietinger 2003] and for the purpose of this book, e-learning can be defined as the 'new' way of distance learning consisting of:

- one or more 'distance learners',
- multimedia and interactive content,
- a software environment working as an interface between learners and their learning objectives (and providing different means to achieve the objectives), and
- one or more 'distance teachers' that assist and guide learners.

At this point, a rather new learning theory comes into play, *exploratory learning* (see e.g. [diSessa et al. 1995] for details). This theory gives first evidences to the field of adaptive e-learning and is based on four principles:

- learning can and should be done at own pace,
- knowledge is rich and multidimensional,
- learners approach the learning task in very diverse ways, and
- learning does not have to be forced.

But how should it be accomplished from the technological point of view? Well, if 'exploring' some topic, it is wished to 'discover' something interesting or new within the topic. Thus, successful discovery can be seen as the ultimate goal of exploration. As depicted in [Lennon and

Maurer 2003], there are hundreds of variants of constructivism. One of them, *discovery learning*, is an inquiry-based learning paradigm that is based on constructivist principles. In discovery learning, students should discover new knowledge and skills based on their prior knowledge and experiences. Unfortunately, as also depicted in [Lennon and Maurer 2003], applying this paradigm is almost impractical, because it is too complex in many traditional class settings. Therefore, Lennon and Maurer propose the use of *managed learning environments*, which have one virtual learning environment as its central component. This virtual learning environment is defined as any environment where students can learn using online resources and serves as a portal to other components that facilitate other services, such as curriculum management, courseware delivery, assessment, communication, tutor support and progress tracking.

One of the most interesting features of such systems is that through this new paradigm a shift is undertaken from 'learning and teaching' to 'learners as teachers'. [Lennon and Maurer 2003] underline this assumption stipulating that

"as all good teachers know, teaching is one of the very best ways to learn".

Further, in [Lennon and Maurer 2003] the following example is given about the use of discovery learning:

"the teacher will usually give an introductory, motivating, talk to fire the learners' imaginations and get them started. The learners may then search the Web, read appropriate material, use email to clarify points and produce a multimedia report".

Naturally, the success of the application of this paradigm is not guaranteed. But the point is not to completely leave learners alone exploring an unknown and fuzzy information space. In accordance to [Dietinger and Maurer 1997],

"the learners have to be guided".

[Dietinger and Maurer 1997] explain further:

"We need 'guidance without dictatorship' but must avoid the 'lost in hyperspace' problem".

Accordingly, there is neither a best teaching nor a best way to learn. Furthermore, most of the paradigms claim that learning should be based

on the particular interests of learners. In terms of these observations one adaptive e-learning system has to fulfil the needs of many actors, different teachers and different learners in different situations, i.e. like a *social system*. The field of Systems Theory can help to understand this perspective, as follows. [Banathy and Jenlink 2003] stated that the models of social systems are built by the relational organisation of the fundaments that represent the context, the content, and the process of such systems. For that purpose, [Banathy 1992] defined three models that represent

"(a) systems–environment relationships, (b) the functions/structure of social systems, and (c) the processes/behavior of systems through time. The models are some kind of 'lenses', which can be used to look at educational systems and understand, describe, and analyze them as open, dynamic, and complex social systems".

The systems–environment model describes an educational system in the context of its social environment. It comprises a set of questions to help users to assess, on the one hand, the social responsiveness of the system and on the other hand, the appropriateness of this responsiveness towards the system. The functions/structure model focuses on the state of the educational system at a certain point of time, like a 'screenshot' of the system. [Banathy and Jenlink 2003]

Among others, it enables to choose the components of the system that are able to carry out the main functions as well as to formulate the relationships of the structural components of the system. Here again, the model contains a set of questions to assess the functional and structural adequacy of the system. The process/behavioural model concentrates on what the system does through time, like a 'movie' of the system. It helps to understand its behaviour as a changing and living social system, i.e. it describes how it processes and transforms input for internal use, how it processes and assesses the output, and how the system might be adjusted if a redesign is needed. Thus, this model comprises a set of questions to evaluate the processes inside the system. For more details, please refer to [Banathy and Jenlink 2003].

With respect to the corroboration of the hypothesis given at the beginning of this section, which formulates E-Learning as *'technology-based distance education'*, it can be stated at this point that it is insufficient. In order to corroborate the hypothesis, it must be changed and at least a new perspective enclosing modern notions is needed. For that purpose, the term 'online' is suitable. Being *online* refers to the status of being connected to a computer network or available through a communication system (see e.g. [Merrian-Webster 2007]). Thus, it is possible at present to think about e-learning in terms of *'online distance learning'*. In other words, e-learning is like the new variant of distance learning, where participants of instructional activities interact with each other or with a computer system by means of electronic devices that are interconnected with the aid of communication channels. Thereby, *interactors* are teachers and learners, *electronic devices* are e.g. laptops, personal computers (PCs), personal digital assistants (PDAs) or mobile phones, and *communication channels* are e.g. wireless hotspots, satellite or telephone cable.

For the context of this book, it is relevant to having shown so far that a deep investigation on adaptive e-learning systems implies to understand fundamental issues coming from related research fields. The next section summarises how the confluence of these different streams have determined the evolution of adaptive e-learning systems over many decades. The three general topics treated so far (adaptation, learning & teaching and e-learning) let assume at this point that (a) technology, i.e. the use of computers, can significantly enhance learning, (b) the advantages of technology-based instruction are not limited by learning goals, tasks or subjects, and (c) teaching through technology is applicable to a variety of learner types.

2.2.4 Technological Roots of Adaptive E-Learning

As mentioned in the last section and in accordance to [García-Barrios et al. 2002], the knowledge transfer process within the context of technology-based teaching and learning environments can be interpreted as a holistic phenomenon composed of two related streams: the *teaching process* (concerning knowledge generation and delivery) and the *learning process* (concerning

knowledge acquisition). Furthermore, e-learning embraces more than simply reading online lessons. It is a large and complex field of research comprising a variety of learning and teaching paradigms, for example: *constructivist, serial, symmetric* [Jain et al. 2002], *cognitive, face-to-face* [Pivec 2000], *discovery, managed learning* [Lennon and Maurer 2003].

E-learning paradigms and implementations have brought many advantages to technology-based distance education. It is now possible to identify, analyse, track and monitor relevant aspects of instruction, such as different velocities in the learning process, distinct learning paths, or in general, different strategies of learning. Social aspects and problems, which are common in conventional face-to-face learning, such as censorship of information or racism, can be regulated or partially solved through mechanisms of e-learning. [García-Barrios et al. 2004b]

Given that the objectives of technology-based educational environments and their impact on individuals are linked with complex and context-dependent constraints and conditions, the author of this book is aware of the fact that a single system solution may only be valid for specific subfields of e-learning. Experiences regarding the observation of user behaviour within the learning process have confirmed that learners tend to stick to distinct learning methods [Pivec 2000].

As stated in [Jain et al. 2002], e-learning, even if standardised, tends to produce asymmetrical learning, as its tools reach out to a dispersed audience where individuals may arrive at different stages at different times, even if along a common learning trajectory. These are two of the reasons for considering adaptivity to be one of the key issues in modern e-learning environments. Many currently available solutions are not able to fulfil all the conditions needed to solve the main problems and to achieve most of the aims of adaptivity and personalisation[6]. Furthermore, they do not consider important pedagogical features in any depth. According to [ADL 2007], the value of personalised instruction is measurable by means of its effectiveness, e.g. a learner in a classroom

---

[6] The terminological and practical differences between adaptivity and personalisation will be described in the next chapter.

setting asks on average 0.1 questions an hour, whereas in an individual tutoring setting, a learner may ask or is required to answer about 120 questions an hour. Thus, the achievement of individually tutored learners' performance, as measured by test scores, may significantly exceed that of classroom colleagues (for details see [ADL 2007] and [Bloom 1984]).

In addition, it is worth to state at this point that (in some cases) holding an e-learning course as a completely virtual course might not reach the pedagogical advantages or objectives as scheduling real-world lessons, i.e. a combination of e-learning with traditional face-to-face meetings allows a controlled regulation of symmetrical and asymmetrical learning. This technique is known as *hybrid* or *blended learning*: it comprises the balanced utilisation of synchron/asynchron, online/offline as well as virtual/non-virtual knowledge transfer phases. [García-Barrios et al. 2004b]

Taking into account the depicted aspects, a more extensive solution is needed, which allows the binding of effective modern technologies and solution approaches in order to enhance the adaptation of knowledge provisioning and to increase the effectiveness of personalisation. Hence, the need emerges to search for and define new useful parameters, which enable a deeper insight into learners' behaviour during the learning process. And, as shown through the reminder of this section, this need is not a completely new issue of research, neither from a technological point of view, nor from a didactical one.

From a historical point of view on adaptive learning theories and according to [Park and Lee 2003], three main theoretical approaches to adaptive e-learning systems can be identified: the *Macro-adaptive Approach*, the *Aptitude-Treatment Interaction* approach, and the *Micro-adaptive Approach*. In [Kaenampornpan and O'Neill 2004] it is shown that *adaptive instructional design* can be tracked back to the beginning of the 20th century. Though, the development of technology-based systems does not start before the 1960s.

A brief summary of the main distinctions among these three theories can be given as follows ([Mödritscher et al. 2004c], [García-Barrios 2006b]):

- In the macro-adaptive (MacroAd) approach, instructional alternatives are selected on behalf of learning goals as well as general abilities and achievement levels of learners, determined prior to instruction.

- In the aptitude-treatment interaction (ATI) approach, instructional strategies are adapted according to specific learner traits, such as intellectual abilities, cognitive or learning styles, expertise level, and achievement motivation.

- The micro-adaptive (MicroAd) approach is based on the diagnostics of the specific learning needs during instruction, and consequently, provides instructional prescriptions for these needs.

Thus, pure didactical methods as well as technology-based instructional systems can be analysed by means of these theoretical approaches. Based on them, the reminder of this section shows some examples of *systems for adaptive instruction*, i.e. 'adaptive e-learning systems'.

In some early MacroAd projects (such as the Burke, Dalton and Winnetka plans), learners were allowed to go through the materials at their own pace (see [Park and Lee 2003] and [Reiser 1987]). MacroAd instruction often included elements like explaining or presenting specific information, asking questions to monitor the learning process, and providing feedback for learners. In 1963, the Keller Plan system enabled personalisation and offered features, such as required mastery to proceed to the next learning unit, and the usage of workbooks.

CMI (*Computer-managed Instructional*) systems follow also the MacroAd theory and have functions to diagnose learning needs and prescribe instructional activities according to these needs. To give an example, the PLM (Plato Learning Management) system follows CMI. It is important at this point to emphasise that the MacroAd-based functions of CMI (i.e. prescription of instruction) and the features of MicroAd systems (i.e. prediction of student learning needs) are distinct. To show this aspect, consider the PLM system, which is able to provide tests on different levels of instruction. Thus, the didactical intention of PLM is that depending on the embedded processes of 'test–evaluation–assignment', specific didactical instructional prescriptions are provided, such as repeating the test or the whole unit, offering additional units or activities, or selecting the time to study and proceed at the own pace. ([Park and Lee 2003], [García-Barrios et al. 2004b])

CMI systems are also called *Computer-Assisted Instructional* (CAI) systems and, consequently, follow the MacroAd theory. In literature, the PLM and CLASS systems belong to this category (see e.g. [Crowell and Traegde 1967]). But also the SOLO system interesting within this category, as it explicitly fosters 'creativity' to the CAI functions by means of individual exploration. As explained in [Dwyer 1970], the SOLO system is based on the 'dual-solo' learning scheme for practical flight instruction. Thus, training starts with a student-with-instructor task-repetitive phase and shifts gradually to student-solo environments where the learner is responsible for making all decisions. Dwyer also points out that the 'solo' work to be efficient must take place within a highly structured and well-prepared complete environment. Further, he identified at that time that a

> *"solo mode is a critical need in next-generation education, but that it only makes sense if coupled with the proper complete solo environment".*

Other examples for MacroAd systems are socalled *Mastery Learning Systems*, such as the IPI (Individually Prescribed Instructional) system developed in 1964, IGE (Individually Guided Education) in 1965, and ALEM (Adaptive Learning Environments Model) in the early 1980s. In mastery learning systems, adaptation is limited to the *time* variable, because learners achieve their goals by getting sufficient time and materials for the learning process according to the results of formative and summative examinations. [Park and Lee 2003]

The ATI approach, indeed, seems to represent more a user modelling technique than a sum of didactical functions. [Cronbach 1957] suggested that facilitating individualised instruction for many different types of learners would require many different environments adapted to specific traits of the each individual learner, such as *cognitive* (e.g. verbal or mathematical abilities, mental speed, cognitive/learning styles, and prior knowledge), *conative* and *affective aptitudes* (e.g. motivational traits like anxiety or interests, and volitional or action-control traits such as self-efficacy).

As stated in [Park and Lee 2003], several researchers consider the ATI approach to be very problematic, because of the following reasons: (a) the assumed optimal abilities may not be exclusive, and thus, one ability may be used as effectively as another ability in the same context; (b) abilities may vary with the learning progress, and thus, it becomes more or less relevant for one lesson than for another, and (c) validated ATIs for a particular context may not be applicable to other areas, and thus, vary for different kinds of content. Because of these limitations of the ATI model, researchers have developed MicroAd solution approaches using 'on-task' measures. [Park and Lee 2003] stipulates further, that these systems were developed through several attempts beginning with *Programmed Instruction* (PI) to the application of *Artificial Intelligence* (AI) methods for the development of *Intelligent Tutoring Systems* (ITSs).

Intelligent Tutoring Systems embrace simultaneously many aspects: they represent internally the learning content, implement the didactical goals and model the learner's knowledge (and lack of knowledge). For this purpose, an ITS consists mostly of the following components: (i) the *expertise module* evaluates the learner's performance and generates the corresponding learning assets, (ii) the *student modelling module* assesses the learner's current knowledge state and assumes mental understanding and strategies, and (iii) the *tutoring module* selects the learning material and decides how and when to deliver it. In ITSs, AI methods are mostly used in order to represent knowledge structures, and natural language dialogues in order to adapt the contents to the learner and to allow a more flexible interaction with the system. [Park and Lee 2003] [García-Barrios 2006b]

Following the rules of MicroAd theories, *Adaptive Hypermedia Systems* (AHSs) were born in the mid 1990s ([Beaumont 1994], [Brusilovsky

2000]). Since then, AHSs constitute an emerging and new direction of research, and due to the fact that adaptive systems have become an interesting issue for e-learning environments, the need to define models for *Adaptive Educational Hypermedia Systems* (AEHSs) arose. Among others, [Brusilovsky 2001] identifies a taxonomy of adaptive hypermedia technologies classes of technologies (see Figure 2.2).

In general, AEHSs apply these technologies on the educational domain. Accordingly, AHSs and AEHSs apply mainly two forms of adaptation:

- adaptation of the page content, also known as 'adaptive presentation' or 'content-level adaptation', and

- adaptation of the behaviour of hyperlinks, also called 'link-level adaptation' or 'adaptive navigation support'.

According to [Stewart et al. 2006], AEHS aim at granting each learner a tailored lesson accord-ing to the individual educational needs. This goal is related with and responds to the 'grand research challenge in information systems – number three' (see [CRA 2002] for details), which claims to 'provide a teacher for every learner'.

All these observations regarding AHSs are interesting from the 'teaching process' point of view, but again, what about the 'learning process'? [Shapiro and Niederhauser 2003] identified features unique to hypertext, which influence the reading behaviour during *Hypertext-Assisted Learning* (HAL), such as its non-linear structure, flexibility of information access or a high degree of learner control. These features lead to many meta-cognitive demands on hypertext readers, such as cognitive load or resource allocation. In accordance to [Shapiro and Niederhauser 2003], during a study of navigation patterns, researchers observed subjects reading hypertext and identified six different strategies: skimming, checking, reading, responding, studying, and reviewing. Thus, the need to navigate through hy-

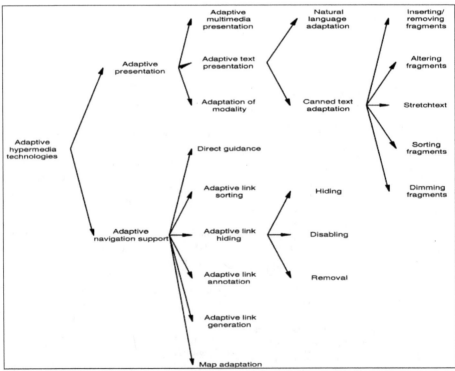

Figure 2.2: The Taxonomy of Adaptive Hypermedia Technologies. [Brusilovsky 2001]

pertext is a prevailing indicator, which differentiates reading and learning with hypertext from reading and learning with traditional printed text. Further, personal strategies for hyper-navigation depend on interest, motivation, and intrinsic or extrinsic goals of the reader. As shown in chapter 5, the research project AdeLE (Adaptive e-Learning through Eye-tracking) examines, among others, these issues with the aid of eye-tracking (or gaze-tracking) and content-tracking technologies.

Indeed, ITSs and AHSs are different research fields. Though, as stated in [Brusilovsky 1999], both represent the main influential streams on modern adaptive e-learning systems. In [Brusilovsky and Peylo 2003], an analysis of the technologies of ITSs and AHSs led to the introduction of a notion that integrates both fields: *Adaptive and intelligent Web-based Educational Systems* (AIWBES). On the basis of this new notion, Brusilovsky and Peylo identified additional new technologies for learner-adapting systems.

As a result, the taxonomy of adaptive hypermedia technologies shown in Figure 2.2 can be expanded to the scope of classic AIWBESs, as shown below in Figure 2.3. To those already defined for AHSs, three ITS-based technologies are 'added':

- Curriculum Sequencing,

- Intelligent Solution Analysis, and

- Interactive Problem Solving Support.

As explained by [Brusilovsky and Peylo 2003], curriculum sequencing aims at providing to a learner the best individual sequence of knowledge assets to learn and of learning activities, thus, helping the learner to find a personalised *optimal path* through learning materials. Intelligent solution analysis aims at responding adaptively to learners' answers to educational problems; this technology aims at giving feedback (if necessary) about errors as well as detecting those pieces of knowledge that could be responsible for the error (knowledge diagnosis).

Finally, [Brusilovsky and Peylo 2003] stipulate that interactive problem solving aims at tutoring a learner with intelligent help while problem solving; on each step of the solving task, the system may e.g. give hints to master the step or demonstrate a partial solution to the learner. Thus, this point of view on AIWBESs can be used to assume that the technologies of AHSs concentrate on content-node-analysis, whereas those in ITSs seem to be more oriented towards assessment- assistance-analysis. This assumption is analysed more in detail in section 3.1.3.

The analysis in [Brusilovsky and Peylo 2003] goes further than the consideration of AHSs and ITSs. It identifies (between the end of the 1990s

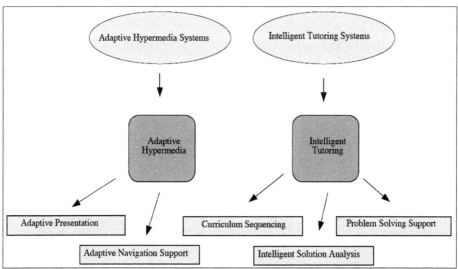

Figure 2.3: Classic Adaptive Technologies for AIWBESs and their Origins. [Brusilovsky and Peylo 2003]

and beginning of 2000s) additional influences from the research fields of Machine Learning and Data Mining (ML&DM), Information Retrieval (IR) as well as Computer-supported Collaborative Learning (CSCL). Taking into account that these three areas have represented interesting targets for applying specific AI techniques, Brusilovsky and Peylo claim that intelligent techniques applied on ML&DM, IR and CSCL have helped the following groups of adaptive technologies to emerge in the last years (see also Figure 2.4):

- Adaptive Information Filtering,

- Intelligent (Class) Monitoring, and

- Intelligent Collaborative Learning.

Adaptive Information Filtering (AIF) embraces technologies that aim at finding only those items in a large pool of learning texts that are significant to the interests of a learner. There are two main AIF technologies: *content-based* (focusing on document contents) and *collaborative filtering* (ignoring the content, but trying to match groups of users interested in similar documents). Intelligent class Monitoring (IcM) utilises AI techniques to help teachers that have no visual face-to-face contact with their remote learners and therefore, they can not detect, track and support troubled learners. [Brusilovsky and Peylo 2003]

Intelligent Collaborative Learning (ICL) is based on a similar goal as IcM. Particularly in Web–based educational systems, collaboration support is highly important, because otherwise, the learners do not meet face-to-face. Currently, within ICL, three adaptive and intelligent technologies can be identified: adaptive group formation and peer help, adaptive collaboration support, and virtual students. The didactical goals of ICL and IcM may look similar, but the adaptive aid has different targets, ICL helps learners, IcM helps teachers. Please refer to [Brusilovsky and Peylo 2003] for more detailed descriptions and some examples of existing systems supporting the presented AIWBES technologies.

On another side, neither the field of AHSs nor the one of ITSs give concrete answers to how knowledge assets or domains can be (automatically or semi-automatically) extracted from learning materials and organised within the systems.

As stated in [Schmidt 2005], *Knowledge Management Systems* (KMS) and E-Learning Systems both address the same basic problem regarding the topic of instruction: facilitating learning in organisations. Both disciplines approach the problem with two different paradigms, resulting in two different types of systems, because KMSs concentrate on requirements of industry, and most e-learning systems are limited to academic institutions. The critical aspect within the context

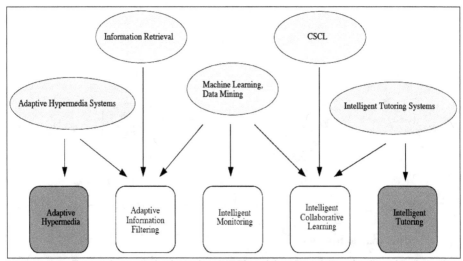

Figure 2.4: Classic Adaptive Technologies for AIWBESs and their Origins. [Brusilovsky and Peylo 2003]

of this book is that highly relevant issues concerning some relevant sub-disciplines of KM, such as *Knowledge Organisation*, are sometimes ignored when designing adaptive systems, e.g. semantic classification schemes or ontologies, which enable an abstraction of the practical meaning and intention of applicable (or newly acquired) knowledge. And this point of view leads to the investigation of the 'modern' field of the *Semantic Web*, which, in combination with *service-oriented programming paradigms*, enables an enhanced *machine-machine-interaction* view on modern Web-based distributed systems (see chapter 4 for more details).

Considering the above depicted aspects, a major challenge arises in order to integrate these approaches into flexible solution frameworks for efficient, service-based and *context-dependent* delivery of learning assets as well as generation of on-demand learning activities. The term 'context' is limited at this point to the field of *Context-Aware Systems* (CAS), which play an increasing role within modern adaptive environments. The connotation of the term 'context' in association with semantic systems is defined further in this book when dealing with semantic modelling of knowledge (see section 4.3.8).

Within the scope of CASs, a context describes the situational state of system users regarding the devices they use to interact with the system and the environment they are at a given moment in time (see e.g. [Kinshuk et al. 2001]). At present, with the penetration of mobile devices (such as PDAs, notebooks or smart phones) in everyday life, pervasive systems are becoming increasingly popular. *Pervasive*, a term firstly introduced by Mark Weiser in 1991 (see [Weiser 1991]), describes the fact of seamlessly integrating electronic devices into users' everyday life. Pervasive systems are also called *ubiquitous*. Thus, following [Schilit et al. 1994], the term *context-aware* is defined as

"systems [that] adapt according to the location of use, the collection of nearby people, hosts, and accessible devices, as well as to changes to such things over time".

Hence, the notion of *everyday computing*, as stated in [Abowd et al. 2000], connotes a paradigm of usage in which users interact constantly

(in an implicit or explicit manner) with pervasive (ubiquitous) devices based on ever-changing user contexts.

Further, taking into account the present role of CASs, adaptive e-learning systems should not only *personalise course content according to users' traits* but also (form a more general point of view) *adapt their interaction interfaces to the device 'in context'*. For additional different definitions of the term context, refer to [Kinshuk et al. 2001]. Though, a last remark within this scope. [Schmidt 2005] makes a distinction between the following types of learning processes regarding knowledge management and e-learning within context-aware corporate learning systems: *course-steered*, *self-steered* and *context-steered* learning. From this perspective, personal competencies represent the semantic glue between the learning resources and the user's context.

So far, several terminological issues, some learning and teaching schools as well as technologies and systems have been presented. The next section summarises the most relevant findings.

2.3 Summary and Conclusions

This chapter aimed at presenting a review on Adaptive E-Learning (AEL) systems from a *general point of view*. Three questions have been presented in the introductory part: What means 'Adaptive' and what means 'E-Learning'? Is AEL just a buzzword? What is new in AEL?

In order to clarify denotations and connotations related with these terms, sub-chapter 2.1 treated the concept of 'adaptation' in a general context, section 2.2.3 focused on 'e-learning', and section 2.2.4 investigated if there are other fields with the same or similar aims as AEL. Setting the focus of investigation on the last decade leads to the following conclusions: (a) Adaptive Hypermedia Systems represent the 'guiding line' for current AEL systems, but (b) the concept of adaptation is not really 'tangible'. Particularly regarding the second point, literature shows some weak points in the utilisation of the term 'adaptation'. Therefore, a journey back to the roots of the topic has been undertaken in order to (a) examine the validity and utilisation of terminology at present, and (b) investigate the novelties and benefits (if any) of currently used technologies.

In concrete, the last decade indicates that there exist two types of adaptation-pertinent systems: *adaptive* and *adaptable systems*. In the former, adaptations are initiated by the system itself without user's intervention, whereas if the system adapts its behaviour according to parameters that were changed directly by the user, it is adaptable. Following this finding, it could be then stated that an adaptable system is just 'configurable' by the user and an adaptive one is something like 'self-configurable'. Section 2.1 in this chapter has shown that the topic of adaptation is more complex:

- First, an adaptive system adapts *towards different targets in the environment*, not solely to users (the next chapter shows the difference).

- Next, the terms 'adaptation' and 'adaptive system' implies issues, such as *sensitivity, responsiveness, stability, fitness* or *predictability*, which are relevant to highly relevant to measure the success of the system.

- Further, well-established research fields outside the scope of adaptive e-learning systems (but dealing with the broader scope of adaptive systems) deal with aspects, such as *generation procedures, population of adaptable entities, self-organisation, feedback, content-based adaptation, reception zone*, or *adaptive control*, showing e.g. that adaptation implies more than solely the selection of one in a set of adaptable entities.

- Adaptive systems comprise *generative tasks* that are able to dynamically create new solutions in order to ensure the fitness of the system within its stimulative environment. This aspect requires the incorporation of sophisticated mechanisms in the system, as done through techniques from the field of Artificial Intelligence in e.g. Intelligent Tutoring Systems (long before adaptive e-learning was born).

These aspects have not been deeply treated during the last decade within the context of adaptive e-learning. Section 2.2 shows further that *user-adapting systems for instruction* are known since the mid of last century, but not as 'adaptive e-learning systems'. Consequently, the objectives of adaptive e-learning have their origin in longer existing disciplines. From the didactical and psychological point of view, the study of these 'older' fields has already generated successful results (which have been often overlooked or ignored since the last decade).

According to [Weibelzahl 2002], the *evaluation of current adaptive systems* represents a research and technological challenge. For example, the lack of consistency in adaptational procedures is a major threat to usability issues ([Benyon 1993], [Edmonds 1987], [Thimbleby 1990], [Woods 1993]). Weibelzahl brings this aspect to a practical context with the following scenario: if an adaptive system modifies its behaviour over a period of time (e.g. internally by changing parameters in user models or externally by tailoring the user interface) but is not able to reproduce exactly the adaptational procedures undertaken in the past, then this lack of controlled flexibility in the system will be the price that a user must pay. So why do current systems not cover issues like 'stability of responsiveness' or 'self-regulated control of generation procedures'?

Within this context, a challenge is given by the need for developing software systems that permit a self-adaptability of their components (in order to promote e.g. self-regulated fitness) as well as an evaluation of their adaptive behaviour (in order to achieve e.g. self-feedback for optimisation of adaptivity). The implementation of adaptive behaviour should not be reduced to a static internal functionality. This means that setting the focus firstly on the adaptable design of the entire system and then on the adaptive behaviour of parts of it, could foster and support the efforts of moving towards efficient solutions of *longer-living adaptive systems*.

Regarding the question *Is AEL just a buzzword?*.. well, it depends, because what is a buzzword? If buzzwords are *"stock phrases that have become nonsense through endless repetition"* [WordNet 2007], then the term Adaptive E-Learning may a new word, but not a buzzword. If buzzword refers to a term that *"has become fashionable and popular"* [Oxford 2007] or is a *"voguish"* [Merrian-Webster 2007] word, then AEL is one. Though, more relevant is the fact that in the scientific communities a buzzword arises often to describe a new concept that denotes the same as already existing concepts, but embraces more complex or newer connotations. In this latter case: *yes, adaptive e-learning is a buzzword and it is ok*, because as shown in section 2.1, it can be seen as 'the' new term that integrates all features of many older disciplines. In essence, with the 'boom' of the Internet (and implicitly, of the Web) in the last decade, AEL represents the point of convergence of longer existing disciplines.

Regarding the question *What is new?* it can be stated that, indeed, there exist novelties regarding the concept. Consider the following two observations regarding 'distance learning' and 'ICT delivers learning'. Firstly, [Nirmalani and Stock 2003] stipulated in relation with distance learning:

> "*Much of the early research in distance education since the 1960s has focused on comparisons between delivery media such as television, video, or computer and traditional face-to-face teaching. Other research compared the effectiveness of one distance delivery medium over another. [...] Recent trends in distance education research indicate a preponderance of studies focused on understanding pedagogical issues in the computer-mediated communication (CMC) environment. [...] What is of significance is that new methods are being explored for understanding interaction and the learning process [...]*"

And second, according to [Hasebrook and Maurer 2004], the following benefits of using *Information and Communication Technologies* (ICT) to deliver learning have to be underlined:

> "*the choice of learning style; customized and personalized learning materials and services; individualized tracking and recording of learning processes; self-assessment and monitoring of learner performance; interactive communications between participants and influences in the learning process; interactive access to educational resources*".

Both statements have been made in the era of AEL. For the context at hand, the key thoughts behind the aforementioned statements are *media delivery, understanding of interaction* and *understanding of learning processes*. These key thoughts

establish the novelties regarding the technological streams of the last decades. As shown at the end of section 2.2.4, emerging technologies on media and modern interaction devices make possible to enhance the knowledge transfer process (e.g. new user-tailoring methods for content delivery, improved tracking of learning behaviour and performance, efficient collaboration tools) as well as to move towards a really transparent access to distributed educational resources (e.g. 'location transparency', whereby a user accesses resources and it does not matter where they are).

To conclude this chapter, a last aspect ought to be clarified: *where is the technique or system that shows the success of AEL, if any?* Indeed, there seems to be one field that contributed with a special system. According to [Lockee et al. 2003], Programmed Instruction (PI) was the first empirically determined field that played an important role in the convergence of computer and educational sciences. As indicated in the previous section, PI follows micro-adaptive techniques utilising on-task measurements for adaptations. As stated in [Lockee et al. 2003], perhaps the most productive and longer standing example of computer-based PI is the PLATO (*Programmed Logic for Automatic Teaching Operation*) system (see also [Denenberg 1978]). Further, [Lockee et al. 2003] states that

> "*the revolution of the desktop computer, coupled with the diffusion of the Internet on a global scale, has provided access to unlimited learning resources and programs through distance education*".

PLATO was created as a *Programmed Instruction Project* at the University of Illinois in the 1950s and has been developed and exploited since then following the continuous evolution of the computer and offering a range of computer-based curriculum that is incomparable, state [Lockee et al. 2003]. PLATO has evolved into a Web-based learning system, which offers a variety of instructional programs to learners of all ages and distinct life experiences. According to [Foshay 1998], although the primary design goal behind PLATO has changed to suit more the constructivist philosophy, it still maintains some of its micro-adaptive foundations (following the ideas of PI). Further, [Lockee et al. 2003] claim that, while other networked system types have expanded to support a great variety of instructional theories, PI still represents an effective and empirically validated possibility for the design and implementation of mediated instruction.

Against the conceptual background given in this chapter, the next one examines the topic of personalisation from a technological point of view.

3 Personalisation and User Modelling

The environment is everything that isn't me.

(Albert Einstein)

This chapter focuses on both main topics of this book: *Personalisation in Adaptive E-Learning* and *Multi-purpose User Modelling*. As shown in the previous chapter, (a) the general conceptual basis for the topic of personalisation-pertinent systems is given by the field of adaptive systems, and thus, if focusing on the field of e-learning, then the specific conceptual basis for personalisation and learner modelling is represented by the field of adaptive e-learning, (b) adaptive systems consist of several specialised components (e.g. an adaptive control engine, distinct sensors, feedback chains as well as modelling components for knowledge domains, instructional process, electronic devices or 'users'), and therefore, (c) one of the main functions for user-adapting systems is defined by its capacity of knowing its environment, i.e. *user-adapting systems need an internal representation of those entities in its environment that provide (directly or indirectly) relevant stimuli as triggers for adaptational procedures.* The present chapter covers the technological background of the three aforementioned aspects, but focuses on the third one. In the first part (sub-chapter 3.1), while the topic of personalisation is investigated in the overall context of adaptive e-learning systems, the focus is maintained on the role of user modelling components in such systems. This first part comprises (a) the presentation of main aspects regarding the distinctions between adaptation and personalisation, (b) an overview into the applicability of adaptive e-learning, (c) the analysis of the most relevant components and functions in existing models and architectures, and (d) an overview on benefits and success stories of applying adaptive e-learning. The second part of this chapter (sub-chapter 3.2) concentrates on the topics of User Profiling and User Modelling. It includes (a) differences between both concepts, (b) basic principles in context, (c) available techniques, and (d) a review on existing solutions. Due to the fact that the methods and goals of user modelling components are equally applicable in different areas, the general field of User Profiling and Modelling as well as the specific field of Learner Modelling are investigated. This chapter shows that design and implementation of user modelling components for adaptive e-learning systems imply the fulfilment of distinct requirements from additional components in the system, i.e. the set of functions in user modelling components depends on the variety of adaptational purposes at hand.

3.1 Personalisation in Adaptive E-Learning

From a general point of view, the terms *Adaptive Systems* and *Personalisation* represent two relevant concepts in several research fields as well as for the design and development of commercial products. From the point of view of Computer Science, and according to [García-Barrios et al. 2005], the topicality of both terms can be explained as a consequence of the impulse given by the broad dissemination of the World Wide Web (WWW) during the last decades.

For example, although the Web has been the 'Big Thing' since the beginning of the 1990's, still at the turn of the century it was difficult to find accurate information in an ocean of billions of Web pages. As stated in [deBra et al. 2004], adaptive hypermedia architectures should lead to *adaptive Web-based systems* as the 'Next Big Thing'. And indeed, at present personalised search services like Google[7] or Yahoo[8] dominate the market of adaptive Web-based systems. Fur-

---

[7] http://www.google.com/ig, http://mail.google.com
[8] http://my.yahoo.com, http://groups.yahoo.com

ther, almost all users of adaptive systems are conscious of the increasing amount of well-tailored information they may access for their particular needs. Moreover, they are also aware of the fact that, in the majority of cases, they have to pay the price or hazard the consequences of delivering personal data to ensure those services.

In order to corroborate these statements, just consider that in accordance with several surveys, such as [ChoiceStream 2004], most of the users are willing to do so. Specifically, the high value of personalisation and adaptive systems can be identified in various application fields, as for example in e-commerce and adaptive e-learning ([Hof et al. 1998], [Kobsa 2001], [Cooperstein et al. 1999]). No doubt, rather than a *trend*, there is a *need* for personalisation-pertinent systems in several situations of modern life.

This section addresses the relationships and distinctions between the terms *adaptation* and *personalisation*, from a terminological as well as from a practical point of view. It gives an overview on some general models and systems, whereby the attention is concentrated on the functions and methods of user modelling components. The aim is to bridge the gap between a general view on adaptive e-learning systems and a specific view on user modelling components and techniques.

3.1.1 A User-centred Point of View

As mentioned and showed in the previous chapter, the topic of user-adapting systems (i.e. the field embracing the general concepts of *Adaptation* and *Personalisation*) is neither a new technique in the field of Computer Science nor a new issue of research in other fields (e.g. Cybernetics, Evolutionary Research, Biology or Climatology). In the last two decades, extensive research work took place to investigate critical issues regarding both topics. And within that context, also some new definitions arose.

Among others, researchers have been contributing to (re)define terms in different ways, sometimes as an attempt to reinvent the wheel or just to give the impression of talking about something different or innovative. For example, consider 'relatively new' software solutions of adaptive systems for which the terms 'customisation or personalisation' are used, respectively, with

the same denotation and connotation as 'adaptability or adaptivity'.

In order to better expose the aim of this chapter consider a first attempt towards a (deliberately abstract) formulation for shifting from the general view on adaptation (as depicted in the previous chapter) to the specific topic of personalisation [García-Barrios et al. 2005]: *as a response to some environmental 'stimuli', a system 'alters something' in such a way that the 'result' of the alteration corresponds to the 'most suitable solution' in order to fulfil some 'specific needs'*. Within the context of this book, the specific needs relate to the *goals of the person interacting with the system*.

[García-Barrios et al. 2005] argue that this 'first general formulation' applies to both, adaptation-pertinent and personalisation-pertinent systems, as it describes at the same time their general purpose. Hence, the need arises to find a model, which meets the following three purposes (which in turn represent the motivation for this chapter):

- semantically establish a terminological distinction between the topics,

- systematically describe both topics, and

- methodically define participatory components, functions and interactions.

The first purpose is met - according to the findings of the previous chapter, and within the context of computer-based adaptive systems - through the ensuing definition:

Personalisation is an adaptation towards a known system user and triggered off by a certain stimulus related to a specific individual goal; thereby, the adaptor (adaptive component) alters some known adaptable entity or entities and must interact either with the user directly or with the user's internal representation (user model) in order to generate the best suitable solution to satisfy a user's need or interaction goal.

This connotation is sufficient for the scope of this book and leads, among others, to the following main characteristics of personalisation-pertinent systems [García-Barrios et al. 2005]:

- Personalisation processes imply system's knowledge about the specific users interacting with it.

- A component responsible of controlling adaptational procedures is needed. This component is usually known as adaptive or personalisation engine.

- The adaptive capacity of the system depends on the set of adaptable entities, which in turn defines the adaptability of the system.

- A personalisation process begins with a stimulus and ends with a best-suitable response.

Considering the point of view depicted so far and taking into account the realisation of computer-based user-adapting systems, the following different types of personalisation can be identified [García-Barrios et al. 2005]:

- explicit or implicit,

- perceivable or hidden,

- predictive or deterministic,

- controlled or uncontrolled, and

- individual or stereotyped.

An *explicit personalisation* takes place when an adaptation is undertaken according to the parameters defined in a concrete internal representation of system users (i.e. with the direct aid of 'user models'). For example, a user-adapting e-learning system provides explicitly personalised learning content by using a specific learner model, as in the case of delivering more textual or more graphical content according to the cognitive style of the learner interacting with the system. As an opposite to this type of personalisation, an *implicit personalisation* is about adaptation resulting from a certain context (i.e. environmental situation) and without using a user model, e.g. when adapting the content's layout to the display options of an output device.

Personalisation is called *perceivable*, if a user recognises the result of user-adapting process. For instance, a user-adapting e-learning system may show, hide or modify control elements on the user interface. Specifically, consider the case of forcing the visual presence of a tree-view showing the course structure to users internally modelled as wholists[9]. *Hidden personalisation*, on the other side, does not affect the user interface or the presented content at all, e.g. if it is intended to update or optimise a user model according to some internal adaptation rule. Thus, the user may not 'see' the result of this personalisation step.

Predictive personalisation comprises some 'work in advance'. For example if the best instructions to go on through the course contents are generated for a learner, this personalisation step is not only hidden, but also predictive (this example refers to the technique of path-prediction in the field of Artificial Intelligence, as shown later in this chapter, section 3.1.2). Also in association with a time-dependent adaptational computation and response, *deterministic personalisation* takes place within the current adaptation process, e.g. if the system chooses or calculates a personalised instruction and this is aggregated and displayed 'right now', i.e. as an immediate response to the user interaction.

Controlled personalisation puts the user in the centre of all considerations and so (a) makes the system scrutable for a user as well as (b) describes also the idea that the user may take control of adaptation processes at any time. Consider for example the case for which a user-adapting e-learning system can not decide the aggregation of a certain instruction and thus, the learner is asked to choose the appropriate instruction among a set of delivered (known) suggestions. In contrary to this aspect, *uncontrolled personalisation* does not allow (or even does not necessitate) the user to influence any adaptational issues, such as parameters, initiations, decisions, executions, optimisations or results.

Finally, on the one hand, a so called *individual personalisation* comprises a personalisation procedure towards 'one specific person'. An on the other hand, *stereotyped personalisation* allows 'personalisation towards groups or anonymous users', e.g. an adaptation towards the 'customers' of a company aims at 'all users known as customers'. Other valid denotations for this last type of personalisation could be *group-based* or *role-based personalisation*.

---

[9] The term 'wholist' describes a specific cognitive trait of users that need always an overview of the information space (for details on this cognitive model please refer to section 5.1.2).

According to [Feeney and Hood 1977], a major theme running through their experiments during the 1970s was that *'people are different'*. People, specifically computer users, think distinctly and react differently to similar informational stimuli. Thus, Feeney and Hood claimed that an information system, which is able to deal with its users on an individual basis (or treating them as members of small populations), can be said to be truly adaptive. Approximately since that time, the terms related with the stem *adapt* have been used to express any adaptation process towards the needs and traits of computer users.

As shown in this section, the author of this book aims at distinguishing between the cocepts of *personalisation* (meaning a specific adaptation based on traits directly related to the user) and *general adaptation* (meaning an adaptation towards other entities in the environment of the adaptive system, which do not pertain to the user). Further, taken for granted that always 'a person interacts with a computer', the term 'implicit personalisation' has been introduced in this section and refers to those adaptations resulting from user-independent stimuli, usually caused by additional sensors coupled with the interacting devices. As an example for this last observation consider modern computers, such as the 2007 versions of Apple's MacBook[10] laptops, which adapt automatically the brightness of their display according to the stimuli from their integrated ambient light sensors.

For the reminder of this chapter, the scope of investigation is mostly focused on (but not restricted to) adaptation-based techniques and their applicability in explicit, perceivable, deterministic, controlled and individual personalisation within adaptive e-learning systems. Therefore, in order to prepare the transition to the technological aspects of user modelling in such systems, the next section gives a more tangible 'big picture' of the topic through the presentation of practical scenarios.

3.1.2 Applying Adaptive E-Learning

Although some models and systems are mentioned in this section, the aim is not to explain their technical details. Instead, the goal is to give an insight into some of the main practical uses, methods and purposes of well-known solutions.

As stipulated in [Brusilovsky 2001], the notorious growth of research activity in the field of Adaptive Hypermedia Systems (AHSs) since 1996 has two main reasons. Firstly, the continuous and rapid increase of the use of the World Wide Web (WWW). And second, the number of concrete achievements and the consolidation of research experience in the parent fields of AHSs, which are Hypertext and User Modelling according to Brusilovsky. Further, [Brusilovsky 2001] identifies the following six main types of AHSs:

> *educational hypermedia, on-line information systems, on-line help systems, information retrieval hypermedia, institutional hypermedia, and systems for managing personalized views in information spaces. [Brusilovsky 2001]*

To give a concrete view on sample applications of modern adaptive e-learning systems, and under the consideration that AHSs represent one of the most relevant research areas at present, this section focuses on Adaptive Educational Hypermedia Systems (AEHSs). A simple classification schema for adaptive methods, as shown in [Specht and Burgos 2006], may consist of three basic dimensions that describe..

(i) the adaptable entities, such as learning content, media selection, instructional paths or annotations,

(ii) the underlying learner-related personal traits, like previous knowledge, individual preferences or interests, and

(iii) the reasons for personalisation, e.g. an underlying preference model, compensation of knowledge deficits or increasing ergonomic efficiency.

Respectively, such a schema can provide answers to the following three questions:

(i) Which parts of an educational system or process are personalisable?

(ii) Which features does the system need and use to personalise?

(iii) Which are the reasons for the system to trigger and execute a personalisation?

---

[10] See e.g. http://www.apple.com for details.

Adding some examples for adaptive methods to Specht and Burgos' three dimensional schema leads to a specific classification, such as the one depicted in Table 3.1 (see next page), where different examples for each dimension are grouped within four categories: adaptive sequencing, incremental interfaces, adaptive presentation and adaptive navigation support. The latter two categories represent the two main categories in Brusilovsky's taxonomy of AH technologies (see Figure 2.2 in section 2.2.4).

So far, it is identifiable that whenever dealing with the topic of A(E)HSs, the focus is set on the user-tailored transfer of hypermedia in Web-based environments. Considering that this book concentrates on the knowledge transfer process (which is a bidirectional relation consisting of teaching as well as learning streams) the practical application of Adaptive E-Learning (AEL) can be explained as follows. Starting from AEHS, the concept of hypermedia builds the central entity for the adaptability of a system, and consequently, it has direct and high influence on its adaptive capacity. Based on the definitions from the parents of Hypertext (see e.g. [Bush 1945] or [Nelson 1974]), Nora Koch describes the concept of Hypermedia as a

"network of nodes connected by links. A node is a unit that contains text and/or multimedia elements [...]. A link is usually directed and connects two nodes [...]. A link is associated with a specific part of the content of the source node [...] called an anchor. It is the linking capability which allows a non-linear organisation of the text or multimedia content". [Koch 2000]

Hypermedia[11], if utilised to build a coherent interconnected structure, defines an explorable information space (*Hyperspace*) from which knowledge assets can be extracted. Thus, the power of the usage of hypermedia in AEHSs has to do with the way it enables to publish learning content on the Web. Despite of the distinctions made in [Nelson 1970] between hypertext, hypergrams, hypermaps and other related notions, consider the following way in which Nelson describes the usage of 'discrete hypertext' and compares it to writing on paper:

"Hypertext means forms of writing which branch or perform on request [...]. In ordinary writing the author may break sequence for footnotes or insets, but the use of print on paper makes some basic sequence essential. The computer display screen, however, permits footnotes on footnotes on footnotes, and pathways of any structure the author wants to create. Discrete [...] hypertexts consist of separate pieces of text connected by links. [...] An asterisk or other key in the text means not an ordinary footnote, but a jump—to an entirely new presentation on the screen. Such jumpable interconnections become part of the writing, entering into the prose medium itself as a new way to provide explanations and details to the seeker." [Nelson 1970]

Based on these findings stated, the practical application of AEHSs depends on the way in which didactical principles can be used on hyperspaces. This simple (but original, and thus, correct) point of view allows then to state that both, hypernodes and hyperlinks represent the central personalisable entities in AEHSs. And precisely these two entities are the basis of all subsequent levels of adaptation, as also identified in [Brusilovsky 2001] regarding AHSs: adaptive presentation (i.e. personalising the content of hypernodes) and adaptive navigation support (i.e. personalising structures through hyperlinks).

Furthermore, the function of AEHSs is based on a recurrent two-step principle:

(i) 'calculate the next hypernode according to the hyperlink behind the anchor activated by the user', and

(ii) 'personalise the presentation of this next hypernode as well as the presentation of its corresponding hyperlinks'.

Explicitly, these two steps comprise the overall functionality of such systems, i.e. how they are implemented in order to satisfy the teaching side of the knowledge transfer process. But implicitly, due to the fact that AEHSs are interactive systems, the learning side of the process is also satisfied, through the interactions of learners, i.e. 'responses and requests' by activating hyper-

---

[11] For further definitions of Hypermedia refer to e.g. [Lennon and Maurer 1994] or [Wardrip-Fruin 2004].

links. Moreover, these two ('hypermedia-based') steps allow specific points of view that provide specific answers to the main questions around the topic of this section: *Applying AEL*. These main questions are derived from Specht and Burgos's classification of adaptive methods in AEHS (see Table 3.1) and concentrate on the context of personalisation: What to adapt? Adapt to what? Why adapt? How to adapt?

Hence, starting with a simple taxonomical view on 'what can be personalised?', the first level is defined by *presentation of media*, and in turn, the next level consist of the adaptable entities derived from media, *content nodes* and *node linkages*.

The next point of view regards 'how to adapt?' and does not focus on computer-aided human strategies, such as questionnaires, behaviour tracking or diagnostics. Rather, the technological possibilities for interactions in user interfaces (UIs) are meant, because they define the limitations of adapting the presentation of media and provide the aids for the generation of the above mentioned computerised human-techniques.

Based on this, and coming back to the context of Web-based AEHSs, an additional possibility is given to enable personalisation: interactive UI components, also called 'widgets', such as buttons, alert and prompt boxes, menus and so forth. These components can be personalised not only by placing them at a certain position and changing its graphical presentation, they can also be adapted to activate navigational events (in analogy to hyperlinks) or as '*process commands*'. This last aspect is possible because a UI widget consists always of two parts, a human-readable and a machine-readable one; respectively, these parts represent layout and behaviour.

In [Koch 2000] three main issues on AHSs are mentioned: content, structure and presentation. Regarding presentation, [Koch 2000] states that it is related to the layout of content (i.e. visualisation) as well as to the description of the *interactive elements*. According to Nora de Koch, interactive elements enable (a) to access other elements, i.e. navigation support, (b) to show passive elements, i.e. display hidden elements, or (c) to control non-static elements, e.g. activate multimedia.

Further, the technological possibilities behind the presentation of media are also a matter of *composition granularity*, especially within the context of Web-based AEHSs. In general, interfaces between humans and machines are defined by input and output devices. In terms of UIs, a computer display screen has always represented the mainly utilised interaction device. The notion of computer display screen, as also used in [Nelson 1970] (see previous page), is intentionally used as basis in the following observations, because talking about Web pages or browser windows may blur the meaning of *how to adapt 'what' is delivered to the user*.

In essence, one hyperlink leads to one hyper-

What?	To what?	Why?	How?
Adaptive Sequencing			
Content, learning activities	Learner tested knowledge, navigation history	Compensation of deficits, encouraging	Tests, tracking
Incremental Interface			
Interface complexity, number of functionalities	Tasks, skills, domain knowledge	Usability	User tracking, questionnaires
Adaptive Presentation			
Selection of media	Knowledge, preferences, goals	Compensation of deficits	Diagnostic
Adaptive navigation support			
Hyperlinks, restriction of navigational freedom	Knowledge, background, preferences	Adaptation to zone of proximal development	Diagnostics, tests

Table 3.1: Examples of Classified Adaptive Methods [Specht and Burgos 2006]

node, and usually, one hypernode stands for one Web page, which represents the content of the hypernode. Nonetheless, for example the increasing use HTML frame sets, inline frames or innerHTML[12], and the possibilities offered by DOM[13] changing techniques in combination with the trendy utilisation of multiple hidden requests to interchange parts of the currently displayed content (AJAX[14]), make it difficult for a Web user to perceive the boundaries within single Web pages (i.e. single hypernodes). Therefore the notion of *Web screen* is used in this book within the context of A(E)HSs and enables to define the previously mentioned 'composition granularity' as the degree to which one Web screen contains hypermedia elements. As a consequence of this issue, firstly, the 'what to adapt?' taxonomy must be complemented by the *interaction through process commands* as well as by the *composition of hyperspaces and Web screens*.

Finally, to meet the aims of this section, the findings and statements given so far (valid in principle for AEHSs) can be translated to the context of AEL and lead to the graph shown in Figure 3.1, which describes the application of AEL systems by means of a <u>S</u>imple Taxonomy for <u>P</u>ersonalisation <u>i</u>n Adaptive <u>E</u>-<u>L</u>earning Systems (SPIEL). Three categories of 'what to adapt' can be identified and are explained as follows:

- Presentation: This notion is derived from the models of A(E)HSs presented in this section, concerning hypernodes and hyperlinks. In AEL systems the nodes are usually called Learning Objects (see e.g. [Atif et al. 2003]), and the linkages refer to intra- and inter-object relationships.

- Interaction: Interaction represents the means to adapt access, presence and activation of

media (like interactive elements in Koch's model), as well as to trigger the behaviour of system components through embedded process commands. Thus, a process command is a piece of software that is able to change the behaviour of both, UI elements and tasks of the adaptive engine.

- Composition: In the same way in which hyperspaces are composable in AHSs, nodes and linkages enable the generation and manipulation of coherent AEL spaces. From the point of view on user interfaces, one or more portions of these spaces can be composed and transferred to the screen of the user. In general, composition comprises the structuring of nodes, linkages, hyperspaces, interactive elements and process commands on the user's screen. In terms of general adaptation, this category comprises the scalability of

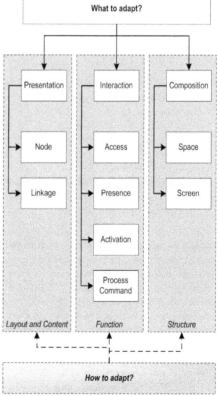

Figure 3.1: <u>S</u>imple Taxonomy for <u>P</u>ersonalisation <u>i</u>n Adaptive <u>E</u>-<u>L</u>earning Systems (SPIEL).

---

[12] For details on the HyperText Markup Language, the "publishing language of the World Wide Web", see *http://www.w3.org/html* or *http://www.w3.org/markup*
[13] See *http://www.w3.org/DOM* for details on the Document Object Model, a platform- and language-independent interface to modify content, structure and style of Web documents.
[14] For details on AJAX, shorthand for *Asynchronous JavaScript + XML* (first coined by J.J. Garrett, 2005), see *http://adaptivepath.com/publications/essays/archives/00038 5.php*

adaptable structures, and thus, includes 'decomposition'.

The SPIEL model illustrated in Figure 3.1 also comprises Specht and Burgos' dimension of 'how to adapt' but from the technological point of view. Thus, personalisation can be achieved in AEL systems through the targeted modification of the following aspects:

- Layout and content, in order to personalise the presentation of media.

- Function, i.e. tailoring program pieces or machine-readable content so that the tasks (behaviours) behind each expected user interaction can be altered. Though, the notion of adapting through functions has many sides, because the resulting changes from machine-readable code can lead to a further adaptation of media presentations, of compositions (and decompositions), of interactive elements, and also of the behaviour of other system components (such as rules in the adaptive engine or states in user models). Hence, this way of personalisation is very powerful, but it depends highly on which possibilities are given by the underlying technologies.

- Structure, in order to define and administer data structures. This notion comprises two points of view. On the one hand, for the UI it expresses the way in which a screen is composed (or parts of it). On the other side, it comprises the possibilities given by AEL systems to generate or adapt the description of data structures, such as models (usually graphs) for domain knowledge or interrelated chunks of user information.

As already mentioned, 'how to adapt' is related with technologies, and thus, with the techniques used to start, process and end an adaptational procedure. A step towards this technological point of view is given e.g. by P. Brusilovsky's taxonomy of adaptive technologies (see Figure 2.2 in section 2.2.4), or by the techniques describing the main models used to categorise adaptive systems in section 2.2.4 (ATI, MicroAd and MacroAd approaches). A concrete and complete answer to this question is beyond the scope of this book, because it implies a deep study of all existing processes concerning the behaviour of

adaptive engines. Though, the main issues on this topic are included in this document through the examination of adaptive models and systems (see next section) and through the techniques of User Profiling and Modelling (see next sub-chapter 3.2).

For the remaining two dimensions, regarding 'adapt to what?' and 'why to adapt?' the following can be stated. First, 'adapt to what?' involves that the system knows its environment, i.e. it depends on the functionality and sensitivity of its sensoring components. Sensoring components are mainly input devices that are specialised to collect data about environmental entities, such as user-related data as well as surrounding 'context data' (see e.g. the topic of context-aware systems introduced in the previous chapter).

As already explained in the previous chapter, this collected data about the environment can be stored and interpreted in internal modelling components. Due to the fact that this section focuses on explicit, perceivable, deterministic, controlled and individual personalisation, the answer to 'adapt to what?' is the set of all learner-related data that is available from the User Modelling Component (UMC) of the system.

To give some examples, based on [Eklund and Zeiliger 1996] and [Brusilovsy 1996a], the following main categories of what is represented in user models are identifiable: user *goals*, user *tasks*, user *preferences*, user *interests*, user *knowledge* (as part of the domain knowledge presented in media), user *background knowledge* (as portion of the general domain knowledge), and user *experience* (regarding e.g. the use of the system of the media).

Speaking concretely about AEL, 'adapt to what?' is tightly related with 'why to adapt?' and therefore, 'what' is hold in the UMC depends on the 'purposes' of user modelling. According to [Koch 2000], some examples of these purposes are to provide..

- topic-based assistance of users while learning,

- learner-tailored information,

- adaptation of the personal UI,

- provision of help to find information,

- feedback about personal knowledge progress,

- support of cooperative work, and

- assistance in using the system.

These purposes, despite of shifting the point of view from the dimension 'adapt to what?' to 'why to adapt', imply both, learning and teaching goals in knowledge transfer processes. The area of didactical goals behind AEL systems was already introduced through the presentation of teaching and learning styles in the sections 2.2.2 and 2.2.4. Hence, on the one hand, 'why to adapt' is tightly coupled with the didactical strategies of the teachers (which are mostly coded as rules in the adaptive engine of AEL systems), and on the other hand, 'why to adapt' depends on the information hold in the UMC.

In general, applying AEL is a combination of the use of the four described dimensions[15] in order to improve the knowledge transfer process. In other words, to apply AEL involves the attempt to personalise and improve learning processes for learners through the available didactical means provided by the system to the teachers. Regarding learning content, AEL is then the set of technology-based aids for teachers in order to personalise didactical media and tasks for 'online' learners.

Yet, a transition is now needed from viewing on the applicability of AEL to a point of view on the architecture and components of existing solutions, as given in the following section. After the investigation of AEL models and systems, the focus can be set on the specific techniques of UMCs, which are presented in sub-chapter 3.2.

3.1.3 Identifying Components, Functions and Interactions

The title of this section already depicts its aim, namely to identify the most relevant components of adaptive e-learning (AEL) systems as well as their functions and the way they interact with each other. Thereby, the focus is set on the investigation of theoretical as well as process-driven and architectural models. An additional aim is to meet the remaining requirements on the necessity of finding a general personalisation model[16], which should clarify the technical background of AEL systems. To retain the methodology of the previous chapter, firstly, a perspective on general adaptive systems is given and then, the scope is shifted and focused on the topic of AEL systems.

As stated in section 2.1.3, in 1962, John Holland defined a full mathematical theory for adaptation-pertinent automata, which describes in detail the functions and interactions of the main participatory components in General Adaptive Systems (GASs[17]). Particularly remarkable in this theory is the fact of taking the natural *generic character* of GASs for granted. Firstly is to be mentioned that according to [Holland 1962], an adaptive system consists of specialised procedures ('generation procedures') that are able to generate specialised adaptation resolvers ('population of programs'). Further, in Holland's theory an adaptive system is able to alter generation procedures, which in turn corresponds to successive alterations in the population of resolvers.

Hence, because these resolvers act upon changes in the environment in order to produce successful adaptational solutions, the environment of the system can be treated as a 'population of problems' and allows to arbitrarily replace individuals and populations everywhere within the theory. Another highly important issue is that Holland's theory considers also success and optimisation measures through the assignment of a numerical reward (also called 'activation') to a generation procedure according to its produced solution to given problem. Thereby, given that different systems can produce distinct solutions to the same problem, the efficiency of a set of systems can be compared and evaluated. [Holland 1962]

As will be shown along the reminder of this section, most adaptive systems at present do not comprise this generic nature, as it was formulated previously. Still, the problem is not only the difficulty to inject a 'generic intelligence' into the system, rather the universality and generality of Holland's theory seems to be hard to put in prac-

---

[15] (i) What can be adapted? (ii) Adapt to what? (iii) How to adapt? and (iv) Why to adapt?

[16] See fourth paragraph in section 3.1.1.
[17] The GAS acronym is introduced here to emphasise on a general point of view on adaptive systems.

tice. Nevertheless, [Holland 1962] also identifies this problem and distinguishes between complete and incomplete models of his theory. Complete models correspond to artificial systems where the 'rules of the game' are limited, whereas the properties and specifications of incomplete models (such as natural systems) comprise an unlimited number of factors, and thus, the theory can not handle all of them. The following paragraphs will clarify the meaning and intention of this observation.

Based on the statements depicted so far, [Holland 1962] also indicates that

> "following Turing, the set of all effectively defined procedures will be identified with the set of all programs of some suitably specified universal computer".

And thus, when focusing on 'unrestricted adaptability', i.e. when it is assumed that nothing is known about the environment, an adaptive system should

> "be able initially to generate any of the programs of some universal computer. The process of adaptation can then be viewed as a modification of the generation process as information about the environment accumulates. This suggests that adaptive systems be studied in terms of associated classes of generation procedures - the associated class in each case being the repertoire of the adaptive system".
> [Holland 1962]

Traducing both observations into the present and taking into account the findings from the previous chapter, this first general point of view allows the following observation: modern adaptive e-learning systems have a 'limited' behaviour because of their specialised adaptive repertory. As will be shown in this section, they behave mostly 'selective' and do not have a generic character, because they calculate or choose the most suitable solution (if any) within their repertory and do rarely generate new ones.

As such, using Holland's terminology, they are explained with the aid of complete models with already known rules. Nonetheless, given the fact that some functions in modern adaptive e-learning systems have been directly influenced by the field of AI (*Artificial Intelligence*, see section 2.1), some system units show the ability to

self-organise themselves following an adaptive plan in order to find the most suitable solutions for personalisation problems.

Hence, in analogy to *self-organising systems* (see e.g. [Ashby 1962], [Lendaris 1964] or [Holland 1992]) and following Holland's theory, a GAS consists of many components within the following categories: (a) *environment* of the system, (b) *adaptable structures* in the system, (c) components to *control* the adaptive plan, and (d) *measurement* of success (e.g. components to optimise the fitness or the performance of structures, or to give feedback on how well problems were solved). Though, this model does not explicitly include an internal component for modelling the environment. Implicitly, the system 'knows' the origin of environmental stimuli and learns about them through the different classes of generation procedures.

Since mid of the 1990's, two highly important process-oriented models regarding *User-Adapting Systems* (UASs) can be identified, the first defined by R. Oppermann in 1994 and the second by A. Jameson in 2001 (see [Oppermann 1994], [Jameson 2001] and [Jameson 2003]). Basically, Oppermann distinguishes in his model between afferential, inferential and efferential components. As the system is assumed to be able to observe the behaviours of its users, an *afferential component* is responsible to collect the data gained form observations. These data (e.g. mouse movements, key strokes, click interactions, commands in shells) represents then the input for adaptational processes.

Further, *inferential components* represent the 'intelligent' engine in the system and thus, they are able to infer user traits from the raw afferential data. At the end of this process chain, *efferential components* act as decision-making units that are able to control and specify how the system's behaviour should be altered in order to meet a specific adaptational goal, e.g. the activation of other internal units, some modifications in the user interfaces, the delivery of specific messages for the user, and so forth (for more details please refer to [Oppermann 1994]).

Anthony Jameson's general model for UASs, illustrated in Figure 3.2, distinguishes between two processes: one related to the acquisition and a second to the application of a *user model*. Thus, mapping Jameson's model to Oppermann's, the acquisition process is an afference and the application of the user model is an efference. In Figure 3.2, the ovals on the bottom represent input from or output to the environment to which is to adapt, the rectangles relate to both previously mentioned processes, and the cylinder on the top represents the stored data. Further, the dotted arrows represent information usage and the solid arrows the production of results. On the one hand, these dotted arrows can be seen as inferential components from Oppermann's point of view. And on the other hand, in relation with Holland's Theory, they embrace the functions of 'population of programs', the resolvers [Jameson 2003]. On another side, focusing on educational systems, it is interesting to observe that whereas many CAI models only supported controlled personalisation features until the late 1980's (see e.g. [Kalmey and Niccolai 1981]), Crowell and Traegde defined long before, in 1967, a simple interaction-based model for what they describe as a self-regulating instructional system with *hierarchically-organised control* (see Figure 3.3).

According to [Crowell and Traegde 1967], the main actors in such systems (*Teacher, Computer* and *Student*) communicate in terms of so-called *Interaction Events*, which are part of the program code and thereby tightly coupled to the predefined *Behavioural Objectives* and *Curriculum Units*. The way in which the system adapts its behaviour to the learning states of users is given by *Meta-control* and *Sub-control Strategies*. Programmed meta-control strategies assume the didactical roles of the teachers and can be super-

vised by programmed sub-control units. It is up to the teachers to take entire control of the system or to delegate the functionality to the Sub-control Strategies unit. Concluding, despite of the fact that this model does not utilise an internal learner model, it fully implements personalisation in terms of the definition given in the previous section, because it (constantly) knows about the user through the Interaction Events and their corresponding Behavioural Objectives, i.e. the system applies *implicit personalisation*.

In 1974 - and still within the context of CAI systems - John Self uses (perhaps for the first time) the notion of a *student model*. As an expert in the field of Artificial Intelligence (AI), his subsequent remarkable achievements made him the world's expert in the field of Student Modelling, later also referred to as *Learner Modelling*. In chapter 2 it has been identified that the field of ITSs, Intelligent Tutoring Systems, arose through the influences of CAI and AI also at that time (for more details on Self's work see e.g. [Self 1974], [Self 1977a], [Self 1977b] and [Self 1988]). Generally speaking, ITS models consist of three main components: *expertise module, student modelling module* and *tutoring module* (see section 2.2.4).

Nonetheless, [Brusilovsky 1994] stipulates that surveys beginning in 1984 have identified an additional component for ITS models, namely an *interface module* (see Figure 3.4 on the next page). This module is in charge of dealing with the knowledge of students' interactions, and thus embraces not only afferential and efferential

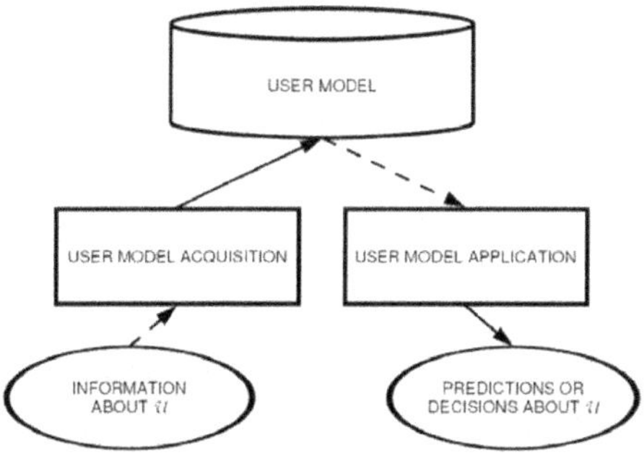

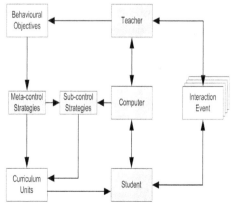

Figure 3.2: Anthony Jameson's Model for the Processing in a User-adaptive System. [Jameson 2003]

Figure 3.3: Crowell&Traegde's Model for CAI Systems. [Crowell and Traegde 1967]

functions but inferential as well. At present, due to the prevailing architecture types of component-based distributed software systems (e.g. client-server or peer-to-peer architecture models) and because of the currently most used programming paradigms (e.g. object-oriented or service-oriented), the Interface Module comprises mostly just dialoguing functions and is called *User Interface* (see Figure 3.5).

As shown in [Martens 2003], modern ITSs are strictly modular and incorporate still both streams of the knowledge transfer process, the teaching process within the *Expert Knowledge* and the *Pedagogical Knowledge Model* as well as the learning process within the *Learner Model* component. Further, inferential functions are embedded into modern ITSs by means of so-called 'adaptation components', e.g. *feedback/correction, presentation* and *diagnosis/evaluation* (see Figure 3.5). These *adaptation components* build the intelligent core of the system and may interact with all other modules in the system.

So far, the investigation line GAS-UAS-CAI-ITS (subsequently named *ITS-movement*) has focused on process-driven models and indicated clearly that a *user model* represents an essential component within the chain of personalisation processes, but interestingly, it lacks of emphasis on learning materials. The next investigation line presented in this section examines again technology-based models but those concentrating on content issues.

Nonetheless, having yet a process-oriented point of view on personalisation, this opportunity is taken to give a brief overview of relevant issues concerning pure didactical models for *in-structional design*. This (brief) shift from technology-based analysis to a pure didactical one is not a focus in this section, but represents an interesting thematic transition into the subject of the next investigation line, as shown in the next two paragraphs.

A comparative analysis of more than sixty(!) different models of instructional design is given in [Andrews and Goodson 1995]. The aim of this analysis was to provide a list of the most common and relevant tasks during the development of instructional models, from a didactical point of view. As a result, based on G.L. Gropper's findings of 1977, [Andrews and Goodson 1995] added four new common tasks to the ten encountered by Gropper, as follows:

(1) formulation of goals from observation,

(2) development of pre- and post-test goals,

(3) analysis of skill- & learning-based goals,

(4) identification of goal sequences as learning facilitators,

(5) characterisation of learner populations,

(6) formulation of instructional strategies that meet the domain and learners' needs,

(7) selection of strategy-oriented media,

(8) development of strategic courseware,

(9) empirical tryout, diagnosis and revision of courseware,

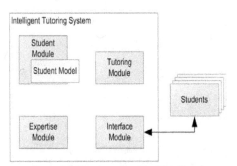

Figure 3.4: Structure of an Ideal Intelligent Tutoring System. [Brusilovsky 1994]

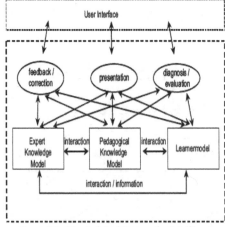

Figure 3.5: Classical Architecture for Intelligent Tutoring Systems according to [Martens 2003]

(10) development of procedures to maintain the instructional program,

(11) assessment of need, problem identification, occupational analysis, competence, or training requirements,

(12) consideration of alternative solutions,

(13) description of entire environment and formulation of constraints, and

(14) costing instructional programs.

In addition, Andrews and Goodson have identified twenty different 'dimensions' that can be used in model schemata. These dimensions are a sort of guidelines to consider when designing a model by means of other existing ones (see [Andrews and Goodson 1995] for details). Though, of highest relevance for this section is the task list, which describes in general the process of instructional design as well as provides indicators and requirements for relevant components in an adaptive instructional environment.

Hence, point 1 indicates the need of an interaction model, point 2 requires an evaluation unit, point 3, 4 and 6 can be coded in a pedagogical or tutoring component, point 5 requires a learner model, and so forth. So far, this is nothing new regarding the first investigation line. But considering the points 7, 8 and 9, this didactical point of view indicates clearly a strong need for components dealing with the creation and authoring of course materials. Why the ITS-movement does not show explicitly any indicators of these components will be clarified in the following pages through the analysis of a second investigation line of models, which (a) has strongly influenced current adaptive e-learning solutions and (b) is mainly component- and content-based. Yet, to not leave the reader with this open issue, a first clue for an answer is given in advance: the design of ITSs has pure computational roots coming from the field of AI, whereas the design of AEHSs is mainly content-based (specifically, hypermedia-based). Thus, based on the findings depicted so far, the following investigation line focuses on component-based models, starting from the field of UASs.

According to [Weibelzahl 2002], Benyon and Murray introduced in 1993 a component-based architecture for UASs, which contains three main

parts: a User Model, a Domain Model and an *Interaction Model*. In addition, Benyon and Murray's proposed architecture (see Figure 3.6) shows relevant internal sub-models within the previously mentioned. Hence, the *Student Model* in the User Model represents all system's assumptions about the learner's knowledge in a certain domain. The *Profile Model* holds information about the learner's general knowledge and interests. Cognitive and affective characteristics of the learner are stored in the *Psychological Model*. Further, the Domain Model is the basis for all inferences and adaptations, which are described by means of different 'levels', three of them being the Task, Logical and Physical Levels. It can be then said that this Domain component represents (apart from the learner) the 'rest of the world', i.e. it comprises all *information about the environment and the system* that is needed for inferential procedures. As depicted in Figure 3.6, communication between learners and system is handled by the Interaction Model, which contains also inferential functions, e.g. to infer user properties or for evaluation purposes [Benyon and Murray 1993]. Remarkable of this model, as stated in [Weibelzahl 2002], is the fact that the architecture can be used to 'evaluate' the different states of UASs. Though, evaluations within this context are just (superficial) 'observable snapshots', because the model does not describe the

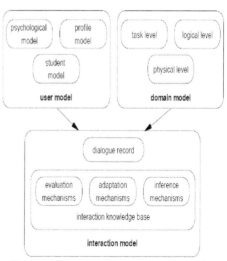

Figure 3.6: Benyon and Murray's Architecture for Adaptive Systems, according to [Weibelzahl 2002]

processes and interactions of components, and thus, the reasons that lead to the current status remain always hidden, [Weibelzahl 2002].

In 2003, Shute and Towle claimed that the premise, which underlines *adaptive e-learning*, is based on the successful findings coming from the ATI theory regarding the mapping of aptitudes or trait complexes to instructional sequences in learning environments. The main idea behind the Aptitude-Treatment-Interaction approach has been already introduced in section 2.2.4. As stipulated in [Shute and Towle 2003],

> *"the goal of adaptive e-learning is aligned with exemplary instruction: delivering the right content, to the right person, at the proper time, in the most appropriate way – 'any time, any place, any path, any pace".*

In the line of this definition, [Shute and Towle 2003] identified the following components of an adaptive e-learning system: (a) *Content Model*, (b) *Learner Model*, (c) *Instructional Model*, and (d) *Adaptive Engine*.

The Content Model may be interpreted as a *knowledge map*, i.e. a hierarchical structure of the domain topics (holding what is to be instructed and assessed) and the corresponding references to learning materials. The Learner Model is, on the one hand, like a particular instance of the content model, and therefore, it holds the current

knowledge status of the learner. On the other hand, stores additional individual traits used by the system. The information in the Instructional Model serves as basis for all internal decisions regarding a personalised content presentation and concerning the necessity of a teacher's intervention. All the information from these models can be accessed by the Adaptive Engine, which delivers the adapted learning content respecting a set of predefined adaptation rules. [Shute and Towle 2003]

Section 2.2.4 of this book has given an overview on the evolution of adaptive e-learning systems based on following three main theories: Aptitude-Treatment-Interaction, Micro-Adaptation and Macro-Adaptation. [Shute and Towle 2003] integrate also the latter two approaches into their analytical point of view on adaptive e-learning systems. At this point it must be emphasised again that the basic idea behind this point of view is to identify 'different ways of assessment' in personalised environments.

In the classical way to deliver content by means of inferential knowledge rules (as shown in the bottom part of Figure 3.7), the system focuses on the identification of weak areas, i.e. which knowledge gap exists between the personal knowledge in the Learner Model and the knowledge structures in the Content Model. This methodology, where Instructional Rules are used

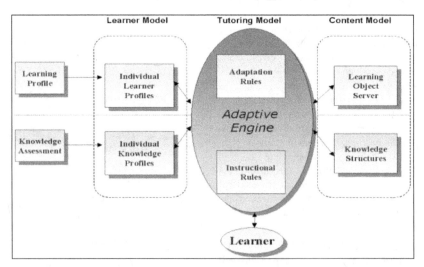

Figure 3.7: Assessment in an Adaptive Learning Management Framework. [Shute and Towle 2003]

to predict which the next most suitable unit is in the individual learning path, corresponds to Micro-Adaptation. In order to provide information about 'how' to present the next unit, the modules in the upper part of Figure 3.7 implement the theory of Macro-Adaptation, i.e. additional inferential rules are needed in order to consider all other relevant learner traits, e.g. cognitive or learning style, affective or emotional aspects, or everything else that seems relevant to compose the best learning objects in the repertory of the adaptive system.

Hence, among others, Shute and Towle's model indicates that temporal aspects, e.g. the current learner knowledge status, can and should be monitored to provide effective personalisation. [Akhras and Self 2000] consider also these issues as highly important, but from the point of view of constructivist learning. The approach described in [Akhras and Self 2000], called *Situation-based Constructivistic Learning* (SCL), was implemented in the INCENSE system, an intelligent e-learning environment for the domain of Software Engineering. SCL is based on a set of the four constructivist learning processes *cumulativeness*, *constructiveness*, *self-regulatedness*, and *reflectiveness*. In accordance to the affordances of situations regarding these processes (i.e. depending on certain patterns of interaction with the system), an SCL system predicts which learning opportunities can be provided to a specific learner. [Akhras and Self 2000] identify four aspects that coexist in any learning process:

- Context is a physical and social situation in which a learner is engaged in a learning activity, such as physical entities, tools, or other people.

- Activity comprises the aspects regarding the knowledge built by learners through adopted experiences and through their interpretations of own experiences.

- Cognitive structures are related to the influences of previous knowledge on new experiences, thoughts and actions.

- Time-extension, i.e. the periods of time where learners build new knowledge by trying to connect old to new experiences.

In contrast to many models depicted so far, in which a personalisation result is based on *knowledge holes*, the INCENSE system personalises new learning experiences by means of identifying particular situational cases of learners. Hence, not a teacher models instructional steps, rather the learners themselves. According to [Akhras and Self 2000], the following concepts define such a model: (a) processes (i.e. activities), (b) materials used in the processes, (c) process results, (d) data (i.e. content of materials/results), and (e) sequences of processes. Akhras and Self's model focuses strictly on the processes of learning, rather than on the products of learning. And as stated above, some learning activities also imply a specific context: interaction with other people. Consequently, this leads directly to the scope of *cooperation* in learning activities.

According to [Johnson and Johnson 2003], theories based on individual learning claim that instruction should be personalised to specific learner traits, such as aptitude, learning style, motivation or needs. Further, Johnson and Johnson claim that the use of technology in education may be more productive if combined with Cooperative Learning (CoopL), which is

> "the instructional use of small groups so that students work together to maximize their own and each other's learning".

In addition, Johnson and Johnson distinguish between cooperative and *competitive learning*, i.e. learners work together to achieve goals that only a few of them can attain, and consequently, they can only succeed if their colleagues fail to obtain their goals. CoopL is also different from *collaborative learning*, which, according to [Johnson and Johnson 2003], originates from England in the 1970s. Thus, following [Vygotsky 1978], Johnson and Johnson stated that

> "a student's learning is derived from the community of the learner".

Thus, collaborative learning recommends to create groups of learners and to let them create an own culture with their own learning procedures. Further, [Johnson and Johnson 2003] explained that strict collaborative followers see any structure provided by teachers as a manipulation that only creates training and not learning. The motto is then 'working in groups, yes.. but without guidelines!', which vanishes the role of a teacher. At present, both terms (cooperative learning and collaborative learning) are used as interchangeable synonyms. For the scope of this book, the term cooperative is related to 'learning in groups' and therefore, comprises also the terms competitive and collaborative.

At present, system implementations for adaptive e-learning try to solve the big number of problems implied in CoopL. In general terms, these problems can be seen as the set of 'natural group conflicts' and emanate from the differences among individual intentions, beliefs, goals, behaviours, experiences and others. One solution approach in this context is given by [Azevedo-Tedesco 2003], a model to solve conflicts in group interactions through an artificial mediator (see Figure 3.8). The solution proposed by Azevedo-Tedesco fosters the use of adapted dialogues that are controlled by a Mediator and infers solutions from user data in the Maintainer component. Here, the Maintainer stores and updates all user and group models. As for the model in [Azevedo-Tedesco 2003], other modern personalisation systems make use of artificial characters to support users in their interaction goals, e.g. as shown in [Bull et al. 2003].

Concerning group-working, the issues depicted so far show that AEL systems may also need components and functionalities for *stereo-typed personalisation*. Thereby, the question arises about how to assess learning performance in groups. The main didactical strategies and intentions behind assessment methods are either embedded into the functionality of the adaptive system (as previously shown in Shute and Towle's model for AEL) or provided by additional, independent modules, as in the case of the following approaches.

As illustrated in Figure 3.9, [Koyama et al. 2001] provides a Web-based solution through so-called 'judgement' components, which use agent-based modules integrated in personal clients that interact with the learners. In this model, the agents collect learner data and check the relevance of the learning materials. Further, agents control the learning sessions and deliver form-based tests to evaluate the degree of understanding. Although the approach presented in [Koyama et al. 2001] includes an assessment module, this component is tightly-coupled with the system, and hence, could be difficult to modify or exchange in order to meet other application-specific requirements.

From a technological point of view, it is relevant to identify if existing solutions provide flexible and strictly modular approaches for assessment components that allow an easy integration. A solution with an exchangeable assessment module is e.g. described in [Amelung et al. 2006]. An example for an independent stand-

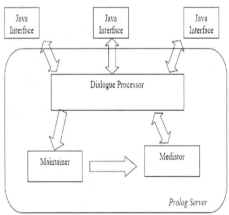

Figure 3.8: Main Components in a Conflict-Resolver System Artificial Mediators. [Azevedo-Tedesco2003]

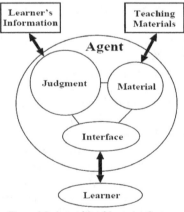

Figure 3.9: Agent-based Learning System integrating Assessment Modules. [Koyama et al. 2001]

alone solution is the automated essay grading system described in e.g. [Williams and Dreher 2004], [Gütl et al. 2005a], [Gütl 2007]. Thus, assessment modules can be identified as potential client applications for user modelling components.

Internet-based technologies as well as Web-based applications represent the hype of the last decade and have been extensively analysed through research and industrial work. Within this context, the next research field investigated is referred to as *Internet-Based Learning* (IBL), a term coined in [Hill et al. 2003]. In the concluding part of their publication, [Hill et al. 2003] give their opinion towards the application of IBL since the mid 1990's, and state that

"Internet remains on the threshold as learning tools. [...] the potential can be lost if steps are not taken to realize the true potential of these information technologies for learning. [...] As we continue to implement and examine the use of the Internet in our learning environments, the factors contributing to their successful implementation will become clearer. Taking the next steps toward the creation of active learning environments using the Internet is just a matter of choice; choosing not to take these next steps will leave the technologies like many other educational technologies before them: great ideas whose true potential was never realized".

In terms of 'why this technology now?' the author of this book agrees with these observations. In many cases, this essential question is answered by just saying that either it is a good 'opportunity' or an interesting 'challenge'. But, what makes this technology so different to others, especially within the context of teaching and learning? Some of the reasons are linked with the influence of the increasing amount of technological achievements of the last decade, which emanate and evolve much faster than their evaluation in concrete application areas. Among others, consider the following issues regarding some achievements since the 1990's: (a) standards evolve continuously and rapidly, e.g. XML[18], RDF[19], OWL[20], Web Services[21], and so forth, (b)

some standardisation communities and technologies concur with each other (e.g. RDF vs. XTM[22]), (c) the application, evaluation and comparative analysis of innovative solution approaches for models and architectures of distributed systems are hard to undertake (e.g. client-server vs. peer-to-peer vs. agent-based models, respectively many existing middleware solutions), or (d) the changes regarding the requirements on the Web as the largest distributed system (mostly) diverge from the goals of currently used software development paradigms (e.g. transparency vs. platform-independence vs. grid computing vs. service-oriented programming[23] vs. Semantic Web[24] vs. Web 2.0 and so on).

These examples show that (within the context of Internet-based technologies) first, the amount of technological achievements is overwhelming, second, one emanating technology is usually derived from an evolving one, and third, apparently the periods of time in which new technologies emanate are shorter than those time phases needed by research and industry to prove that the application of these new technologies is certainly efficient in 'real life'. Using again the terminology in John Holland's Theory and staying within the scope of adaptive e-learning systems, these issues describe the Web as a *rich environment*, because they imply a *complex prediction problem*.

As stipulated in [Holland 1962], the simplest rich environment is given when it is composed of graded sequences of problems. Therefore, solutions are needed that permit the development of systems solving problems at one level of difficulty and enabling their reuse in solving problems at the next level. Indeed, the practical success of AEL systems in the present cannot be measured at all; nonetheless, a large number of solution approaches for some partial problems are known. Some of these solutions have been implemented and evaluated, and show success-

---

18 http://www.w3.org/XML
19 http://www.w3.org/RDF
20 http://www.w3.org/2004/OWL

21 http://www.w3.org/2002/WS
22 http://www.topicmaps.org
23 The implications of these technological issues will be discussed in the next chapter.
24 http://www.w3.org/2001/sw ; the technological aspects of this topic will be also discussed in the next chapter.

ful practical results, as will be show in the next section. But the majority, adopting and adapting [Hill et al. 2003]'s words, *are great ideas whose true potential has not been realised*. Because of this *richness of the Web environment*, only some of the currently used, best-known IBL models for adaptive e-learning systems are presented in the reminder of this section.

Firstly, as already shown in previous sections, the field of Adaptive Hypermedia Systems (AHSs) can be seen as the most representative type of adaptive systems in the last decade. For AHSs, several reference models have been developed and their first implementation area was the field of Web-based instructional systems. The Adaptive Hypermedia Application Model AHAM (see e.g. [deBra et al. 1999], [Wu 2002] or [deBra 2006]) is based on the Dexter's Hypermedia Model (see e.g. [Halasz and Schwartz 1990] or [Halasz and Schwartz 1994]) and is e.g. implemented in the AHA! system (see [deBra and Calvi 1998] or [deBra et al. 2003]).

The AHAM extends Dexter's Model by dividing its Storage Layer in three parts: Teaching, Domain and User Models (see Figure 3.10). The Teaching Model component holds didactical rules and one layer above it, in Dexter's Presentation Specifications module, [deBra et al. 1999] integrate the central inferential component and call it also 'adaptive engine'. As in Dexter's model, the AHAM concentrates mainly on the Storage Layer in order to efficiently manage the informa-

tion of the models. But in contrast to Dexter's model, which deals with nodes and links, AHAM's storage unit is composed of 'concepts and concept relationships', which is the way to represent domain knowledge in the model (see [deBra et al. 1999] for details).

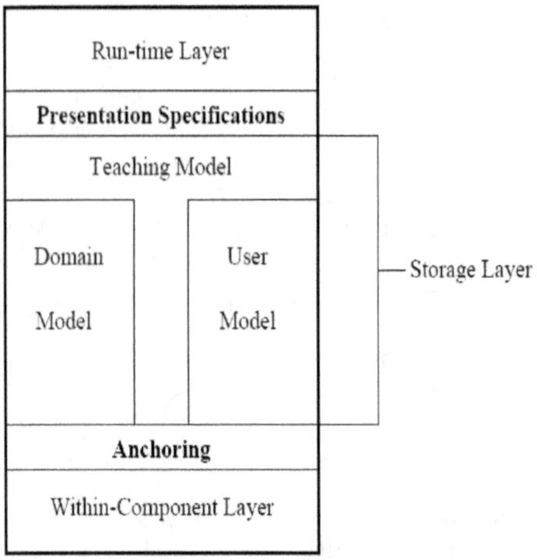

Another relevant AHS-oriented model is presented in [Koch and Wirsing 2002] and is based on the Munich Reference Model for adaptive hypermedia applications (see Figure 3.11). As shown in [Koch and Wirsing 2002], the rules for adaptive behaviour are implemented within the Storage Layer in the Adaptation Meta-Model. Like in Dexter's model – but in contrast to the AHAM - it includes a Web-based Within-Component Layer for the representation of nodes and links.

Many AHS models more have been introduced in the last years. In an attempt to incorporate several well-known models into a single one,

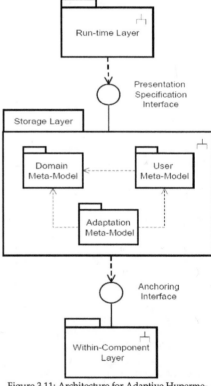

Figure 3.11: Architecture for Adaptive Hypermedia Applications [Koch and Wirsing 2002]

Run-time Layer	
Presentation Specifications	
Teaching Model	
Domain Model	User Model
Anchoring	
Within-Component Layer	

Figure 3.10: The AHAM Model [deBra et al. 1999]

[Ohene-Djan et al. 2003] defined a general, open architecture for adaptive hypermedia applications, as shown in Figure 3.12. The author of this book identifies in this model the first solution approach that explicitly splits the interaction layer of the core system into specialised modules according to the specific role of each user or client application, i.e. the model defines different types of modules that are able to interact with distinct users. Thus, it supports the interaction of a highly relevant actor, which was not identified so far in other models: the 'Hyperpage Designer'.

Further, a more detailed view of the component 'Core of Adaptive Hypermedia Functionality' is depicted in Figure 3.13 (see next page) and shows that the model does not focus on inferential instructional rules, rather it is designed to provide general adaptation functions, and therefore, it provides separated modules for composition and inferential functions (Composer & Engine) and for the storage of rules describing the adaptational processes (Rule Set).

In addition, this model given by [Ohene-Djan et al. 2003] provides another component for the representation of the hypermedia structures and descriptions (HyperLibrary), which can be used within the context of web-based adaptive e-learning system for the representation of the domain knowledge.

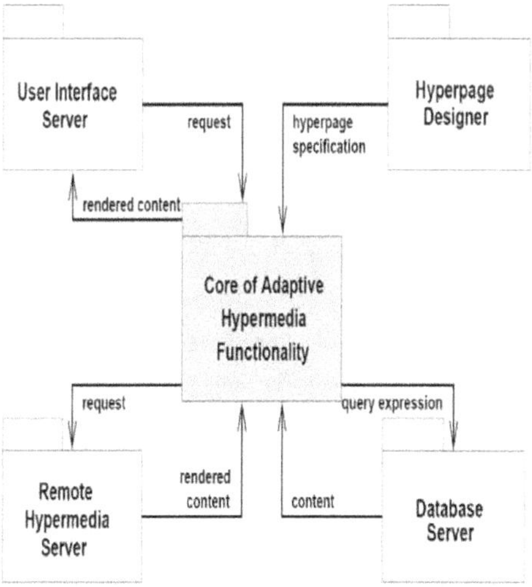

Although the architecture shown in [Ohene-Djan et al. 2003] allows Hyperpage Designers to create and edit *hyperspaces* through a separate interface component, this general model is not defined exclusively to educational hypermedia systems. Thus, the notion of 'designers' includes but is not restricted to 'authors of hypermedia-based learning materials'.

Concretely speaking about AEHS (Adaptive Educational Hypermedia Systems) and as stated in [Brusilovsky 2003], there already exist some tools and systems that support the process of designing and authoring hypermedia for adaptive systems. According to [Brusilovsky 2003], approximately from 1998 until 2003 the focus of research has moved from just innovating adaptive hypermedia technologies to investigating the problems of designing and authoring adaptive courseware. This process is different as for regular educational hypermedia (i.e. non-adaptive). For example, [Brusilovsky 2003] differentiates between the steps for creating regular and adaptive educational hypermedia as shown in Table 3.2. As can be derived from the table, there exist also tendencies to integrate inferential adaptation rules directly into the learning material.

Within the research context of educational software environments, [Brusilovsky 2003] stipulates that the development trends since the be-

Figure 3.12: A General, Open Architecture for Adaptive Hypermedia Systems [Ohene-Djan et al. 2003].

Creation steps for educational hypermedia		R	A
D E S I G N	Design and structure the knowledge space		√
	Design a user model		√
	Design learning goals		√
	Design and structure the hyperspace	√	√
	Design connections between the knowledge space and the hyperspace		√
A U T H O R I N G	Create page content	√	√
	Define links between pages	√	√
	Create descriptions for knowledge elements		√
	Define links between knowledge elements		√
	Define links between knowledge elements and pages with educational material		√

Table 3.2: Creating of Regular and Adaptive Educational Hypermedia. [Brusilovsky 2003]

ginning of the 1990's can be divided into two streams. On the one hand,

"the systems of one of these streams were created by researchers in the area of intelligent tutoring systems (ITS) who were trying to extend traditional student modeling and adaptation approaches developed in this field to ITS with hypermedia components". [Brusilovsky 2003]

And on the other hand,

"the systems of another stream were developed by researchers working on educational hypermedia in an attempt to make their systems adapt to individual students". [Brusilovsky 2003]

This second investigation line, AUS-AEL-IBL-A(E)HS (subsequently called *Hypermedia-movement*) can be interrupted at this point in order to analyse the question stipulated in page 47 regarding the reason why the GAS-UAS-CAI-ITS investigation line (*ITS-movement*) did not provide explicitly some indicators of content management components.

In advance it must be stated that both streams are, naturally, not mutual exclusive; they complement each other and have evolved through mutual efforts regarding research and practical achievements. Nonetheless, the point is that they

focus on different strategies: the ITS-movement concentrates on the internal algorithmic functions, whereas the Hypermedia-movement is content-oriented. The author of this book agrees with Brusilovsky's statements and extends them as follows in order to find at least one of the answers to the question concerning why this difference still exists (after almost two decades!).

An extensive investigation on issues regarding hypermedia spaces is beyond the scope of this book. Though, it is worth mentioning that lots of research efforts with respect to *hypermedia data modelling* were made during the mid 1990's (see e.g. [Duval et al. 1998]). In addition, research literature shows that the power of *metadata*, as a means to describe (or enrich the description of) Web-resources and to foster their reuse, influenced notoriously the Hypermedia-movement (see [Duval 2001]).

Thus, to give an example within the context of this book, consider the last decade, the SCORM era. In accordance to e.g. [Mödritscher et al. 2004a,b], one of the main goals of the Advanced Distributed Learning (ADL) initiative, established in 1997, is to promote cooperation between government, industry and academia to develop standards concerning sharing and reus-

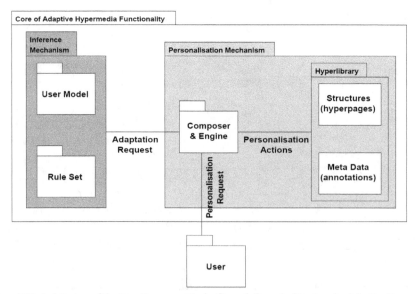

Figure 3.13: Architecture of the Core Component in the General, Open Architecture for Adaptive Hypermedia Systems, as in [Ohene-Djan et al. 2003].

ing e-learning resources. One of the results, the Sharable Content Object Reference Model (SCORM), is a set of specifications adapted from multiple sources that provides a suite of e-learning capabilities to enable interoperability, accessibility and reusability of educational hypermedia. Within the SCORM specification, standards are defined to e.g. describe learning objects (LOM) and learner profiles (IMS LIP), define adaptable sequences of learning resources (IMS Sequencing), express different pedagogies (IMS LD), etc. ([Mödritscher et al. 2004a] [Mödritscher et al. 2004b] [ADL 2007] [Littlejohn and Buckingham 2003] [Atif et al. 2003]).

In summary, the following can be stated. The Hypermedia-movement focuses on the creation and authoring of learning resources, i.e. it aims at generating new didactical means for course authors and teachers e.g. through the utilisation of evolving and enhanced standards. The ITS-movement also considers adaptation through standardised personalisation rules attached to the learning materials. At this point they converge and are still following the same line, i.e. improve the teaching side of the knowledge transfer process. On the other side, both movements work together to enhance the learning side of the knowledge transfer process by fostering personalised Web-based learning environments. The practice of this second aspect takes place through the development of new adaptive e-learning systems or through the integration of adaptivity and adaptability in existing non-adaptive systems. One last remark concerning authoring of learning resources: also in the mid 1980's (but regarding the field of CAI) the need for sharing and reusing learning resources was pointed out, e.g. in [Maurer 1985].

In general, both investigation lines presented so far, the ITS- and Hypermedia-movement, have shown very similar components and functions. The main distinction identified was the way they embed inferential rules in their models, either in the source code of the system or attached to the content. The reminder of this section closes the context of the Hypermedia-movement with some modern AEHS models, and finalises with a brief presentation of some other models that have influence from current trend-setting research fields, such as the Seman-

tic Web and Context-aware Systems; these research fields also contribute to the identification of additional components, which have not been defined in the previous models and represent further afferential and/or efferential modules for a user modelling component.

According to [Henze and Nejdl 2004], a sort of common understanding to explain Adaptive Educational Hypermedia does not really exist, and therefore, the corresponding models and systems are difficult to compare and analyse. To move towards a solution for this problem, Henze and Nejdl elaborated a logic-based definition for the characterisation and analysis of AEHSs, which comprises the following four main components:

• Document Space (DS)

• Observations (OBS)

• User Model (UM)

• Adaptation Component (AC)

In terms of these main components, Henze and Nejdl's model allows the analysis of AEHSs according to the following aspects. The DS defines and holds graphs, which describe (a) the set of learning *resources* (including their *metadata*) as well as (b) the set of corresponding topics from the *domain knowledge*. The OBS component is used to describe how the system monitors the users' interactions. Sentences in the UM describe the information about individual users, groups of users as well as their traits, whereas predicates and rules in the UM are intended to associate traits and users. And finally, the AC comprises the rules that describe the system's adaptive functionality. [Henze and Nejdl 2004]

After analysing distinct models for AEHS, Henze and Nejdl found out that the Document Space plays a decisive role as inferential source of adaptation components, because the way how to adapt is coded there. In relation with this aspect, what Henze and Nejdl call the ‚closed corpus problem' in AEHS was already identified at the beginning of this section as 'limited' or 'selective repertory' of modern adaptive systems. Henze and Nejdl's model shows that the adaptive repertory of AEHSs is mostly fixed at design time, and therefore, subsequent changes are difficult to process, i.e. the 'opening' of the adaptive repertory in order to meet the requirements of the rich Web-environment is difficult to solve without a redesign of the system. The reason for this problem is that a rich environment generates problems dynamically, and such a generic behaviour leads to the need of a 'generic' problem-solving system, or at least of a very dynamic one.

In most AEHSs the adaptive repertory is distributed among two components: on the one hand, the DS represents the static source for adaptive inferences, and on the other hand, the data collected in the OBS component describe the dynamics of the overall system at run-time [Henze and Nejdl 2004]. Both (sub-)repertories are processed by UM and used by AC in order to make the decisions (in AC) respecting the most suitable solution for the current needs of the learner (from OBS) in accordance to persistent traits (from UM). Thus, a step towards generic (or at least very dynamic) solutions in AEL systems could be possible through open, extensible and flexible implementations of OBS, UM and AC components.

So far, it has been shown that personalisation in adaptive e-learning systems is a complex task. This complexity results e.g. from the huge amount of possible relationships among domain knowledge entities, intra-document relations, inter-document dependencies and observed behaviours. Moreover, the consideration of *situational contexts* of system users as a source for a variety of additional relationships increases computational complexity.

Due to the rapid evolution of the *Semantic Web* in the last years and to its remarkable technological achievements, many standardised ways are now available to efficiently manage knowledge- and context-based information delivery. The Enhanced AHAM (EAHAM) version presented in [Kravcik and Gasevic 2006] describes such topics and can be seen as another solution approach for Situation-based Constructivist Learning (in addition to e.g. the already mentioned INCENSE system). The architecture for EAHAM is depicted in Figure 3.14 and shows that the functionality of AHAM's Teaching Model in the Storage Layer is replaced by three models: Adaptation, Activity and Con-text Models.

As stipulated in [Kravcik and Gasevic 2006], a context in EAHAM is interpreted as the set of circumstances in which an activity occurs, a 'setting'. In essence, such a setting (i.e. situational context) represents the sum of all environment traits that complement the user model. Thus, on the one hand, when connecting contexts with Henze and Nejdl's model, these traits should be observable. On the other hand, the Context Model in EAHAM embraces a similar basic functionality as the Domain Model of Benyon and Murray's model, because both are implemented as task-based modules. Another example for a task-based and context-oriented model can be found in [Scharl 2000]. The EAHAM also provides functionality to compute learning processes within the Activity Model. As stated previously in this section, one course within an educational hypermedia space can be described by means of metadata standards; such metadata does not only define relationships to domain knowledge, it may also refer to relationships among the documents. Through this way, as typically done in AEHSs, it is possible to map didactical models into specific structures of learning resources, such as sequences or graphs. Additionally e.g. pre- and post-conditions can be integrated into the resulting nodes of the graphs to stipulate rules for the control and assessment of the mastery progress of learners. These graph-based models representing interrelated tasks are usually called *workflows*. And this way of view-

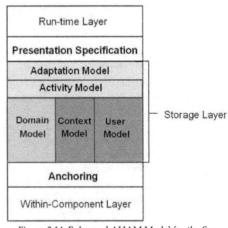

Figure 3.14: Enhanced AHAM Model for the Semantic Web [Kravcik and Gasevic 2006].

ing on the knowledge transfer processes is known as *process-driven* or *workflow-based learning*. Examples for systems of this type are described e.g. in [Sadiq et al. 2002], [Helic et al. 2005] or [Helic 2006]).

Regarding workflow-based learning systems, the most interesting issue within the scope of this book is the fact of moving towards what was called 'long- living adaptive systems' in section 2.2.5 and is known in the field of e-learning as *life-long learning*. Generally speaking, life-long learning implies both, the phases in human life involving 'traditional' educational (e.g. school and university education) and 'work-place' education. This latter phase in human life is the central scope of the field known as *learning-on-the-job*. Precisely this phase in modern life leads to the investigation of e-learning within business process, which can be modelled by means of workflows. As can be seen, at this point the convergence is given between process-driven learning, life-long learning and learning-on-the-job.

As stated in [Helic et al. 2005] and [Helic 2006], firstly, due to the dynamic nature of learning processes they imply an own *lifecycle*. According to [Helic et al. 2005], this lifecycle consists of three phases: (1) a *modelling phase* where particular learning processes can be defined, (2) a *learning phase* concerning the execution of a learning process, and (3) an *observation and improvement phase* where the processes are monitored and personalised. Thereby, in analogy to task-oriented models, personalisation is meant to meet the current needs of learners 'in a particular situation or context'.

From this point of view, different personal roles emerge for distinct *workplace situations* or *educational projects*, which in turn can be modelled through workflows, respectively scenarios. Within the field of Web-based Training (WBT) environments, [Helic et al. 2005] address these aspects and present a solution system (called WBT-Master), which is able to support different didactical strategies meeting the requirements of different educational theories, e.g. collaborative, experiential or project-oriented learning.

Of relevance for the scope of this section is not the lack of explicitly defined, internal adaptive functions or modules within the WBT-Master

system. Instead, the interesting aspect is that the WBT-Master system is composed of many specialised components that can be arranged to meet distinct didactical goals. Thus, it enables a highly adaptable composition of modules in order to map didactical goals into scenario-based learning process lifecycles, and can be seen therefore as a *tool supporting system* [Helic et al. 2005]. The WBT-Master places several tools at the disposal of teachers and learners, such as discussion fori, annotations management, expert knowledge bases, knowledge mining mechanisms, project-based group task management, course library support, and so forth (see [Helic et al. 2005] for details). And again, the tendency (and fact) that current e-learning systems comprise a variety of interdependent components is identifiable. This means that a big set of systems that are capable of providing useful personal information to a user modelling component is identifiable.

On another field, the number of Context-aware Systems (CASs) has increased notoriously in the last decade (see section 2.2.4). An interesting example of the application of CASs is presented in [Specht and Oppermann 1999] and refers to so-called *Nomadic Information Systems* (NISs) [25]. Specht and Oppermann consider such systems as a derivation of *Mobile and Wearable Computing Systems*, because users of NISs carry some interconnected electronic devices with them, which allow the definition of their geographical position (e.g. gained through technologies like GPS, DGPS, or digital compass; see [Specht and Oppermann 1999] for details about these technologies). NISs collect and process information about changes of location, and in turn, the gained data can be used to identify changes in learning situations, scenarios of contexts. Also in association with NISs, the connectivity context of a user is also known as Personal Area Network (PAN). According to [Karypidis and Lalis 2006], a PAN-oriented system concerns the collaborative operation of wearable, portable and stationary devices that compose the current immediate networked environment of its user. Hence, the target environment of adaptive NISs

---

[25] Note: The author of this thesis relates NISs with CASs, because the spatial position of users also belongs to their 'situational context'.

is defined by the adaptability of a user's PAN, where e.g. a learner can be spatially monitored and depending on the type of her current PAN (laptop, mobile phone, PDA, etc.) the presentation or delivery of learning content can be implicitly personalised[26]. Hence, also the influence of current solutions for context- or situation-aware systems indicate that additional sensors for user-related information (i.e. context-sensing devices) represent another set of potential data afferential modules for a user modelling component.

3.1.4 Concluding Remarks on Personalisation in Adaptive E-Learning

The presentation of the models for adaptive systems described in the last section helped to identify the most relevant components and functions under special consideration of the role of User Modelling Components (UMCs) in Adaptive E-Learning (AEL) systems.

The Role of Personalisation in AEL Systems

It is identifiable that UMCs assume distinct functional roles depending on distinct aspects, such as interaction goals, application areas, research fields, or in the case of AEL, didactical paradigms. These distinct roles have been indicated from three points of view, personalisation (i) as a system's architecture, (ii) as a process and (iii) as a goal.

Personalisation as a system's architecture

All depicted models have indicated the need or presence of a single, independent UMC. In spite of the different names for a UMC (e.g. 'student model', 'learner model', 'user profile' or just 'user'), its presence is indispensable due to its highly relevant role as main internal representation of the adaptation target (i.e. the learner). Still, some application areas have also indicated the need of providing implicit personalisation, i.e. adapting to a user but without the utilisation of a UMC. In this latter case, particularly CASs show AEL scenarios where it is needed e.g. to adapt the delivery of learning resources according to the traits of the interaction device, which does not necessarily imply a knowledge of personal traits of the learner.

Personalisation as a process

The models shown so far indicated that in the majority of cases a UMC is an indispensable member of personalisation process chains, but it is also highly dependent on its interactions with adjacent components. For example, starting with the first interaction of a new user, one of the primordial goals of any personalisation system is to acquire as much relevant information as soon as possible, so that each adaptation result meets the individual needs of a specific user with the highest possible accuracy and success.

From this point of view, the functional role of a UMC depends mainly on the following two aspects. On the one side, a UMC depends on its data providers, i.e. the set of sensors ('afferential components') with which it interacts and from which user data is gained; providers of a UMC are e.g. UIs (user interfaces), knowledge modellers or activity engines, but also assessment tools or feedback mechanisms from an adaptive engine. On the other side, a UMC should provide information about individual user traits to its clients ('inferential and efferential components'); in most cases, the main client of a UMC is an adaptor, i.e. the adaptive engine, which knows how to utilise UMC data in order to generate the most suitable personalisation result.

Personalisation as a goal

Indeed, the diversity of existing models is a limitation to analyse the functions of UMCs from a general point of view. Each model and system is mostly defined within a set of constraints that restricts the functional scope of its components, i.e. the requirements on the components vary from one model to another. As shown so far, the different AEL models were created to meet the goals of distinct didactical paradigms. Thus, which information holds a UMC and to which extent it has inferential capabilities, is a matter of the expected practical impact of the system and thus, depends on the aims of the exercise. From this point of view, this sub-chapter represents an attempt to indicate that the functional role of UMCs is not restricted to storage and delivery of data, it also implies the derivation (i.e. inference) of new traits from existing ones.

---

[26] See 'implicit personalisation' in section 3.1.1.

The following two examples clarify this aspect in brief. First, in ITSs and AEHS the UMC should manage a data model about the knowledge state of learners, but the general model of the knowledge domain is hold in the expertise module. Consequently, the UMC may either (a) get, hold and provide ready-to-use data, or (b) get raw data, generate itself individual instances (like partial snapshots) of the general model in an adjacent knowledge domain component, and make them available. Second, SCL and CoopL systems provide personalisation in accordance with individual or group activities. In this case, again, the UMC may either (a) contain static information about individual group memberships, or (b) continuously monitor the interactions of learners (i.e. observed behavioural data), and generate individual learning situations by comparing observation data with models for already known group activities. This last point of view (viewing on personalisation as a goal) describes one of the most critical aspects for the design and implementation of UMCs. An investigation based on the theory of architectural models for AEL systems, as presented in this sub-chapter, provides indicators to identify some duties of a UMC. These duties depend highly on the specific purposes of each model as well as on the aims of the underlying didactical paradigm.

An extensive evaluation of existing running prototypes and commercial products is beyond the scope of this book, because it does not focus on the study of adaptive engines; this book concentrates on the design and development of *User Modelling Components* in Adaptive E-Learning Systems. The reminder of this section discusses some examples of known success stories of applying AEL, however, problematic areas are also mentioned.

Benefits and Drawbacks of Adaptive E-learning

Does adaptive e-learning have a future? Which are the benefits? Where are the examples for success? These and more critical questions arise when we face the reality and observe that (in general!) most learners and teachers do not accept or adopt e-learning as an alternative for effective instruction. Further (as can be concluded from the findings of chapter 2), on the one hand, computer-assisted teaching and learning exists practically as long as the computers themselves are, and on the other side, research and development efforts for systems that deliver learner-tailored instruction have been made since several decades. In short: more than 50 years of history through PI, CMI, CAI, ITS, AHS, and others, and still 'adaptive e-learning' is not mature enough to revolutionise instruction?

Educators

Also in the past century, the evolution of Artificial Intelligence methods and their high impact in the field of Robotics gave the fiction of androids a notion of a nightmare coming true. And this 'menace to humanity' increased as automated machines (not really looking like humans) replaced people at their work-places, and still they do. The same problem exists with e-learning and will remain as long as some people think that intelligent teaching machines (e.g. sophisticated adaptive e-learning environments) could entirely replace teachers in their role. Interpreting adaptive e-learning systems as a solution that solves all didactical problems will lead to the same 'blurred' and dangerous connotation as for robots[27]. As has been stated so far at different points in this book, an adaptive e-learning system is just another tool for teachers to improve the knowledge transfer process through the use of digital media. And as such, these systems cannot be more successful as the degree of efficiency in which they are applied and used by teachers and learners. 'Since ever', humans have invented useful tools but not everyone has the talent or the desire to use it conveniently.

Learners

Further, regarding the scope of technology-assisted teaching, consider the work done and the results achieved in Wallemberg Hall at SCIL (Stanford Center for Innovations in Learning) since 2002 with so-called High-Performance Learning Spaces (HPLSs)[28]. Among others, HPLSs consist of three 12-feet flat-screen dis-

---

[27] Consider that this point of view is the same as for CAI in the 1960s, as can be identified e.g. in [Chapin 1968].

[28] For more details on HPLSs please refer to *http://scil.stanford.edu/news/high_performance_learning.html*

plays, flexible seating for up to 50 students, laptops for all attendants, wireless LAN, a central control for enabling or configuring Internet access, etc. (i.e. a HPLS has almost everything a modern classroom should have in the eyes of technology lovers).

According to [Steinhardt 2005], an evaluation of student performances in HPLSs have shown that teachers with better skills for working with digital media have achieved better didactical results than in traditional classrooms, but teachers that usually avoid additional didactical media have performed worse in HPLSs than in traditional classrooms. The interesting point within this context is that HPLSs are based on the requirements of current 'digital natives'. Marc Prensky is known as the person who firstly coined the terms *digital natives* and *digital immigrants* (see [Prensky 2001] for details). Under the premise that *kids are natives in a place that most adults are immigrants*, [Steinhardt 2005] stated that, at present, most teachers are digital immigrants, because they did not grow up with the same digital tools (if any) as their students. Thus, young people are digital natives looking at and working with new digital media in a way, which is mostly unknown for their teachers (in terms of practical use). Based on this notion of different 'digital generations', an additional relevant reason why technology-based instruction (particularly AEL systems) has failed to find intensive use at present, is the fact that concrete results of using 'new digital media' for teaching have appeared almost too late, namely at a point where a next generation of digital natives has already grown up with newer media.

Individuality

INSPIRE is a prototype of a Web-based AEHS, which was designed to support *web-based blended learning* (i.e. Web-based as well as traditional classroom-based teaching). According to [Papanikolaou et al. 2002], INSPIRE is based on the idea of restricting the domain knowledge at the beginning of learner interaction (which is a proper solution for novices) and enriching it gradually as learning progresses. It personalises each learning instruction by adapting previous learning results to the knowledge level and learning style of users. A first pilot study, con-

ducted with undergraduate students of the Department of Informatics and Telecommunications of the University of Athens attending the course on Computer Architecture, focused particularly on user feedback about the system's design. As stated in [Papanikolaou et al. 2002], most user comments led to improvements of the user interface and *corroborated the assumption that distinct didactical strategies should be applied to distinct personal learning styles*. Another evaluation of INSPIRE (consisting of one expert review with 8 expert-instructors and two group evaluations with 33 learners) showed that 7 among 11 questions regarding the topic ‚Prescriptive Instructional Strategy - Individualised Support' (one of five topics in total) indicated a very high rating[29] of usefulness. The following three highest rated issues concerned 'individualising educational content based on learners' knowledge level' could be identified[30]; firstly,

"the system [...] generates a sequence of lessons for each particular learning goal. Content planning [...] of each lesson depends on learners' progress"

... with a mean rating of 4.7; secondly,

"the domain concepts are gradually presented to learners [..]. The outcome concepts proposed to the learner for study are determined based on learner's knowledge level on the outcome concepts of the previous layers, i.e. in order to go on to the next layer the learner should be 'Advanced' on the outcome concepts of the previous layer. However, the concepts of each subsequent layer enrich those of the previous ones, augmenting the domain presented to the learner. "

... with a mean rating of 4.5; and finally,

"the system provides individual navigation advice following learners' progress, without restricting the educational material and limiting learners' freedom to browsing. [...] Following learner's progress, the system proposes the pages of the Use level (when his/her knowledge level becomes 'Almost Adequate') and finally of the Find level

---

[29] Mean ratings for each question: 1 to 5, where 1 stands for useless, 2 for almost useless, 3 for rather useful, 4 for
useful, and 5 for very useful.
[30] See [Papanikolaou et al. 2002] for more details on the results of this evaluation.

(when his/her knowledge level becomes 'Adequate').'

... with a mean rating of 4.5 (see [Papanikolaou et al. 2002] for details).

Expectations of Learners

Regarding AEHSs, [Brusilovsky 2004a] shows the three main benefits of using the ISIS-Tutor[31] system in comparison to using a non-adaptive system, as follows:

- *"The students are able to achieve the same educational goal almost twice as faster.."*

- *"The number of node visits (navigation overhead) decreased twice."*

- *"The number of attempts per problem to be solved decreased almost 4 times (from 7.7 to 1.4-1.8)."* [Brusilovsky 2004a]

Another system called Interbook, a web-based adaptive textbook application, was evaluated regarding the value of adaptive annotation (see [Brusilovsky 2004a]). The evaluation indicated, among others, that (a) there was no performance difference among both subject groups (with and without the utilisation of adaptive annotation), (b) adaptive annotation fosters non-sequential navigation, and (c) adaptive annotation met the benefit-related expectations of the users(!).

Learning Performance

EDUCE, an adaptive intelligent educational system, as denoted in [Kelly and Tangney 2006], is based on Gardner's theory of Multiple Intelligences (GtMI) for dynamically modelling learning characteristics and for designing instructional material. The results of the empirical study presented in [Kelly and Tangney 2006], which was based on 47 test subjects with an average age of 13, explored two main questions:

"What is the effect of using different adaptive presentation strategies instead of giving the learner complete control over the learning environment?"

and

"What is the impact on learning performance when resources are matched and mismatched with learning preferences?"

Among others, [Kelly and Tangney 2006] analysed the variables (i) 'Presentation Strategy' with the values 'Least Preferred' (i.e. resources the learner least prefers to use) and 'Most Preferred' (i.e. resources most preferred), (ii) 'Activity Level', which shows the percentage of resources used, and (iii) 'Activity Groups', showing 'Low', 'Medium' or 'High' degree of learning activity. The expectations of the evaluators should show that the learning gain would be higher for learners (a) adaptively guided to preferred resources, (b) with adapted access to additional resources, and (c) adaptively guided to resources based on a dynamic model of behaviour. Regarding the mentioned issues, [Kelly and Tangney 2006] found out that nothing could be concluded about the effect of level of choice regarding expectation '(b)' due to insignificant statistical values, but unexpectedly, the results regarding of adaptive presentation strategy (see '(i)' and '(a)')suggest that the learning gain was higher in the least preferred condition than in the most preferred one. Thus, a post-test concerning the issues '(ii)' and '(iii)' was conducted, and showed that

"students with low levels of learning activity can improve their performance when adaptive presentation strategies are in use. This suggests that by promoting a broader range of thinking and encouraging students to transcend habitual preferences, it is possible to increase learning performance for learners who are not inclined to explore the learning environment." [Kelly and Tangney 2006]

Kelly and Tangney stipulated also that the presentation strategy had no impact on high activity learners, because they can automatically adopt distinct ways of thinking if they may access different additional resources.

Further, within the context of learning performance, [Kelly and Tangney 2006] found out that using adaptive presentation strategies for providing learners with not-preferred resources enhances just the results of low activity learners. As a conclusion, Kelly and Tangney stated that these surprising results (somehow in contradiction to traditional GtMI-based approaches) may be explained through the assumption that the best personalised presentations are those which consist of resources that imply a challenge for the

---

[31] See [Brusilovsky and Pesin 1994] for details on the adaptive hypertext learning environment ISIS-Tutor.

learners. This topic has not been evaluated yet, but, considering that 'challenge' in game-based learning is highly motivating for digital natives (as treated in [Prensky 2001] or [Bransford et al. 2000]), and taking into account that the study of Kelly and Tangney was based on boys with an average age of 13 years, the mentioned assumption may find corroboration, and hence:

" [...] maybe that in education too, challenge at the appropriate level is also needed." [Kelly and Tangney 2006]

Investors

As stated in [Gütl et al. 2005b], people have to be supported and assisted during their life-long learning activities. Traditional learning methods and environments do not meet the contemporary needs of our information society any more. As a result, e-learning can be identified as one of the emerging areas in the last few years, as shown by means of concrete numbers in the (IDC 2003) study. Approximately 934 Mio US$ were invested in e-learning worldwide in the year 2003 and the European market was meant to be the best one. Though, a lot of failures and only a few - in most cases locally restricted - success stories can be identified (see also [Baumgartner 2003]).

Practical E-Learning

To conclude this section, a last observation is worth mentioning with respect to the efficient adoption of AEL systems at present. According to [Brusilovsky 2004a], AEL systems are rarely used in *practical e-learning*, because of the prevailing utilisation of 'non-adaptive CMS[32]-styled all-in-one solutions for Web-based e-learning systems (usually called Learning Content Management Systems, LCMSs). An integration of adaptive methods into LCMSs is a hard task (among others, also technology- and cost- dependent). Though, precisely in this context many results have been achieved during the last years, e.g. standardised interoperability of systems (SCORM, IMS, etc.), or a semantically richer description of content through metadata (LOM, RDF, etc.).

---

[32] CMS is usually the shorthand for Content Management System, but as used by [Brusilovsky 2004] within the context of adaptive e-learning it means Course Management System.

Also within this context, in accordance to [Brusilovsky and Miller 2001], solution systems for e-learning are currently dominated by Learning Management Systems (LMSs), which support knowledge transfer processes by providing an integrated set of tools for the management of distinct teaching and learning activities (such as course development, enrolment, progress tracking, discussion fori, collaboration tools, assessment and grading, etc.). Still, LMSs are based on the premise of *one size fits all*, i.e. user-tailored adaptation is not provided, and thus, every learner uses the same tools and materials. AEL systems, in contrast to this premise, follow educational theories claiming that learners should be taught through those didactical strategies that best suit their distinct cognitive traits, i.e. some functions of LMSs should be personalised. From this perspective, AEL systems have already demonstrated positive results, as shown in previous paragraphs (see also [Brusilovsky 1999]).

So far, this book has given a comprehensive overview on theoretical as well as technological issues with respect to AEL systems. As shown, AEL systems aim at helping to facilitate as well as to augment the knowledge transfer processes through the utilisation of personalisation techniques. Among others, the central role of two components of such systems have been identified, namely (a) an *Adaptive Engine* as the central steering module for adaptational procedures, and (b) a *User Modelling Component* as central module to represent and manage the main environmental target of the system, the user. The next sub-chapter deals with the main issues concerning this second component from a technological point of view.

3.2 User Profiling and User Modelling

Adaptive e-learning systems are implemented under the premise of *one size does not fit all*. As already mentioned in previous sections, some educational theories stipulate that different learners follow different strategies during learning and show different preferences in the use of learning materials. Some evaluation results indicate that each individual learning style is identifiable and that adaptation towards a personal style increases the learning performance of some learners (see e.g. [Rasmussen and Davidson-Shivers 1998]). Though, Web-based e-learning shows some critical problems if the corpus of learning materials is too open, such as a cognitive overload or the 'lost in hyperspace' phenomenon. Adaptive E-Learning (AEL) aims to solve these difficulties of content comprehension and disorientation through user-adapting methods that reduce and optimise the needed material repertory.

To *know the user*, AEL systems generate an internal representation for each learner, in which e.g. personal preferences, goals and knowledge are modelled. Since decades, different strategies have been developed to adapt to these learner traits, such as annotating or hiding links, modifying the learning sequence of materials, hiding or tailoring content, etc. (see e.g. [Brusilovsky 2001]). However, the development of AEL systems is a hard task and implies major research questions, as for example how to identify relevant learning traits, how to model the different learners, how to distinguish between internally generated assumptions and facts about the learners, or how should an AEL system adapt what and for which purpose based on which user information.

This sub-chapter focuses on the *User Modelling Components* (UMCs) of adaptive systems and aims at giving a detailed insight into common architectures, main goals, involved processes, used techniques as well as current solutions and trends. Thereby, the sub-chapter comprises main aspects of UMCs as independent parts of AELs systems, but draws in parallel an investigation of main issues regarding general adaptive systems. For this purpose, the sections of this sub-chapter are structured as follows. First, section 3.2.1 clarifies the distinctions between the terms *user profiling* and *user modelling*. Considering a UMC as independent part of an adaptive system, section 3.2.2 gives an introduction into the theoretical fundaments of UMCs, i.e. the *lifecycle of user information* and the *model content*. In order to complete a general view on User Modelling, i.e. what is to model, why to model and how to model, section 3.2.3 introduces into the most relevant *organisation processes* and *traditional implementation techniques* of UMCs. The practical focus is resumed in section 3.2.4 by presenting some existing solutions of *user modelling systems*, which are application- and domain-independent. i.e. they are reusable for distinct application areas, also outside AEL.

3.2.1 User Profiling vs. User Modelling

Generally speaking, the terms *User Profiling* and *User Modelling* are frequently used in literature as synonyms or near synonyms (as stated e.g. in [Kay 2000]), but in some cases a distinction is made (see e.g. [Koch 2000]). Nonetheless, literature shows that user modelling is mostly used, and comprises implicitly profiling techniques. This book follows this last issue and consequently, a *User Modelling Component* (UMC) may comprise both, user profiles and user models. The distinction between both terms is based on the following observations.

According to [Koch 2000], user profiles are the simplest form of user models and are utilised to represent user traits as simple key-value pairs, i.e. one user trait is assigned to a value. Usually, this value belongs to a known range of valid values. This (widely utilised) connotation for user profiling reflects its adjunctive character in practical use, i.e. the knowledge about a user grows by means of adding key-value pairs to profiles. Hence, these key-value pairs in user profiles can be also called *attributes*. Further, attributes can be of different types, which are e.g. directly derived from typical Computer Science language, such as Boolean (with true/false values), number (with e.g. the sub-types integer, dual or float), or character sets (char, string or text). Attributes can also have values of other types (see e.g. [Kay 2000], [Koch 2000] or [Wu et al. 2001]), such as discrete (e.g. 1 and 2 for low and high) or probabilistic (e.g. 0..1 for none..all), or might be transient (i.e. short-termed, e.g. preferred music genre is Latin jazz, interest in mathematics), permanent (i.e. long-termed, e.g. mental impediments, birth day, age, academic degree) or situation-dependent (e.g. preferences for a device, media, colours).

Thus, user profile attributes can be also utilised to describe individual user traits that are specific for AEL systems, like personal data, skills, learning styles or material-related preferences, whereby one specific trait is defined as attribute key with a number of possible values. Regarding this last observation, it is worth mentioning at this point that in practical use, new values are derived from the input of a user profile component and assigned to already known keys, and not the other way (keys assigned to values).

Taking this last aspect into account, a personal attribute may be also represented by a set of values assigned to one key, and therefore, the "key←value" definition should be extended to "key←{values}". Consider the following examples: 'preferred_books' ← { 'bookA', 'bookB', 'bookC' }; 'preferred_colors' ← { 'blue', 'white', 'red', 'black' }; 'enrolled_courses' ← { 'course7', 'course12' }. From this point of view, user profiling is very similar to traditional user management modules in systems that provide methods for individual authentication to enable e.g. controlled access to information (as in Content Management Systems or access-restricted Internet Portals). In such systems, user data is normally collected directly from user feedback (e.g. username, password, email address, and so forth), stored in form of attributes and provided as stored, without modifications.

The observations depicted so far enable then the definition of *user profiling* within the scope of this book:

> *User Profiling* is a technique for the internal representation of users. It enables the composition of key←values pairs (*attributes*) into collections of user data (*user profiles*) for the description of user traits. This technique comprises functions to create and configure attributes and groups of attributes, as well as functions to e.g. initialise, add, store, modify, remove or extract attribute values. In a given adaptive system, the set of available user profiling functions are deployed as an independent unit called *user profiling component*.

Based on this description for user profiles, an investigation on user models can be started. The notion of the concept User Model is explained in [Kay 2000] as an artificial representation of the human part of reality. Within the scope of this book, Kay's definition has been adopted and adapted (as depicted in Figure 3.15). Regarding this figure, all observations and statements in the following paragraphs are based on Kay's original notion for User Modelling and User Models.

Generally speaking, *modelling* implies the task of representing artificially one part of the 'real world' (i.e. the current 'context'), which is a

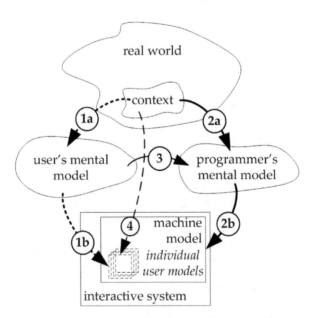

Figure 3.15: General Abstraction for a User Model Definition (adopted from [Kay 2000] and adapted).

process executed in the mind of persons. From the real world and the context, two mental models are of interest, one in the mind of the user (see arc '1a') and the other in the mind of the programmer (arc '2a'). The first step in *user modelling* is then to code an artificial model which fits to the current mental model of users about their real world contexts (see arc '2b' from 'programmer's mental model' to 'machine model'). But naturally, the programmer's model is based on own experiences and learning, and therefore it differs from the user's model. To increase accuracy the programmer should know as much as possible about the user's model (arc '3').

At this point it is worth mentioning that the real target of this first step is not to create one canonical model that fits to all users. Instead, the real target of the programmer is to create a machine model that is able to generate and administer up-to-date individual user models. For this purpose, the programmer implements those user modelling techniques that best fit to the adaptational goals of the interactive system. Hence, after 'the machine runs' the main factors influencing each model in a current real world context are the inputs from the programmer's model (see arc '2b') and from the user's current model states (arc '1b'). Therefore, as very precisely stated in [Kay 2000],

[…] it is the programmer who will define the limits of what can be modelled, how it is represented, how the user can contribute to it and how it can evolve. This is why adaptation should be considered from the very early stages of designing a system.

As can be seen, the second step of *user modelling* comprises mainly the interactions of the user with the system (arc '1b'). Through these interactions, individual user models provide input factors that are utilised by the machine model to create or update individual traits, such as user's preferences, beliefs, knowledge, skills, cognitive styles, goals or interests. Nonetheless, individual user models may need additional relevant user traits that should be gained directly from the real world context (see arc '4'). For example, the need may exist to model the user's learning type by distinguishing between 'imager' and 'verbaliser', i.e. a user that seems to best acquire knowledge through images (figures, diagrams, tables, etc.),

respectively through text. The personalisation goal in this example could be to deliver personalised learning media in terms of mostly image-based, respectively text-based (of course, only if available). This personal trait may differ from the users' beliefs about their learning styles, and therefore, they represent the beliefs of the machine model about individual users. To conclude this example practically, the device in the real world context providing the needed input to the machine model may be an eye-tracking system that monitors the gaze movements of the user on the screen elements (see [García-Barrios 2006e]).

So far, the main issues regarding a definition of the *user modelling* task and of individual *user models* have been discussed. Implicitly, also the functional boundaries of a *user modelling component* (machine model) have been identified. According to e.g. [Gütl et al. 2004] and [García-Barrios 2006b], the field of user modelling may complement and improve user profiling techniques through the application of sophisticated methods, e.g. a user modelling component may organise user knowledge spaces dynamically and autonomously with the aid of ontologies, or provide enriched semantics of profile data through the management of context-specific metadata, or assemble profile data to complex data structures. In essence, user modelling components comprise more enhanced functionalities as user profiling components by means of *reasoning techniques*.

Furthermore, this section has shown that the adaptive capacity of a personalisation system is complex. It is not only limited by the number of adaptable entities (what to adapt) or the goals of an adaptive engine (why to adapt) or the techniques used (how to adapt). In addition, it depends on the sensitivity of the running system, which is defined through the input streams of the user modelling component (see arcs '1b' and '4'). From this point of view, it was also shown that these input streams are captured on distributed sensing components, which are not only implemented by interfaces of the user modelling component, rather they can reside in the current real world context and be represented by specialised devices, e.g. eye-tracker, keyboard, microphones and the like.

3.2.2 Basics of User Modelling

As mentioned in the introductory part of this sub-chapter, this investigation of *user modelling components* (UMCs) is based on the premise that they are independent modules within adaptive systems. Further, taking into account the findings of previous chapters and sections, it is possible to consider a UMC as an autonomous system which is able to serve external systems with specialised *user modelling tasks*. Accordingly, the following aspects are treated in this and the next sections: lifecycle of information in models and content of models, respectively in the next section, internal organisation processes and traditional user modelling techniques.

Two Points of View: from and to the UMC

The first lines of this section have already brought a UMC to the centre of attention whereas other components interacting with it can be seen as providers or consumers of user information depending on the task at hand. In addition, as depicted in the previous section (see also Figure 3.15), highly important duties of a UMC

are to serve as a flexible factory and as an efficient administrator of individual user models. To describe what happens with user information within a UMC the need arises to look from inside out, i.e. the point of view **from the UMC** to external components is needed. These duties as factory and administrator of individual models, to be really successful (i.e. usable and useful enough for adaptational procedures), require first, the acquisition of enough relevant user data to build the best possible, first individual model. From this starting point onwards, the UMC is in charge of the efficient maintenance as well as of the accurate provision of individual user information to its (adjacent) clients.

Thus, in general, the four main global tasks of UMCs can be generalised as follows:

- Acquisition of user information,
- Creation of individual user models,
- Administration of user models, and
- Provision of appropriate user information.

These four global tasks, depicted in Figure 3.16, represent the main aspects regarding *information interchange* and *information processing*. As

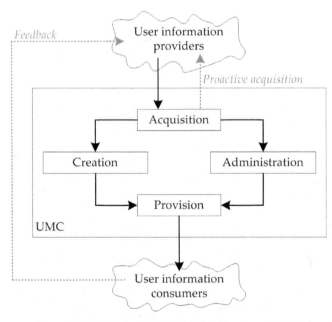

Figure 3.16: Main Tasks of a User Modelling Component (UMC).

stated in the last section, three main ways to acquire information can be identified: interaction, observation and proactive analysis. Accordingly, user information does not enter a UMC just by feeding it from outside; the notions of observation and proactive analysis imply that a UMC can be also able to collect data by itself. For the matters of completeness, consider some examples. On the one hand, an adaptive engine may be able to optimise adaptational procedures, and thus, may give direct feedback to its UMC to adjust the corresponding user model states. Thus, a consumer may also adopt the role of provider alternately. Further, the sensors of the system can be part of the UMC and expand its duties with a fetch-and-provide task by means of observing its environment by itself. On the other hand, the collection of user information might be implemented as an own task in the UMC in terms of proactive analysis, i.e. the UMC might be able to proactively explore an information space and collect the needed data by itself. With these special cases for the tasks of feedback provision and proactive acquisition the point of view stays simple and becomes complete, as shown through the grey dashed arrows in Figure 3.16.

Furthermore, if the point of view is changed, i.e. from the external systems **to the UMC**, the question arises about *which functionality does a UMC place at the disposal of its external clients?* To get an answer, the focus is reduced just to the scope of *information interchange*. A simple way to reflect which basic functions are available to external components is to describe them by means of the *CRUD principle*. The CRUD acronym stands for *create*, *retrieve* (or *read*), *update* and *delete*, and defines the set of basic functions used for data management in persistence layers. Usually, the CRUD principle is well-known as design pattern for distributed object-oriented systems as well as in fields dealing with e.g. file system access, or relational and object-oriented databases (see e.g. [Brandon 2002] and [Yoder et al. 1998]).

Within the scope of this document, the CRUD principle is not used with a direct and exclusive relation to the scope of database management. Rather, it is used to easily address and categorise some relevant functions concerning *information interchange* with UMCs, for example: *creating* a user model (including e.g. initialisation of the model, insertion and addition of new data, creation of stereotypes), *reading* from a user model (including e.g. retrieval, selection of viewing of personal data), *updating* user models (e.g. overwriting existing data, moving information in modelling structures, editing data), and *deleting* functions (e.g. removing data, hiding old data, destroying unused models, archiving data to simulate that a user 'forgets' knowledge). In addition, the notion of CRUD also makes possible to address the management of access rights for different external components, which can be not only be software systems but also humans behind some user interface.

Both presented points of view have been introduced, because it is important to know that the implementation of a UMC as an independent system (or system module) is an *elastic task*. As already shown in the sections 2.2.4, 3.1.2 and 3.1.3, to implement the needed functionality of UMCs depends e.g. on the underlying system goals, the adaptation paradigm and techniques, or the overall system architecture or programming paradigm.

Further, as already indicated in the previous section (see also [Kay 2000]), the goals of a UMC as part of an adaptive system must be considered from the very beginning of the design phase, because the practical limits of what, how and why is modelled are in the hand of the programmers, which in most cases implement the system by means of *elastic users* in their minds.

Lifecycle of User Information

The *global tasks* of UMCs regarding interchange and processing of user information have been defined as: *Acquisition, Creation, Administration* and *Provision* (→ **acap**). As also indicated, this notion leads to the need of services for the consumers of user information. In general, because a UMC can be seen as a (user) data management system, the CRUD principle may be taken to describe the basic *services* of UMCs provided to their consumers: *Create, Retrieve, Update* and *Delete* (→ **crud**).

Thus, the combination of global tasks and basic services of UMCs can be used to define the *lifecycle of user information* in individual user models. That is, information in a UMC can *emerge, stagnate* or *evolve*, and *disappear*. The lifecycle can be explained as follows:

- Emerge

  If an external system wants to add new user information (**c**r**u**d), a process of initialisation must be started in the UMC, i.e. the UMC must first collect the needed user information (**a**c**a**p) and then build a new individual model or a new attribute in an existing model (a**c**a**p**).

- Stagnate or Evolve

  On the one hand, if information already exists, it can be stored and maintained 'as is' (e.g. in simple user profiles). That means that the value is not expected to change over time, it will stagnate rapidly. This type of user traits is known as static, permanent or long-termed (see section 3.2.1). On the other hand, existing information could be 'enriched', e.g. through reasoning methods or semantic post-processing in order to 'increase' its value for the system, i.e. it evolves internally. User information that evolves over time is called dynamic or short-termed, and is used to model e.g. cognitive or behavioural aspects. Within this context consider also the notion of inferred and generic attributes in user models. UMCs may also generate from existing data some new information by itself, e.g. the UMC may infer the level of expertise in a knowledge domain from existing data about learning performance or grading. These aspects imply internal processes of maintenance (a**c**a**p**) to modify (c**r**u**d**) information.

- Disappear

  As in most information systems, information can also be permanently removed in UMCs (cru**d**). Though, consider also some possible special services of UMCs, which could make the impression that information disappears: archiving (i.e. a long-time moving and hiding of data until reactivation) or partial deletion (i.e. making data retrievable or not without permanent deletion from the system).

The topic about the lifecycle of user information in personalisation is considered by the author of this book as highly important, because of the implications of not considering it. For example, consider firstly that privacy and security might be violated if interactive systems do not explicitly inform their users about what happens exactly with accumulated personal data (e.g. who can read the information, why and when will it disappear?). Second, if an adaptive system only supports uncontrolled personalisation (see section 3.1.1), users might wonder why something provided was adapted, and in turn, the acceptability of the system will decrease if not even disappear at all.

Therefore, either the UMC is made directly scrutable for the users or the adaptation decisions should be made visible and understandable (in terms of responding to user questions, such as How and why does information emerge or evolve?, Which is the current state?).[33] Thus, the need arises at this point to investigate in more detail which information can be or is modelled in UMCs, and in addition, for which purpose and how. Due to the fact, that the type of information that is hold and processed in UMCs depends on the application domain, only some aspects can be drawn. The reminder of this section introduces some possible contents of user modelling components.

The Contents of User Models

A user modelling component (UMC) models knowledge about the user of the adaptive system in order to support a better management of personalisation procedures that best match the individual needs of the user. This knowledge can be 'facts' about the user (acquired directly from users' feedback) or 'system beliefs' about the user (acquired from sensors observing the user or inferred through reasoning methods), and is hold within the UMC by means of what was called user information in the previous sections. UMCs provide this information to its consumers in order to satisfy the needs of system users according to the specific application domain.

---

[33] Judy Kay, Alfred Kobsa and Jörg Schreck are some of the experts dealing with such aspects (see e.g. [Kobsa and Schreck 2003], [Kay et al. 2005], [Kay 2006] or [Carmichael et al. 2006]). Some aspects regarding these topics are presented in section 4.2.

According to [Self 1994], within the context of instructional systems, a comprehensive individual user model should include information about *learning goals, learning progress, domain knowledge, preferences, thematic interests* and all additional user information that is significant for the adaptational goals of the system. Further, also within the context of user information, [Sleeman 1985] indicates that learner models can be classified depending on (a) nature of the information, (b) form of the information, and (c) method of interpreting the information. [Brusilovsky 1994] adopted this classification and with respect to the nature of information, identifies that learner models can be divided into two major groups: models containing *course knowledge* and models containing *individual, subject-independent traits*. In a more general context, [Beaumont 1994] identifies four types of possible contents of UMCs: (a) *goals and plans*, (b) *capabilities*, (c) *attitudes*, and (d) *knowledge or beliefs*.

It is impossible to present all possibilities of what can be modelled in a UMC, because the types of information depend also on the technological possibilities to store and manage information, i.e. the range of possibilities vary in its technological 'sophistication and complexity levels'. This issue can be seen from a representational point of view, which in turns can grow in terms of *semantic*, as for Knowledge Organisation (KO). The area of KO is a sub-field of Knowledge Management, where the sophistication as well as the semantic and pragmatic value of digitalised knowledge grows in terms of a data-information-knowledge complexity chain, i.e. the representation of structures vary from low to high level semantics. From this point of view, at lowest level the range of what can be represented in UMCs is as wide as the range of possibilities provided by Data Management, and in addition, those of Information and Knowledge Management (see [García-Barrios 2002] for details in this context). The following paragraphs give further examples of what is or can be modelled in user modelling components.

Personal or *demographic data* is associated to all attributes that are relevant in social-living, such as name, address, age, phone number, place and date of birth, e-mail address, and so forth. This type of information represents usually the starting point to create an individual user model. Related to personal data are *social* or *community traits* in terms of all attributes describing that a user is a member of some group or community (see e.g. [Orwant 1993]), such as affiliations, citizenship, but also hobbies and other interests that can be used to infer *collective information*. Related to personal data are also *physical traits*, which may be gained from direct user feedback or through observation, e.g. physical impediments or usage of prostheses.

Modelling *aptitudes* is an issue that represents a classification of high level aptitudes, such as *cognitive* (e.g. cognitive or learning styles, mathematical abilities, mental speed, prior knowledge), or *affective aptitudes* (e.g. anxiety and interests as motivational traits, or self-efficacy as volitional traits). See [Cronbach 1957] for details.

In [Riding and Cheema 1991], two dimensions for *cognitive styles* of people are suggested. First, a *wholist-analytical* dimension, describing how people process and organise information spaces, i.e. retained overview vs. sequential parts. And second, the *verbaliser-imager* dimension, describing how people represent information in their minds, i.e. word and verbal association vs. mental pictures. According to [Sadler-Smith and Riding 1999], if this dimensions are assigned (e.g. with Riding's computer-based Cognitive Styles Analysis test), then other relevant aptitudes can be derived, e.g. preferred instructional method or media. For example, *preferred instructional methods* can be *collaborative* (role-play, discussions or games), *dependent* (face-to-face lectures or tutorials) or *autonomous* (open learning or computer assisted learning). Regarding preferred media, [Csinger et al. 2004] propose to combine direct feedback with the observation of learners' *media usage behaviour* to recognise and design media presentation (this approach is called intent-based presentation).

Regarding the topic of *learning styles*, e.g. [Felder and Silverman 1988] identify four main dimensions: *sensing-intuitive, visual-verbal, active-reflective and sequential-global*. As stated in [Parvez and Blank 2007], from these styles learning preferences can be inferred, e.g. (a) *active* learners prefer to work in a groups, (b) *reflective* learners analyse situations before trying some action and prefer to work alone, (c) *verbal* learners prefer written or spoken materials, (d) *visual* learners prefer illustrations, e.g. diagrams, maps or flow charts, (e) *sensor* learners prefer knowledge delivery through facts and concrete procedures, (f) *intuitive* learners are innovative and prefer theoretical materials, (g) *global* learners need 'the big picture' and are able to work through large phases, (h) *sequential* learners prefer to work through an ordered and incremental path. Other possible learning style theories as well as their implementation in AEHSs are shown e.g. in [Stach et al. 2004] or [Brusilovsky and Peylo 2003]. As for these mentioned (cognitive and learning) dimensions of styles, other types of preferences can be also inferred from tests or by observing users. For example, [Sadler-Smith and Riding 1999] identify *preferred assessment methods* and divide them into *formal* and *informal* methods, such as tests or essay questions, respectively individual assignments or different question types (e.g. multiple-choice or short-answer). In general terms, the detection of users' *interests* and *preferences* is still a big challenge in the field of adaptive systems and play a central role in e.g. recommendation-based systems, where (usually) individual interests are inferred from group interests through machine learning methods over observation-based data (see next section).

UMCs can contain also user information about *personal knowledge*, which is, naturally, one of the main addressed aspects in adaptive e-learning (see section 4.3.8). Though, it is relevant to state at this point that this topic is also one of the main reasons to integrate knowledge-based reasoning methods in UMCs. For example, if the UMC has already acquired information that a learner has a low knowledge level in some topic, e.g. in 'distributed databases', it can be inferred that the learner has also a low knowledge level in 'replication methods', because an understanding of distributed data is a prerequisite to understand the complexity of replication methods. This a highly important issue, because, as mentioned in section 2.2.4, an indispensable function in Intelligent Tutoring Systems is to model *learner's knowledge* as well as *lack of knowledge*. Without these models, adaptive e-learning systems are not able to successfully adapt instructional steps, actions or learning material, which in turn e.g.

will not contribute to enhance learning performance.

Despite of assessment and grading, the learning performance is also a matter of time, because some systems support learning strategies which are based on efficiency and competitiveness (see Cooperative and Competitive Learning in section 3.1.3). Competitive learning can be measured with the aid of observation or tracking *learning behaviour*, such as *navigation behaviour* in terms of measuring the time learners stay on single lessons or tracking *reading behaviour*. This notion of behaviour tracking is only one of many possibilities to track *personal behaviour*, i.e. this notion gives a time-dependent context to UMCs, which can be modelled through footprints or landmarks defining a chronological order of related user information.

According to [Burton et al. 2003], *mental activities* can be seen as actions (in the sense of e.g. thinking is talking to ourselves), rather than as indicators of present consciousness, and thus, this point of view play the central role for distinguishing between behavioural and cognitive approaches. One possibility for behaviour tracking comes from the field of exploratory learning, where learners are 'free' of exploring a specific space of learning materials (see e.g. [diSessa et al. 1995]). To give an example, [Stelmaszewska et al. 2005] shows how the learner's *information seeking behaviour* can be tracked through the navigational paths of users in the retrieved set of materials. In order to deliver most relevant documents to learners, most successful *navigational patterns* are calculated from the behaviours of all learners. The resulting best matches are the presented to each individual learner by means of 'tips' attached to the search results. On the other side, also the 'behaviour' of documents can be tracked in terms of a chronological presence within the set of 'best matches', 'bad matches', 'too many matches', 'original query', 'reformulated query', etc. Among others, the study presented in [Stelmaszewska et al. 2005] showed that the implementation of a user-adaptive interface to explore digital libraries can help non-experienced learners to apply expert retrieval strategies, because the system provides a proper, timely and contextualised feedback.

A last interesting example of what can be the content of a UMC, is *motivation*. Learner motivation can be measured using behavioural information modelled as transient or permanent attributes (see section 3.2.1), such as mental effort or focus of attention, respectively topics of interest or presence of distraction. These attributes can be associated with other existing traits, e.g. knowledge level or topic complexity. A solution for such a UMC was proposed by [Far and Hashimoto 2000] on the basis of motivation and knowledge states. In this solution approach, an individual motivational state is represented in a graph that holds possible dependencies among motivational attributes and learning tasks. In simple words, if there exists currently a connection from the distraction node to the attention node, this means that the learner's attention is being influenced by distraction.

The variety of possible user traits that can be modelled in UMCs depends on the predefined goals of the overall adaptive system, i.e. *why to model* is a matter of *why to adapt*. Within the scope of Adaptive Hypermedia Systems, [Koch 2000] identifies the following objectives of user modelling: help to learn, support collaboration and assistance, find information, tailor information, improve man-machine communication. In the more general terms of general adaptation, [Finin 1989] identifies four main utilisation purposes for user models; accordingly, user models are units for representation and reasoning that are utilised to (a) understand users' information seeking behaviours, (b) provide help and advice, (c) get feedback from users, and (d) provide output to consumers of user model information.

The aspects depicted so far give a first overview on *what can be modelled* in a UMC, and connected with this, *why are models needed and used*. The full range of attributes and inferred traits that can be modelled in UMCs is a topic treated within the scope of Generic User Modelling Systems (GUMS). Such systems are described in the next section. Further, the lifecycle of user information was already shown. Though, the answer to *how to model user information* is still missing from the technological point of view. An investigation of structural and functional aspects of GUMSs requires basic knowledge about the rele-

vant user modelling techniques, which are introduced in the subsequent section.

3.2.3 User Modelling Techniques

This section aims at presenting an overview on the technological methodologies implemented for user modelling components. Ideally, some kind of 'comprehensive categorisation' would be desirable to assign unique user modelling techniques to unique user modelling tasks. Unfortunately, such an attempt fails, because either single technique are used for very distinct tasks or hybrid solution approaches are required to satisfy the requirements of complex tasks. As stated in [Orwant 1996],

> "no single technique will be best for all, or even a majority, of tasks. Some data streams are temporal, others spatial; some are discrete, others continuous; some are numeric, others symbolic. It will not always be clear which technique is best, or even valid, for a particular domain."

The previous sections have shown that the development of adaptive systems for e-learning is a hard task due to the variety of existent requirements on the systems. For example, the techniques used for the implementation of personalisation engines in Adaptive Educational Hypermedia Systems depend on the tasks and goals at hand, such as adaptive link annotation, adaptive assessment, adaptive guidance, and others (see e.g. section 2.2.4). Consequently, the implementation of a user modelling component depends also on the tasks and goals of the personalisation engine. Further, as introduced in the previous section, different points of view are possible to describe User Modelling, e.g. focusing on 'information processing' to describe internal global tasks and the lifecycle of user information, or focusing on 'information exchange' to define the set of services available to external user information consumers. Thus, the variety of interdependencies within the components of adaptive systems represents the main reason why the development of User Modelling Components (UMCs) is often restricted to the specific needs of the systems in which they are embedded. An attempt to simplify the view on *how to model user information* (in other words: which techniques are used for which tasks) is undertaken in this section following two approaches: (a) a *traditional perspective* concentrates on the 'main global tasks' of UMCs (see Figure 3.16) and assigns commonly used techniques to these tasks, and (b) a *relative perspective* introduces some distinct techniques from other research fields that are utilised to build and manage user models for the distinct purposes of adaptive systems.

Traditional Perspective: The Tasks at Hand

In the previous section, the four main *global tasks* of User Modelling Components (UMCs) were identified: *acquisition of user information, creation of individual user models, administration of user models,* and *provision of appropriate user information.* In literature, these four issues are referred to as *processes* in user models. According to [Kay 2000], for building and maintaining user models the following processes have to occur: elicitation of user information (i.e. acquisition of information from direct feedback), modelling (comprising different techniques, such as observation, stereotyping or knowledge-based reasoning), managing uncertain information, and updating information. Depending on the degree of complexity of the UMC goals, these processes can be, from very simple to extremely complex.

The observations in this section are based on the first presented notion of *global tasks*. Hereby, the introduced 'traditional techniques of user modelling' are mostly related to the following two issues: *initialisation of an individual user model* (i.e. acquisition of user information and creation of individual user models), and *maintenance of an individual user model* (i.e. the methods required to process and administrate user information). The task of providing user information is omitted at this point, because it is interpreted as a pure communication issue in technological sense, and thus, depends on the semantics of the output interface of UMCs, i.e. the consumer component must implement this interface in order to understand the data structures administrated in the UMC.

The *initialisation* of an individual user model is a process that comprises two main techniques: *feedback* and *observation*. Further, the initialisation process ends when the user model reaches the

state of 'being ready', i.e. the information it contains is usable and useful for its consumers. In the context of initialisation, *feedback* denotes that individual user information is gathered and interpreted within the most rapid and efficient one-step-process. *Observation* is a multi-step-process needed when user's feedback is not enough for initialisation and requires e.g. a longer-termed interaction with sensors that track user behaviour, or the support of other system components.

Thus, initialisation of user models can be critical in terms of the so-called *ramp up* or *cold start* problem (see e.g. [Burke 2002] or [Denaux et al. 2005a]), which arises when a user interacts for the first time with the system and there is a *lack of enough information* to initialise an individual user model. According to [Self 1994], a user model can be initialised through (a) *explicit questions*, (b) *initial testing* or (c) *stereotypes*. Explicit questioning can be problematic (for the users) if the number of questions is too large. Therefore, *adaptive questionnaires* can be used to reduce the number of questions (see e.g. [Kurhila et al. 2001]), where the length of questionnaires is optimised through AI techniques that drop-out uninformative questions. Initial testing aims also at providing questionnaires to gather as much useful information as possible, but the ultimate goal is to use this information to infer other user attributes, such as background knowledge or cognitive styles.

In contrast, the main goal of explicit feedback is to acquire as much user attributes as possible for a first usable profile. To control the length of tests, *neighbourhood of knowledge states* may be applied (see [Self 1994]), e.g. if the knowledge domain of question B is a sub-domain of question A, the mastery of domain A implies the mastery of sub-domain B, and thus, a reduction of the length of the test is possible (with the precondition that the test is well constructed). Furthermore, if e.g. only a specific number of questions is tolerable, but more user information is required, then *default assumptions* can be created (see [Koch 2000]), i.e. the UMC compensates its lack of knowledge with the aid of predefined rules (as in the example given above). Once sufficient information is acquired, some other techniques come into play, such as (a) *overlay models* or (b) *stereotypes*.

Within the scope of modelling knowledge of learners in adaptive instructional systems, the *overlay approach,* as coined by Carr and Goldstein in 1977, is based on the idea that an individual model of a learner is a *subset of the expert model* (see [Sison and Shimura 1998]). Usually, an expert model is represented as hierarchical or semantic network of topics. As stated in [Koch 2000], overlay models can be implemented by assigning a numerical value to each knowledge topic in the user model (i.e. in the derived subset), which indicates the degree of confidence for the system's belief that a learner knows the concept.

According to [Rich 1979], a *stereotype* is a collection of most frequent traits of users. For the design of stereotypes, possible types of learners are grouped in set of attributes (stereotypes). For AEL systems, as stated in [Kay 2000a], there exist *fixed and default stereotypes.* If applying the former technique, one (type of) learner is assigned to an existing stereotype according to the measured, inferred or *observed* learning performance. A more flexible approach is possible through default stereotyping, where first, learners are assign to default stereotypes at the beginning of learning sessions, and then, by *observing* patterns of change in individual data streams (e.g. regarding learning performance or interaction behaviour), more individualised stereotypes will replace the initial (default) one.

Further, as for the initialisation of user models, updating user information in UMCs may require acquisition techniques. For example, to update a user model, default stereotyping can be also used, because the changes in observed user information may lead to a replacement or refinement of stereotypes. Thus, a recurrent use of the above depicted techniques may support the *maintenance of individual user models.* Other possibilities to administer the user information in UMCs have been implicitly depicted in this section, such as updating by means of *rule checking* (as for default assumptions) or *knowledge overlays.*

One of the main problems in the field of User Modelling is the *reasoning capacity* of UMCs. For example, if UMCs are implemented as User Profiling Components (UPCs, where no new information attributes need to be generated or trans-

formed), a UMC can be seen as a static repository of user traits. In this case, even overlay techniques can be utilised without breaking the notion of UPCs, because the individual models in the UMC are just static weighted subsets of some (bigger) *canonical models* from which they are derived, i.e. for administration and updating just recurrent comparisons between the models could be enough. Though, problems may arise in this context, because the practical limits of the UMC will depend directly on the structural limits of the canonical models.

The technique of stereotyping (since its first introduction by [Rich 1979] and after some refinements, see e.g. [Chin 1989]) can be considered to be 'the' basic element in many UMCs. As it is important to show that stereotyping is more than just a data-to-stereotype matching approach, some more details must be known about this technique. The aim is to identify the relations and distinctions between stereotyping and (knowledge) reasoning. As stated in [Kay 1994], stereotyping can be seen as a special type of knowledge-based reasoning, which concentrates on reasoning about persons. The underlying principle of stereotyping is: assign default assumptions about the current user until the system has collected better information to assign better assumptions. As can be seen, this principle implies that the system must know why and when to move to a better stereotype, i.e. it must implement some reasoning techniques.

In essence, stereotyping comprises two main phases: (a) *activation* of stereotypes and (b) their *assignment* to or *retraction* from individual models (see Figure 3.17 on next page). For that purpose, some kind of reasoning engine must be implemented, which is in charge of dealing with pre-defined *triggers* and their relation to a known *pool of stereotypes.* Triggers are needed for the activation of stereotypes and can be defined as a set of facts and rules. Thereby, the input of triggers (i.e. 'data about user') feeds specific facts and rules in such a way that specific conclusions can be inferred for the activation of specific stereotypes. Therefore, without triggers an assignment to an individual model is not possible. An example: consider an AEL system that needs to model learners according to their previous knowledge level in a certain domain, e.g. 'novice interface designer' and 'expert interface designer'. Some triggers for this example could be 'has at least one year practical HCI expertise', 'knows at least three usability evaluation methods', 'has completed a study as graphics designer', and so forth. According to the input data (e.g. explicit feedback of the learner or data from a personal knowledge modeller) each trigger shows a validity result in terms of e.g. 'true' or 'false'.

Due to the possible lack of enough information, it may be specified that (a) at least one of the previously mentioned triggers is required to be 'true' for the activation of the novice stereotype, and (b) at least two to activate the expert stereotype. These specifications are referred to as *conclusions* in Figure 3.17. Further, in [Kay 1994] the notion of *essential triggers* is introduced and leads to the definition of *retraction*. For the example at hand, it might be then highly relevant to be an 'expert' that one of the triggers is 'true'. If the system has already assigned the expert stereotype for a specific learner, but after acquiring more detailed information the essential trigger is found to be false, then the stereotype will be deactivated, i.e. the learner might 'fall down' to the novice level. Thus, a recurrent evaluation of the validity of stereotype assignments is also essential within this technique, and enables a refinement of better fitting individual models. Figure 3.17 shows to ways of structuring stereotypes, as a hierarchy or as a set, e.g. moving up and down in a hierarchy might be useful to define levels of detail or granularity for user information.

The simplest way for thinking about the use of stereotypes is considering them as *default collections of user attributes* that can be adopted by an individual model. Further, the level at which reasoning (e.g. inferring assumptions or user traits) is implemented depends on the task at hand, i.e. it might be insufficient to store concrete user attributes in stereotypes, because the information inside should represent a basis for more inferences. For example, the information in a stereotype could be a set of some abstract parameters ('more than 50 learning objects accessed', 'no graphics used', 'all textual contents have been read', 'no use of navigation tree',..) that are used as input for further models ('from the abstract parameters and the good results of the already mastered exams, it can be inferred that the

learner prefers textual presentations of learning materials and is a sequential reader'). As a result, the adaptive engine of the AEL system may decide to deliver more detailed textual materials (if available) and to replace the navigation tree by 'previous and next buttons'.

Hence, the specific content of a stereotype may depend on the adaptation goals of the overall system and can be specified in the design and implementation phase of the UMC. As also stated in [Kay 1994], some of the most relevant technical benefits of this technique are simplicity and efficiency:

"if a large number of user models can have a substantial common part stored once in a stereotype, less storage is needed and if the same reasoning is to be applied to many users, we can devise ways to do this process efficiently."

Nonetheless, as in many other solutions for adaptive e-learning systems, the practical efficiency of solutions is not a matter of the tech-

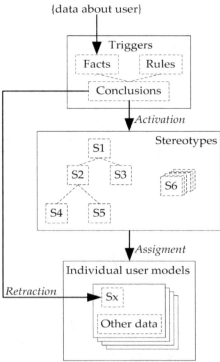

Figure 3.17: Simple Example of the Usage of Stereotypes for Individual User Models.

nique or the interaction with the end-user, it is a matter of setting-up the environment. In [Kobsa 1993] it is recognised that the problem of using stereotypes is the work-load to build and fill the required stereotypes for a specific application or domain. The user model developer must e.g. think about the following tasks: the identification of user sub-groups (to define the types of stereotypes), the identification of key user traits (to define the content of stereotypes), as well as the order or representation of stereotypes (to define the structure of stereotypes). User sub-groups and their corresponding common key traits help to categorise the community of all possible users in stereotypes.

In addition, stereotyping often implies that the supposed members of a sub-group should own some application-relevant skills, i.e. the user model developer should also know about the application domain, a fact that in turn restricts the usability of particular stereotypes to one or a few application domains. Further, key user traits and triggers define the relevance and accuracy with which a learner is supposed to belong to a specific sub-group. Because stereotyping is often used to accelerate the initialisation of user models, the amount of input data must be kept as small as possible to enable a rapid and easy assignment to sub-groups. [Kobsa 1993]

Another critical aspect for this technique is related to its susceptibility to inaccuracy; simply because it is based on beliefs of the system about users (see [Kobsa 1993]). This is why considerable efforts have been made since the early 1990's to make user models scrutable, i.e. to allow users to inspect and modify their personal models (see e.g. [Kay 1990] or [Orwant 1991]). Though, scrutability in this context is a hard task, because the assumptions of the system are difficult to translate from a machine-readable (e.g. formal definitions in modal logic) into a human-readable language (see e.g. [Kobsa 1990b]). Some effects of interactions between stereotypes (e.g. skimming effect, chorus effect of dark horse effect) are shown in [Lock and Kudenko 2006]. Further, hybrid solutions combining overlay models and stereotyping are also possible, such as in the WHURLE system (see [Zakaria et al. 2002] or [Zakaria et al. 2003]).

So far, the description of the stereotyping technique has been presented in more detail to point out that each technique may involve implicitly the usage of other techniques. Thus, implementing the use of stereotypes may be a simple technique, but can also be combined with rule-based reasoning or other techniques to fulfil more complex tasks. Further, it is highly important to emphasise again that the input data of UMCs must not be gathered directly from the user or delivered by other software components, it can be part of a proactive task of the UMC to collect automatically some data related to specific users and infer some assumptions or create internal information spaces autonomously.

Therefore, other highly relevant techniques are identifiable, e.g. inducing and compiling knowledge through Machine Learning methods, or deriving personal traits indirectly from the identification of common patterns with other users, etc. Finally, the ultimate goal of user modelling is to respond to as much questions about specific users as possible. Ideally, the user information in individual models should match the characteristics of the real-world user, but this task implies to ask the user continuously about the certainty and accuracy of the stored data. And this last issue represents one of the most critical issues about the techniques of user modelling: without disturbing the user to gather as much information as possible, how to assure the highest confidence in self-generated assumptions?

Relative Perspectives: Technologies at Hand

The spectrum is broad: different components interacting with a User Modelling Component (UMC) can ask different questions about a specific user for different purposes: Does the user dislike this information?, Does the user like this product?, Does the user like article A more than article B?, Is the user a good learner?, Is the user interested in receiving emails about our newest products?, What is the user's ZIP code?, and so forth. A detailed description of all available techniques that can be used in the field of User Modelling runs out of the scope this book. Nonetheless, the reminder of this section gives some examples of when some techniques are used and which is the main idea behind the techniques.

Mainly, the techniques presented here are based upon algorithmic solutions from other research fields, such as collaborative or content-based filtering, which are commonly used in or derived from the fields of Information Retrieval, Recommender Systems, or Machine Learning ([Mackinnon and Wilson 1996], [Burke 2002]). Further, techniques from Artificial Intelligence to manage uncertainty of assumptions can be used (see e.g. [Kay 2000]). The use of 'apparently' complex techniques must not be inefficient: e.g. most of the inferences performed by some user modelling systems are very quickly done when *cheap techniques* are used [Orwant 1996]. In [Orwant 1996], cheap learning techniques used within the DOPPELGÄNGER user modelling system are presented: Linear Prediction (to model recurrent events), Beta Distribution (to model interests), Markov Models (to model location), Hidden Markov Models (to categorise behaviour), Cluster Analysis (for constructing communities).

Linear prediction is commonly used to predict cyclical behaviours. In DOPPELGÄNGER, linear prediction is used to predict the next point in time when users are going to read their newspaper and for how long they will read. Linear prediction is a robust and fast technique to predict occurrences and durations of events, but the results might not be accurate enough if the predictions must be made too far into the future.

With *Beta distribution*, the predictions are proportional to the number of recollected past observations. For the calculation of interests or preferences, firstly, only two numbers are required, e.g. a quantity of 'hits' and a quantity of 'misses', and then these numbers are used to generate both, an 'estimation' and a 'confidence'. In DOPPELGÄNGER the numbers mean the number of newspaper articles viewed or not viewed by a user from among all shown articles, i.e. if the particular topics of each article are known, it is assumed that a 'clicked' (i.e. 'visited') article means an interest on the topics of the article. Further, as also stated in [Orwant 1996], there exist problems with this approach: (a) because all observations are treated equally,

they all affect estimations in the same degree, and desirable would be to give more recent responses a higher relevance; (b) the topics have a all-or-nothing notion (i.e. it is just define if a topic is assigned to an article or not), but in real life one article may contain just a mediocre explanation of one topic, whereas another article could be a simple and precise explanation of three topics; (c) user feedback has also a 'all-or-nothing' notion, and intermediate stages are more natural, e.g. 'liked it not much' or 'found it awesome'; (d) the granularity of rating an article is not given at topic level, thus, if articles are good examples of many topics, how then to identify exactly which topic was meant to be 'liked'? [Orwant 1996]

Another technique, which not only expresses how long an event is expected to last, but also which event comes next, is the *Markov model*. The DOPPELGÄNGER system uses Markov models for the prediction of changes of users' location. Hereby, the system tracks information about location movements of users and models the locations as a set of 'states'. Based on these states, it calculates a 'matrix of transition probabilities', i.e. predictions are hold in this matrix expressing which state is going to be reached next with the highest probability. In terms of DOPPEL-GÄNGER: the states of a personal Markov model changes according to the locomotion of a user, and further, the accuracy of predictions increases, the higher the frequency of tracked locomotions and the longer the observation period. [Orwant 1996]

The more distinct user behaviours must be observed and predicted, the more distinct Markov models are needed. In addition, this technique requires to maintaining single Markov models for each user. Therefore, if some relevant Markov models already exist and can be categorised in a collection of representative models, a mechanism is required to just select the model with the best predictions. This can be realised through *Hidden Markov Models* to categorise different behaviours. For more details about Markov models see e.g. [Orwant 1996], [Hafri et al. 2003a], [Hafri et al. 2003b].

In order to solve the problem of initialisation of an individual user model for a new user, [Orwant 1996] uses a technique called *Cluster Analysis* to dynamically construct 'communities' of users with similar traits. This technique involves an iterative analysis over all known user models searching for relevant patterns of similarities among the traits of users. Thus, existing 'clusters' are used to initialise new individual user model with the attributes in the cluster that best fits to the input data of a new user (see [Orwant 1996] for details).

Other techniques can be used to infer user traits from observation. For example, *Plan Recognition* may be used to describe learner intentions or desires, whereby a 'plan' is represented as a sequence of actions to achieve a specific goal, i.e. the fundament is given by observing certain actions of system users. Such systems try to determine all possible learner plans, which are valid concerning the observed actions. This calculated set of plans can be decreased by taking new leaner actions into consideration. Within the context of plan recognition and *Path Prediction* different algorithms and types of techniques can be used, e.g. Bayesian networks [Li and Ji 2005], plan libraries or plan construction [Kobsa 1993], sequential and non-sequential patterns [Mobasher et al. 2002], Keyhole recognition or intended recognition [Cohen et al. 1981], prediction trees [Chiu and Webb 1998]. For more details see e.g. [Carberry 2001], [Doane and Sohn 2000], [Sison and Shimura 1998], [Danine et al. 2006], [Desmarais and Gagnon 2006]. Also, predictive statistical models can be also used to anticipate certain aspects of human behaviour, e.g. goals, actions and preferences [Zukerman and Albrecht 2001]. Further, [Claypool et al. 2001] claim that implicit indicators extracted during browsing a Web document space can be as predictive of interest levels as explicit ratings.

It is worth mentioning at this point again that the techniques used for User Modelling depend also on the required form of the individual user models, which in turn depends on the task at hand. For example it may be enough to model users as sets or clouds of knowledge topics represented as graphs or as a single reference to a sub-model of a canonical big knowledge structure. Or it may be needed to infer user interests indirectly from a big repository of data items (e.g. documents or products), whereby the internal representation of interests is a dynamically generated sub-set of data item descriptions.

This last issue is highly relevant in user models of Recommender Systems, where the techniques used to provide personalisation (originated from the field of Machine Learning) can be divided into a five categories (for details refer to [Burke 2000] and [Burke 2002]): (a) *collaborative*, (b) *content-based*, (c) *demographic-based*, (d) *utility-based*, and (e) *knowledge-based techniques*. Regarding the application of Machine Learning techniques in User Modelling, the distinct forms that a user model may take are variable and depend on the purposes for which user models are needed.

As stated in [Webb et al. 2001], user models may describe (a) some cognitive processes underlying sequences of user actions, (b) differences among user skills and expert skills, (c) behavioural patterns and preferences, or (c) static user traits. Therefore, four major requirements for the use of Machine Learning techniques can be identified: the need for large data sets, the need for labelled data, concept drift, and computational complexity (for more details see [Carpenter and Grossberg 1988], [Bloedorn et al. 1996], [Pazzani and Billsus 1997], [Chiu and Webb 1999], [Webb et al. 2001]).

Finally, other techniques for user modelling include: probabilistic methods on weighted graphs of XML documents ([Cannataro and Pugliese 2000], [Cannataro et al. 2001]), rule-based data mining [Adomavicius and Tuzhilin 2001], user exploration based control [Kinshuk and Lin 2003], vector space models [Çetintemel et al. 2000], fuzzy inferencing ([Arabshahi et al. 1993], [Lascio et al. 1999], [John and Mooney 2001], [Sevarac 2006]), Gaussian curves [Specht and Kahabka 2000], ontology-based personalisation ([Pretschner and Gauch 1999], [Middleton et al. 2004], [Weißenberg et al. 2004], [Denaux et al. 2005a], [Kay and Lum 2005]), information retrieval and filtering ([Mackinnon and Wilson 1996], [Freyne and Smyth 2006]). For comparisons among some of these techniques see e.g. [Jameson 1996], [Gonzalez et al. 2006].

The next section presents an overview on application- and domain-independent UMCs, also known as *Generic User Modelling Systems*.

3.2.4 User Modelling Systems

Th considering of user modelling components (UMCs) as independent modules in adaptive systems implies considering them also as stand-alone (sub-)systems. If they have to satisfy the needs of different types of users, applications or systems interacting with them, UMCs have to be domain independent, extensible (to support pluggable additional functionalities), and should include interfaces that support open standards (see e.g. [Fink 2003]). Further, as stated in [Kay 2000], they should be reusable e.g. in terms of 'a database of user models'. At present, some of the existing UMCs following these issues are called *Generic User Modelling Systems* (GUMS), a term introduced by Alfred Kobsa as a result of his works on *user modelling shell systems*. According to [Kobsa 2001], the term 'generic' denotes and focuses on being 'application-independent'.

Evolution of GUMS

At the end of the 1970's, user models had been developed in the application fields of Dialog Systems and Human Computer Interaction (HCI). For example, the dialog system GRUNDY utilised a stereotype-based UMC to provide personalised recommendations of books (see e.g. [Rich 1979]). In the field of HCI, until the mid of the 1980's user models had been small applications that were embedded in systems for the purpose of personalising user interfaces (see [Murray 1987]). Subsequently, as indicated in [Kobsa 2001], a number of adaptive systems emerged in distinct application fields, which collected different types of user information. As said, UMCs of that time were embedded solutions, and thus, entirely application dependent, and consequently, not reusable.

Thereafter, in 1986, Finin and Drager published 'GUMS$_1$: General User Modeling System', a first description of a 'domain independent' UMC for the creation and maintenance of long-term, stereotype-based individual user models. According to [Kobsa 2001], GUMS$_1$ defines the starting point for subsequent *Generic User Modelling Systems* (GUMS). Please note that GUMS$_1$ is the name of the system in [Finin and Drager 1986] and the 'G' in the acronym stands for 'gen-

eral' (Finin's connotation for 'domain independent'), whereas GUMS is the acronym introduced by Kobsa's for those types of user modelling systems that are 'generic', i.e. according to Kobsa, 'application-independent'. Though, Alfred Kobsa worked until the mid 1990's on what he coined as *user modelling shell systems*, which were application-independent shell-based UMCs. The term 'shell' is a result of the influences of the field of Artificial Intelligence (AI), more precisely the field of Expert Systems, as can be concluded from [Kobsa 1995]:

"until recently, hardly any software tools have been available that facilitate the development of user modeling systems. [...] developers had to program the required user modeling functionality from scratch. [...] It therefore makes sense to condense basic abilities which most user modeling components must have into 'empty', application-independent user modeling shell systems. At the time when an application system is developed, a shell system must be selected and filled with user modeling knowledge pertaining to the particular application domain. At runtime, the 'filled' shell system will then perform the role of the user modeling system in this application [...]. A similar line of research can be found in the field of expert systems, in which experience gained from individual expert systems lead to the development of expert system shells."

Until the mid of the 1990's, according to [Fink 2003], explanations for what these *GUMS* are and for their specific functionalities were a matter of intuition as well as of experience from the study of user-adapting applications of that time (see also [Kobsa 2001]). Thus, as can be followed from the influence of the Expert Systems field and as stipulated in [Fink 2003], the development of UMCs was considered a *knowledge processing task*, because of the very strong affinity between the research areas of AI and User Modelling. This is the reason why the main features of user modelling systems of the 1990's (with few exceptions) can be summarised as follows:

"- generality including domain independence [...], - expressiveness (i.e. maintaining as many types of assumptions about users' propositional attitudes as possible), and especially

- strong representational and inferential capabilities." [Fink 2003]

Another notion for application-independent UMCs found in literature is that referring to *user modelling servers*. Since the mid 1990's, the distinct networking possibilities of remote computers (e.g. local area networks LANs, wide area networks WANs, etc.), the increasing application of Internet-based technologies as well as the international spreading of the WWW as the main 'channel and platform' for the publication and interchange of information, reinforced the implementation of several software solutions following the *Client-Server* architecture, which is based on request-response chains of messages among distributed processes. Thus, with the experiences gained from user modelling shell systems it became popular to construct UMCs as single and independent servers within distributed systems. Finally, at present, the most relevant technological influence on the architectures of UMCs is represented by Semantic Web and Web Service-based approaches, which enable current applications to improve not only human-machine interaction but also machine-machine interactions. Then, it would not be surprising at all, if currently developed or next solutions of Web-based, independent UMCs became the attribute to be *user modelling services*.

The reminder of this section gives an overview over some relevant solutions for independent UMCs. For this purpose, the author of this book utilises the term *user modelling systems* (UMSs), which denotes a grouping of the previously introduced types of systems. Hence, within the scope of this book, a UMS is independent in the sense of the two notions behind GUMS₁ and GUMS, namely domain- and application-independent. Based on the analysis of user modelling systems in [Kobsa 2001], the reminder of this section presents a review on systems focusing on UMSs *derived from research work*, though, it makes then reference to *commercial* systems and comes back to the context of adaptive e-learning.

User Modelling Systems

The next pages of this section present the user modelling systems *GRUNDY, GUMS₁, BGP-MS,*

DOPPELGÄNGER, umt, Protum, um, Personis, LDAP-UMS, CUMULATE and *PLUS*. They can be considered as the most relevant results from research work since the early 1980's.

As stated in [Rich 1979 and 1983], the *GRUNDY* system generates recommendations of novels by comparing individual stereotype-based user models to the models of the books known to the system. GRUNDY selects the best match and provides a short description of a book including an explanation of the reasons leading to the set of recommendations. The system also asks the user to rate the recommendation, and if the user is not satisfied, it asks why not. From the acquired feedback, GRUNDY updates both, stereotypes and individual user models. The core of the system is represented by a module that is able to build a personal 'User Synopsis' (USS), i.e. an individual user model. A USS is constructed by combining user's direct feedback information, inferences from user interactions and stereotype predictions made by GRUNDY. Further, the information in the USS is used to steer the rest of the GRUNDY system.

According to [Kass and Finin 1988], user modelling systems must track users' behaviour and infer implicitly something about personal attitudes, because mostly (but depending on the situational context of the application domain) it is not 'socially acceptable' to ask users explicitly about personal attitudes. For example, the *Real-Estate Agent* and *HAM-ANS* systems acquires explicitly some needed information about users' preferences and makes deeper inferences about other preferences. Real-Estate Agent and HAM-ANS are systems in the domains of apartment and hotel room rentals (for details see e.g. [Morik and Rollinger 1985] and [Hoeppner et al. 1983]). Kass and Finin collected and extended some findings from [Rich 1979], [Rich 1983], [Sparck-Jones 1984] and [Finin and Drager 1986] to stipulate six main dimensions to describe user models, as depicted in Figure 3.18 (on next page). Thus, according to this categorisation, a *generic user model* consists of models for six *classes of users*. This categorisation is relevant for the context of this book, because it introduces the term 'agent' as a dimension of user models (see fifth dimension in Figure 3.18). The term 'agent' in Kass and Finin's categorisation of user models refers to general models of individual entities that may have no relation to the modelling system. As stated in [Kass and Finin 1988],

> "imagine a futuristic data base query system: not only do humans communicate with the system to obtain information, but other software systems, or even other computer systems might query the data base as well. The individuals using the data base might be quite diverse. Rather than force all users to conform to interaction requirements imposed by the system, the system strives to communicate with them at their own level. Such a system will need to model both people and machines. A second situation is when a person uses an application such as an advisory system on behalf of another individual; the advisor in this case may be required to concurrently model both individuals."

The number of agents underlines two relevant aspects, first that the UMS may model one or more agents, and further that the modelling of one agent may depend on the models of other agents. This notion of agents (in 1988) might be one of the first hints towards a *multi-purpose* UMS, because in order to provide usable and useful personalised information to different applications or systems, the UMS may need to model e.g. the traits, goals and behaviour of such applications or systems in terms of *agent models*. As depicted in previous sections, information providers and users of UMCs can be humans or adjacent software units, i.e. also devices may interact with UMCs, such as a mobile phone defining the current context of a user in a Context-Aware System. As an 'agent', this mobile phone is one among thousands of possible types of mo-

bile phones. Therefore, it would not make much sense to collect data about each personal phone in each individual user model. Hence, by modelling 'phones' as agent models, is would be only necessary to build a 'linkage' from each user model to the 'personal' phone model.

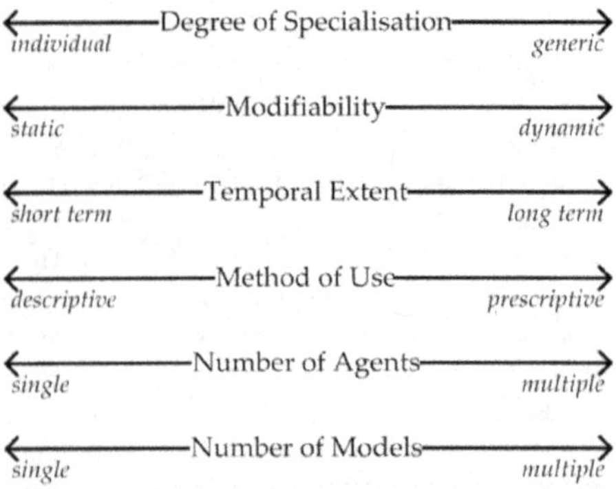

According to [Finin and Drager 1986], *GUMS₁* aims at providing stereotypes in the form of a tree hierarchy (see Figure 3.19). Each stereotype describes facts and rules for the UMS to reason about individual users. From first user information a default stereotype must be assigned. Further, rules might be utilised to derive new user information. If one fact in a stereotype contradicts other facts, *GUMS₁* pops to the next higher stereotypes in the hierarchy that does not include the 'problematic' fact. Therefore, a 'good' set of facts and rules is required to shift the current user up and down the branches until the best suitable stereotype is found. Further, the more

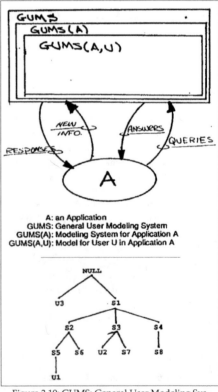

A: an Application
GUMS: General User Modeling System
GUMS(A): Modeling System for Application A
GUMS(A,U): Model for User U in Application A

Figure 3.19: GUMS: General User Modeling System [Finin and Drager 1986].

←————Degree of Specialisation————→
individual   *generic*

←————Modifiability————→
static   *dynamic*

←————Temporal Extent————→
short term   *long term*

←————Method of Use————→
descriptive   *prescriptive*

←————Number of Agents————→
single   *multiple*

←————Number of Models————→
single   *multiple*

Figure 3.18: Dimensions of a User Model according to [Kass and Finin 1988].

user information is available, the more stereotypes may be activated moving through distinct branches down the tree; in GRUNDY, the more information is known, the more different stereotypes are run concurrently [Kass and Finin 1988].

Sleeman's **UMFE** system is able to represent assumptions about a user's knowledge expertise in a specific domain ([Wahlster and Kobsa 1986] [Wahlster and Kobsa 1989]). In concrete, the User Modelling Front-End component UMFE expresses personalised explanations of expert systems by collecting the inference chains from such components and adapting them to the user's level of expertise. According to [Sleeman 1985], the main goal of UMFE is to ensure that an explanation of a knowledge topic was composed only of those concepts known by the user (indicated through explicit feedback or inferred by the system). As explained in [Wahlster and Kobsa 1989], these concepts known by the user are modelled in UMFE vía an overlay technique. On the one side, the concepts are represented in a hierarchy. On the other side, each user model entry is linked to a node in the hierarchy. Each entry contains a simple description of the knowledge state of the user ('known', 'not-known', 'no-information') and a numerical value on the linkage representing the certainty of the assumption, e.g. for a user 'vgarcia' an entry could exist in the form 'adaptive systems'←'known, 60'.

A first description of the **BGP-MS** user modelling shell system (see Figure 3.20) was first published in 1990 by Alfred Kobsa (see e.g. [Kobsa 1990a]). BGP-MS is a customisable, application-independent system, which interoperates with other systems vía inter-process communication [Kobsa and Pohl 1995]. In essence, UMS-specific functionality can be adapted to specific application domains by selecting the needed components of BGP-MS and filling them with domain-specific user modelling knowledge. Thus, BGP-MS is composed of several units with different tasks, e.g. one specific unit is in just charge of activation and deactivation of assigned stereotypes. Further, this UMS includes a suite of knowledge representation mechanisms to represent (a) its assumptions about the user, (b) the domain knowledge of the user, and (c) its general knowledge about the application domain. As shown in Figure 3.20, four main units can communicate with other applications over a Functional Interface. Also within BGP-MS, a Representation System (RS) is the core unit that is utilised to create stereotypes and individual user models. The knowledge in RS can be adapted by a user model developer through the Graphical Interface.

The **DOPPELGÄNGER** UMS was a result of the 1993 Master Thesis of Jon Orwant at Massachusetts Institute of Technology (MIT) (see [Orwant 1993]). As Orwant stated, his work on the DOPPELGÄNGER UMS was heavily influence by the field of Machine Learning (ML). Thus, the main goal of developing this UMS was to investigate how ML techniques could find usage in UMSs. DOPPELGÄNGER is a User Modelling Shell [Orwant 1995], which is able to detect patterns over actions tracked from users. The system aimed originally at providing personalised daily newspapers. The architecture of the system consists of two levels, the sensors and the server level for external applications (see Figure 3.21). Thus, this UMS acquires user information from sensors (hardware or software components), in-

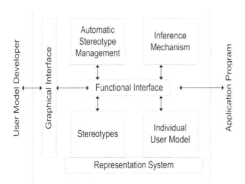

Figure 3.20: Internal View of the BGP-MS User Modelling Shell System [Blank 1996].

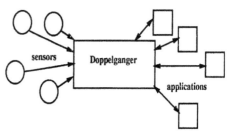

Figure 3.21: The Doppelgänger Architecture [Orwant 1993].

fers additional information and places the results at the disposal of external applications. Each sensor can be seen as a specialised unit, because each one implemented specific techniques in order to extract information from user activities, e.g. one sensor collects data about frequency and duration of computer use whereas others collect data about the physical user's location vía wall-mounted motion sensors. According to [Kobsa 2001], user model developers have access to the ML techniques in order to generalise and extrapolate sensor data (e.g. linear prediction, Markov models or unsupervised clustering for the creation of stereotypes). Further, this UMS was scrutable for users, which could inspect and adjust their own models.

An additional interesting aspect of this UMS is the possibility to run many DOPPELGÄNGER systems at different places and let them exchange their information about a user or a *community*, which is something similar to a stereotype, e.g. communities are teachers, adults, students, children or artists [Fröschl 1995]. Orwant distinguishes communities from stereotypes in that the membership to a community is not binary ('is a member' or 'is not a member'), as for the assignment of common stereotypes. And the Machine Learning Methods? According to [Orwant 1995], a community 'tolerates' members through distinct degrees of memberships, and thus, showing a 'conformity' of a certain user to the community. Therefore, a community is the combination of its constituent individual models, i.e. communities represent the average of their member traits. Further, users may belong at the same time to many different communities [Orwant 1995]. So, each community 'votes' if an information is required for a certain user model; the relevance or weight of the votes is proportional to the similarities among the communities and the specific user model, i.e. the degree of membership.

In the shell system **umt** ('user modeling tool'), user model developers may freely compose stereotypes in hierarchies, where an inheritance of user information to sub-stereotypes is also allowed. Further, *umt* supports also the definition of rules for inferences and contradiction detection. An individual user model is selected by comparing possible stereotypes with preferences. Thereby, *umt* assumptions are rated higher than

stereotypes' inferences. As *umt* is able to record the calculation steps of possible models, they can be also revised or evaluated. [Kobsa 2001] [Brajnik and Tasso 1994]

Protum, the 'PROlog based Tool for User Modeling' combines the mechanisms of $GUMS_1$ and *umt*; it is based on Prolog, a logical programming language. Protum supports the management of stereotype hierarchies (including multiple inheritance), a rule-based resolution of contradictions, and the multiple activation of stereotypes. In order to resolve conflicts between assumptions of two activated stereotypes, Protum uses further inferences on the recorded activation rate of stereotypes, e.g. from two stereotypes A and B with activation rates of 75% and 95%, the assumption of B is chosen to infer a particular user information. [Fröschl 1995] [Vergara 1994]

um was first presented in [Kay 1990]. As stated in [Cook and Kay 1994a] (which is a short Version of [Cook and Kay 1994b]), the very first aim for the design of *um* was to make individual user models *accessible* to the users they model through so-called viewer programs. Later, in [Kay 1995], *um* is presented in more detail within the context of creating user modelling shells as a toolkit that provides support for a variety of co-operative 'agents', e.g. in a recommender application a user model developer may cooperate with end-users in order to refine their individual user models. According to [Kobsa 2001], *um* places a set of tools to the disposal of user model developers to represent assumptions about user traits (e.g. knowledge, beliefs, preferences) in key-value pairs. Therefore, from a strict point of view, *um* is a library of user modelling functions and not an independent UMS.

Each piece of information is part of a '*um component*'. A component comprises a list of recorded data: the assumption about a specific user trait, the evidences for its truth or falsehood, and for each evidence, a timestamp, its source and its type. The five types of evidence used in *um* define the possible types of *components*: observation, stereotype activation, rule invocation, user input and told to the user. [Kay 1995]

Personis focuses on the following main issues of UMSs: user's privacy as well as control and scrutiny of own individual models [Kay et al. 2002]. Personis is based on Kay's *um* toolkit, and therefore, it also aims at making the models *accessible* to the users, i.e. *scrutable*. Thus, if a scrutable user model can be controlled by the user it models, own estimations about traits can be entered in the system in order to alter system's assumptions. Personis had been developed to collaborate with adaptive hypertext applications (AHAs), as shown in Figure 3.22 (*Collaboration Architecture of Personis*, CAP). The CAP is divided into four main parts: (a) a centralised *User Model Server* as core and unique provider of user information, (b) a *Generic Scrutiny Interface* (GSI) as application-independent point of access for all users, (c) a set of *AHAs*, and (d) application-dependent *Views*, which represent the set of those components in the server that are needed for each application. [Kay et al. 2002]

Furthermore, each adaptive hypertext application is complemented by an own scrutiny interface associated with its specific functionality, i.e. each user is given the possibility to scrutinise the adaptivity within the context of the components used by each AHA. In other words, the user model developers for each AHA define the corresponding components needed by each application and describe them in form of a specific View. On the other hand, as stated in [Kay et al. 2002], user's privacy is implemented in Personis by allowing each user to stipulate which components and information of the private user model may be accessed by which AHA, and further.

This *access control information* is also stored within the individual user model and is implemented in the User Model Server by means of application-dependent *Resolvers*. As each resolver associates certain components with certain views, this mechanism allows different applications to interpret the user information behind the same components in a different way.

Storage and management of user information in **LDAP-ums**, as described in [Fink 2003], is based on the Lightweight Directory Access Protocol (LDAP). As shown in Figure 3.23, different *User Modelling Components* (UMCs) can be attached to each LPAD-based user modelling server. These UMCs are considered to be 'internal clients' of the server. Further, LDAP directories enable synchronised and transparent access to distributed user data in a network of LDAP servers. As depicted in Figure 3.23, these directories can be structured in hierarchies (where hierarchy nodes can be further intra-linked vía 'referrals', as softlinks or shortcuts in common file systems). [Kobsa and Fink 2006]

The core of the architecture depicted in Figure 3.24 (on next page) is represented by a *Directory Component* (DC). This DC is further divided in three functional modules: (a) *Communication*, (b) *Representation* and (c) *Scheduler*. The DC and the UMCs communicate via CORBA (*Common Object Request Broker Architecture*) and LDAP. In order to access user data, the Communication module is in charge of the LDAP communication between the external clients of the server and the UMCs. Thus, each UMC can perform specific

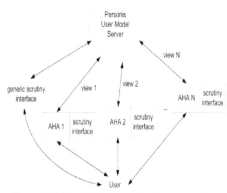

Figure 3.22: The Collaboration Architecture of Personis [Kay et al. 2002].

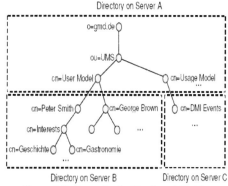

Figure 3.23: Example of a Distributed LDAP-Directory. [Kobsa and Fink 2006]

user modelling tasks for specific external clients, e.g. domain-based inferences for application-related (i.e. client-related) user information. The management and interpretation of the directory contents is a task of the Representation module. Both, the tasks of the UMCs and of the Representation module are implemented as services of distributed objects that can be composed in different ways. Therefore, a CORBA-based Scheduler is used as mediator and main interface between the different modules and components of the server. [Fink 2003]

For LDAP-ums, three different components are proposed by [Fink 2003]: (a) a statistics-based *User Learning Component* that extracts user interests and preferences from the user's data usage; (b) a similarity-based *Mentor Learning Component* that fills the missing values in incomplete individual user models from complete similar models; and (c) a rule-based *Domain Inference Component* that infers individual interests and preferences from the user's explicit feedback or from implicit knowledge in other components [Kobsa and Fink 2006]. Finally, it is also worth mentioning that for external clients, a very relevant advantage of this LDAP-based server architecture is

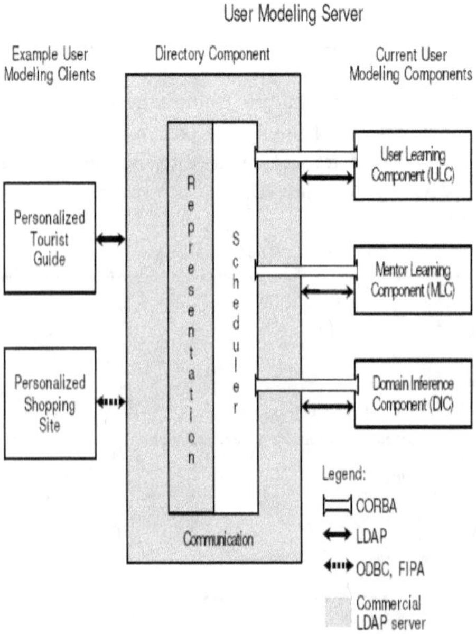

given by the fact that it allows a synchronised and transparent access to user data highly distributed over a cluster of servers on different platforms or machines.

As Personis, *CUMULATE* can be seen as a learner modelling server for distributed AEL environments. CUMULATE, as one of the central units in a distributed AEL system, can collect different events of single learners interacting with different distributed e-learning sub-systems [Brusilovsky 2004b]. CUMULATE represents the main modelling component for each of these sub-systems, because it can store and manage user-specific activities as well as infer additional individual learning-relevant traits. As shown in [Brusilovsky 2004b], to serve each sub-system with accurate application-specific user information, CUMULATE implements *inference agents* (for each sub-system) that are specialised to control the workflow of single events, and accordingly, can update single inferences for specific individual user models. In [Brusilovsky 2004b], a first version of CUMULATE is used within the distributed AEL system 'Knowledge Tree'. In [Brusilovsky et al. 2005], an enhanced version of the Knowledge Tree is presented, called 'ADAPT²', which supports *ontology-based* interoperability between self-contained adaptive Web-based systems. In Adapt², CUMULATE's functionality seems to be extended to interact with ontology-based servers.

I-HELP is a large-scale multi-agent and multi-user system for learners and experts that receive and give peer help in a synchronous or asynchronous way [Greer et al. 1998]. According to [Vassileva et al. 2003], I-HELP evolved from the centralised *PHelpS* system, a peer-help system for workplace training. I-HELP is based on individual user model peers, which are able to find expert peers that are online and can provide help in the task at hand. In addition to these user agents and expert agents, I-HELP provides other types of proactive and independent agents, e.g., *application agents* (in order to represent application variety and independence) or *diagnostic agents* (in charge of creating and maintaining the states of individual knowledge expertise). For example, *personal agents* may search for other agents to get resources that could help them for a specific purpose; hereby, the resources could be Web

Figure 3.24: Architecture of Fink's LDAP-based User Modelling Server. [Kobsa and Fink 2006]

pages known by a *specialised application agent* in a certain knowledge domain, or postings from an I-Help *discussion forum application agent*. [Vassileva et al. 2003]

The I-HELP system is also presented in [Niu et al. 2003a] and [Niu et al. 2003b] as a *Multi-Agent Portfolio Management System* (see Figure 3.25). The multi-agent paradigm of I-HELP might further be seen as a possibility to support learning in Nomadic Information Systems (see 2.1.3 and 3.1.3). In addition to the centralised services and the reactive character of the UMSs presented so far, *agent-based UMSs* as I-HELP provide distributed, mobile and partitioned individual user models that are able to pro-act as negotiators with distinct agents and as mediators for distinct purposes. For details refer to [Niu et al. 2003a] and [Niu et al. 2003b].

Another example for an AEL system using agent technology was developed within the TILE project (see [Kinshuk et al. 2001]). In TILE, an ITS (Intelligent Tutoring System) was developed, which follows a client-server architecture, but parts of them act as mobile agents. As stated in

[Kinshuk et al. 2001], specialised mobile agents (a) interact with a client-side inference engine in order to fetch user data relying on the individual student model at client-side, (b) move to the server-side to perform specific tasks (e.g. updating partial individual student models or group models), (c) collect personalised information at serve-side, (d) return to the client-side, and (d) update the client-side individual student models. The TILE solution approach has been mentioned at this point due to its analogy to the use of agent technology through peers in I-HELP, but does not represent an application- and domain-independent UMS.

PLUS stands for <u>P</u>ervasive <u>L</u>ifelong <u>U</u>ser-models *that are <u>S</u>crutable,* firstly presented in [Kay 2006]. According to Judy Kay, PLUS is more a

> *"vision for achieving long term models that are associated with a person, rather than an application."* [Kay 2006]

Thus, this 'vision' should not be presented at this point in this document, because it is not a system. Though, in the opinion of the author of

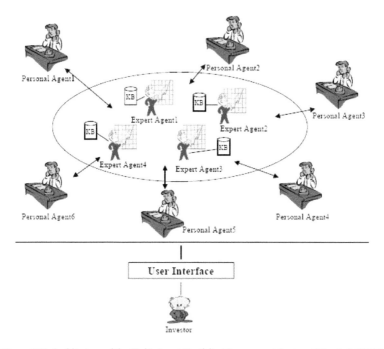

Figure 3.25: Architecture of the Multi-Agent Portfolio Management System. [Niu et al. 2003b]

this book, PLUS gives several reasons to consider it within this section, because it (a) is a very concrete view on personalisation systems, which (b) represents the results of Kay's work and experiences in the field of UMSs, and much more relevant for this book, it (c) is based on a framework that explains clearly how existing applications fit into this 'vision'. The PLUS framework is shown in Figure 3.26.

The basic idea behind PLUS focuses on the *flow of user information*, and according to [Kay 2006], is based on the simple notion of *accretion/resolution*: accretion representing the operation of acquiring user information, and resolution being the delivery of useful individual user information. In the PLUS framework, a user model is a long-term repository containing individual user information, so-called *evidence about people*. The lowest layer in Figure 3.26 contains the User Model Ontology, which is the basis for scrutability as it explains the meaning of modelled traits. Evidence Sources are providers of user information and contribute to reasoning about the user (arrows pointing at the user model), e.g.: *Sensors* observing users unobtrusively; *User (Model) Interfaces* tracking human-machine interactions; non-adaptive *Applications* in daily use (e.g. emails clients, Web browsers) that might either deliver individual information directly to a user model or indirectly providing electronic traces (e.g. log data) which can be mined and collected; or (typically internal) reasoning sources, i.e. software tools supporting

Stereotypes and *Knowledge-Based Inference*.

As stated in [Kay 2006], the implementation of user interfaces that support the 'simple' notions of accretion/resolution is not an easy task, because their usage must effectively meet substantial usability requirements, such as learnability ('easy to learn for novice users') and memorability ('easy to remember for casual users').

In general terms, **commercial user modelling systems** concentrate on most relevant practical requirements from the industry sector, such as integration and consistency of existing user information, special care of critical issues about user's privacy and company's security, as well as overall fault tolerance, performance, availability and scalability of the system. Detailed reviews on commercial user modelling systems can be found e.g. in [Fink and Kobsa 2000] as well as summaries in [Kobsa 2001], [Fink 2003] and [Kobsa and Schreck 2003]. These publications present for example the following systems: *GroupLens* (*MovieLens*), *LikeMinds*, *Personalization Server*, *Frontmind* and *Learn Sesame*. Further details about general UMSs can be found e.g. in [Fink and Kobsa 2000], [Kobsa 2001], [Fink 2003], [Schreck 2003] and [Fröschl 2005].

So far, representative existing solutions for application- and domain-independent UMSs have been presented. Finally, some observations on the topic of **learner modelling systems** are given In essence, UMSs aim at satisfying the needs of distinct user information consumers: (a)

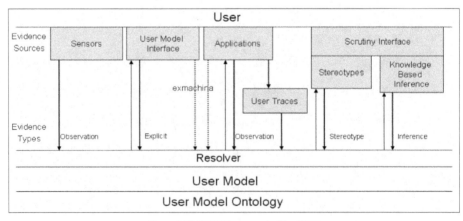

Figure 3.26: PLUS Framework. [Kay 2006]

generally speaking, humans, specifically in AEL, learners, an (b) other software systems or components of other systems.

In the last section, some main techniques for user modelling were introduced. As shown, these techniques are distinctly applied in UMSs depending on the intended usage and expected benefit as well as on the distinct user modelling task at hand. In general, the main idea behind a UMS is to provide a framework of flexible and expandable set of tools to reach its ideal goal of serving as a *multi-purpose system*. Though, the core functionalities of the UMSs presented in the last section rely mostly on one or few of the available techniques, a fact that must be also considered to show that the scope of being multi-purpose is also restricted in this sense. In summary, most UMSs make use of (a) overlay models or/and stereotypes, and (b) a set of distinct reasoning techniques for the distinct purposes of inducing additional and more complex assumptions about user traits (and naturally, for managing the uncertainty of these assumptions).

Based on these observations and in a strict sense, talking about a *Learner Modelling Systems* with the notion of UMSs in the background might be misunderstood, because learner modelling systems are (a) mostly *embedded as fixed components* in AEL systems (see e.g. sub-chapters 2.2 and 3.1), and (b) *neither application- nor domain-independent*. Nonetheless, the author of this book mentions Learner Modelling Systems at this point to indicate that *AEL can also be seen as a field of independent (sub-)applications areas*, i.e. the number and type of possible clients of a learner modelling system in a specific AEL environment is variable and depends (among others) on the didactical tasks and goals at hand (as shown in sub-chapters 2.2 and 3.1).

Thus, components in AEL environments with distinct functionalities and goals may be providers and/or consumers of distinct user information: assessment data, discussion forum postings, chats logs, data usage logs, attendance records, quiz results, personal data, skills, competences, knowledge topics, learning styles, interaction logs, presentation preferences, and many others. Taking into account these statements, an investi-

gation on *learner modelling systems* leads to a consideration of the UMSs presented in this section.

3.2.5 Summary

An essential issue within the context of this chapter was to show the distinctions between User Profiling and User Modelling. In concrete, user modelling components comprise more enhanced functionalities as user profiling components by means of *reasoning techniques*. Further, the more an adaptive system 'knows' about its environment, the more efficiently it will personalise, i.e. the more relevant knowledge about the user is internally represented in the individual user model, the more accurate information a user modelling component can deliver to its information consumers.

Sensing hardware devices as well as software-based graphical user interfaces represent main means to gain knowledge about the user, i.e. *interaction* and *observation* represent modalities of highest relevance concerning the acquisition of information. Another modality is *proactive analysis*, where the user model component autonomously analyses external data collections which belong to users or have been used by them. This last issue is highly important personalisation comprises very critical aspects concerning privacy of user as well.

With respect to the duties of user modelling components, four main *global tasks* have been identified: acquisition of user information, creation of individual user models, administration of user models, and provision of appropriate user information. Thus, two perspectives can be derived to analyse the techniques of user modelling, one depending on the *interchange and* a second depending on the *processing of individual information*. Furthermore, the variety of user modelling techniques is directly proportional to the variety of inter-dependencies within the components of adaptive systems. Therefore, developing user modelling components is often restricted to the specific needs of the systems in which they are embedded.

The *initialisation of an individual user model* comprises two main techniques: *feedback* and *observation*. Accordingly, a user model can be initialised through *explicit questions*, *initial testing* or *stereotypes*. Once sufficient information is acquired, other techniques come into play, e.g. *overlay models*. One problem of stereotyping is the work-load to build required stereotypes for a specific application or domain. Another problem is the susceptibility to inaccuracy, because stereotyping is based on beliefs of the system about users. Therefore, scrutability methods are needed to allow users to inspect and modify their personal models.

Finally, a very critical aspect of designing and developing user modelling systems that aim at being application- and domain-independent is to reach a high level of abstraction and flexibility in order to support a high degree of multi-purposefulness. Some user modelling techniques are based upon algorithmic solutions, such as collaborative or content-based filtering, which are commonly used e.g. in the fields of information retrieval, recommender systems or machine learning. Moreover, techniques from the field of artificial intelligence might be applied to manage uncertainty of assumptions, e.g. linear Prediction (to model recurrent events), beta distribution (to model interests), Markov models (to model location), Hidden Markov models (to categorise behaviour), cluster analysis (for constructing communities), plan recognition methods (to infer user traits from observation), and others.

The next section considers the main findings gained so far and discusses most relevant state-of-the-art challenges for the development of user modelling systems. Some insights into future trends regarding personalisation in adaptive e-learning are given as well.

3.3 Meeting the Challenges

This chapter has presented the main issues concerning the functions and roles of User Modelling Components and Systems (UMCSs) within and beyond the context of Adaptive E-Learning (AEL). Thereby it has been identified that UMCSs should be flexible and reusable in order to efficiently serve their potentially very distinct information consumers. Because of this need for multi-purpose efficiency, different techniques for modelling users have been introduced. On the one hand, applying traditional techniques allows

user model developers to implement effective solutions that are proven to fit very well in many known situations.

On the other hand, more sophisticated techniques might be applied to satisfy complex needs of software components interacting with UMCSs. Further, although more than five decades ago adaptive instructional systems were already developed with components that hold and managed individual user information, the research field known as User Modelling (UM) exists since approximately three decades. It may be then assumed that UMCSs embraces a comprehensive and concrete core of technological methods and strategies with proven practical success. Still, UM represents something like a tennis court on which constantly changing players train with self-defined rules and a wide range of freedoms. Indeed, this is not a negative statement, on the contrary and as shown in this chapter, it indicates that the field evolves as a central part of many other application areas. Though, UM keeps often in the background of other research fields finding mention as by-product of other systems.

The reminder of this chapter underlines some aspects that have direct impact on the present evolution of UMCSs and thus, represent *relevant challenges* for UM developers at present.

Research Challenges

The UM research field is based on distinct epistemological assumptions and has been exploring distinct dimensions of personalisation problems. UM experts have addressed highly relevant distinctions underlying different solution approaches. According to [Fischer 1999], solution approaches from the UM research field

> "seem to be appealing, natural, theoretically justifiable, desirable, and needed (e.g., reuse can be justified by the fact that complex systems develop faster if they can build on stable subsystems). But in reality, progress in these areas has been slow and difficult, and success stories are rare."

Additionally, [Fischer 1999] stipulates that the UM research area has also created challenging research problems, such as how to:

> "(1) integrate different modelling techniques;

(2) capture the larger (often unarticulated) context and what users are doing (especially beyond the direct interaction with the computer system);

(3) identify user goals from low-level interactions;

(4) reduce information overload by making information relevant to the task at hand;

(5) support differential descriptions by relating new information to known information and concepts; and

(6) reach a better balance for task distributions between systems and users." [Fischer 1999]

These represent some of the challenges UM researchers and developers have to face. Further, they address not only technological problems (which is the most-suitable modelling technique for the task at hand?), but also e.g. psychological (which are the interaction goals of users?) and representational problems (how to make context and meaning form the real world 'machine-readable', and vice versa?).

Adaptive E-Learning vs. Multi-Purposefulness

The ultimate aim of AEL is to deliver optimised facilities of e-learning (see sections 3.1.2 and 3.1.3). In summary, AEL solutions concentrate on personalising the following aspects: (a) learning materials (through adaptation of presentation), (b) interactive elements (through adaptation of functionality), and (c) fine-grained elements in human-machine interfaces into usable and useful composites (through adaptation of their structures and relationships). Further, AEL focuses on providing components for the modelling of (a) content, (b) knowledge, (c) instructional activities, and (d) users, i.e. learners. As shown in chapter 3.2, the complexity of developing UMCSs increases with the generality of their duties, and consequently with the diversity of techniques needed to implement distinct tasks. Within this context and at the beginning of the design of UMCSs, it is highly relevant to stipulate the distinctions among the duties of adaptors in Adaptive Systems (usually called adaptive or personalisation engines) and the UMCSs. User models must often contain or represent at least parts of other models, e.g. knowledge models, content models, activity or interaction models. In addition, UMCSs must often make use of reasoning methods, as typically applied also for adaptors.

Essentially these finding make it difficult to stipulate the functional boundaries of UMCSs. Where do the duties of UMCSs end, if they must also process accurate answers to questions that may fall into the duties of adaptors or other modelling components? For example: How should the next instruction look like for this learner? Which is the best learning path for this learner? Or, is an assessment needed for this learner at this point in time? An answer to such questions is mostly simple: these questions do not pertain to the repertory of answers of UMCSs, because they have to do with direct decisions on adaptable entities of the system and not with 'knowing the user'. This is the reason for having made a distinction (in section 3.2.2) between information processing and information interchange, which resulted in (a) global tasks (acquisition and provision of user information as well as creation and administration of user models) in UMCSs and (b) CRUD functions (create, retrieve, update and delete) for information consumers. These issues represent a simple ways to define the set of services of UMCSs.

General Services

As stated in [Kobsa 1995], the following frequently-found services of UMCSs can be identified:

- representation of assumptions about user traits in individual user models (e.g. assumptions about knowledge, misconceptions, preferences, plans, goals, tasks and abilities),

- representation of relevant common traits of users as members of the community using a specific application (e.g. vía stereotypes),

- classification of users as members of these communities (i.e. group modelling),

- integration of the typical traits of user groups into individual user models,

- recording behavioural aspects of users, in particular, their past system interactions,

- inferring further information based on an individual interaction history,

- generalisation of interaction histories into stereotypes,

- inferring additional assumptions based on initial ones;

- maintaining information consistency in user models,

- providing current assumptions (including their justifications),

- evaluation of the topicality of user models,

- comparison of model contents with given standards.

Modelling What and How?

It is known that e.g. the modelling of cognitive styles and learning styles is a hard task, i.e. how to process information vs. how to convey information (see e.g. [Sadler-Smith and Riding 1999] [Cristea and DeBra 2002] [deAssis et al. 2005] [Brown et al. 2006] [Müller and Wiesinger 2006]). Also the modelling of personal behaviour is not trivial. Behaviour brings the time dimension into individual user models: (a) the present implies observing behaviour in real-time, (b) the past implies the analysis of the history of behavioural data, and (c) predictive techniques are required to assume future behaviour. To give examples regarding 'past user traces', the challenges are e.g. how to model user trails through contents, how to model internal system footprints from adaptors, how to model bookmarks or landmarks made by learners, how to model learning progress monitored by the system (see e.g. [Kokkinaki 1997] [Burton et al. 2003] [Nakakoji et al. 2003]).

At present, the Web represents a medium for publishing and communicating in a very 'free' manner. The content stored on the Web is not only data (or resources), it contains also semantic descriptions on data (e.g. metadata) that help to understand the meaning of messages conveyed in the data. As shown in section 2.2.4, the influences of e.g. Semantic Web, Knowledge Management and Context-aware Systems open new challenges and opportunities for user model de-

velopers. In order to avoid disturbing new users with explicit feedback methods to acquire personal information, the Web as an every-day usage medium and its huge amount of 'visited documents' allows to indirectly infer personal traits from Web Usage. Thus, the Web is not only an incredibly big source of information about user behaviour and interests, it represents also the biggest collectively shared knowledge repository. Also in this context, Social Tagging is contributing at present to complement Web data with non-standardised semantic descriptors (i.e. community-based vocabularies). Hereby, machine learning techniques are used to infer individual information about users from similarities with community information, i.e. to infer *MY-Web-usage* from *OUR-Web-usage*. [Henze et al. 2004], [Denaux et al. 2005b], [Liu et al. 2006], [Terrase et al. 2006]).

Distributed Systems vs. Distributed Services

Further, the Web represents a 'Distributed System'. As such, it comprises more than distributed documents; e.g. applications, storage devices, connection channels are also distributed. Thus, from the technological point of view, it comprises also the typical challenges of distributed systems: heterogeneity, openness, security, scalability, failure handling, concurrency, and transparency (see e.g. [Coulouris et al. 2005], [Tanenbaum and vanSteen 2007]). Among others, the openness and scalability of the Web has enabled that distinct application areas could settle down successfully on this technology; these application areas represent a wide spectrum of possibilities to apply UM and adaptive tools. User modelling servers, personalisation engines and recommender tools are just general examples of commonly used tools in modern Web-based applications.

The need for a multi-purpose functionality in UMCSs is clear when viewing at a small part of this spectrum: adaptive user interfaces [Langley 1999] [Thevenin and Coutaz 1999], context-aware systems [Assad et al. 2006], context management [Zimmermann et al. 2005], digital libraries [Hicks and Tochtermann 2001], e-commerce [Ardissono et al. 2002] [Perugini and Ramakrishnan 2003], information retrieval [Grcar et al. 2001], mobile computing [Holmén 2000] [Lankhorst et al. 2002] [Kinshuk and Goh 2003], museums [Zimermann et al. 2003] [Filippini-Fantoni et al. 2005], multimedia retrieval [Li et al. 2001], multi-model adaptive systems [Conlan et al. 2003], news [Billsus and Pazzani 2000] [Shepherd et al. 2002], ubiquitous systems [Carmichael et al. 2005].

Interoperability and Standards

Another challenge in the context of a distributed system is 'interoperability'. Within the context of this section, this issue is restricted to the scope of information interchange, which in turn leads to point out the numerous efforts in the field of standardisation. In the case of user modelling, not only global standards regarding privacy and security issues are worth mentioning (like P3P[34]), also concrete efforts are made to standardise privacy and security in UM (see e.g. [Schreck 2003]). Within the area of ubiquitous user modelling also standardisation efforts have shown remarkable results (e.g. UserML and GUMO; see [Heckmann 2005]). Further, various organisations and consortia, such as Dublin Core (DC) Metadata Initiative, Institute of Electrical and Electronics Engineers (IEEE), Instructional Management System Global Learning Consortium (IMS GLC), Advanced Distributed Learning Initiative (ADL), work in the area of standards for adaptive e-learning (see e.g. [Paramythis and Loidl-Reisinger 2003]).

The most relevant standards regarding the content of user and learner models are vCard, eduPerson, ULF, GESTALT, PAPI Learner (simply known as PAPI) and IMS LIP[35]. According to [Fröschl 2005], PAPI Learner shows advantages in the context of adaptive e-learning, e.g. it fulfils the need of detailed information about the individual learner activities (the reason is that it comprises fundamental ideas from Intelligent Tutoring Systems, where learner performance is one of the most relevant adaptational targets).

Nonetheless, PAPI Learner lacks of topicality in the evolution of its specification. Therefore, it might be advisable to use IMS LIP, because the community behind the evaluation and development is bigger and increases constantly. Furthermore, IMS LIP is based on PAPI Learner, shows good usability and provides an extensible structure. See also e.g. [Alrifai et al. 2006], [Dichev and Dicheva 2006] [Stewart et al. 2006].

Data Acquisition: Intrusiveness and Scrutability

All statements given so far take for granted that the individual user information is already in the system and must just be transferred through some communication channel, perhaps also decrypted for ensuring privacy and security, and in addition, standardised to enable interoperability and ease of use. But how is the data acquired from the user's point of view? Through the 'sensoring devices' of the system. So what about feedback and willingness of users (i.e. asking users if they agree to mechanisms tracking their behaviour)? (see e.g. [Dreher und Maurer 2006], [Kobsa and Teltzrow 2006]) Or an even more delicate issue: what about the intrusiveness and vulnerability of such devices?

In concrete, mouse and keyboard are usual sensoring devices of an adaptive system.. even more, users are aware that their clicks and keystrokes might be monitored. Other alternatives for sensoring devices include e.g. microphones, cameras, or eye-tracking systems. The typical aim of utilising these modalities is to gain more insights into cognitive, emotional or affective traits of users (see e.g. [Jacob 1995], [Picard 1997], [Salvucci and Goldberg 2000], [Hudlicka and McNeese 2002], [Bianchi-Berthouze and Lissetti 2002], [García-Barrios et al. 2004b], [Alepsis et al. 2006], [Merten and Conati 2006]).

Thus, from the point of view of user information acquisition and adaptational goals as well as considering some findings coming from other research fields, developers of adaptive systems should, on the one hand, enable the scrutability of the user model information to give learners a view on monitored information as well as control over adaptational parameters, and on the other

---

[34] P3P: Platform for Privacy Preferences. See http://www.w3.org/P3P.
[35] See also for http://www.imc.org/pdi (vCard), http://www.fdgroup.com/gestalt (Gestalt), http://edutool.com/papi (PAPI), http://www.imsproject.org/profiles (IMS LIP).

hand, consider privacy and security issues e.g. due to the intrusive character of devices such as an eye-tracker system (see e.g. [Ivory and Hearst 2001], [Waern and Rudstrom 2001], [Goldberg et al. 2002], [Czarkowski and Kay 2003]).

The need of making a user model scrutable implies the stipulation of some requirements for scrutability and visualisation. According to [Uther 2001], the following motivations for scrutable user models are identifiable: (a) access to and control over personal information (as a moral and legal issue because of dealing with private information), (b) programmer accountability (e.g. allowing users to scrutinise the system-generated assumptions about their beliefs and actions encourages the programmers to be more careful about their techniques), (c) correctness and validation of the model (i.e. scrutability involves users in resolving the uncertainty of assumptions), (d) machine predictability (users scrutinising system's beliefs can increase their understanding of the system, and in turn, improve the quality of interactivity), (e) aid to reflective learning (with respect to thinking and learning about learning, an available learner model aids meta-cognition, see e.g. [Kay and Crawford 1993], [Paiva et al. 1995], [Bull 1997]).

Some requirements on the visualisation of user models are [Uther 2001]: (a) overview (reflecting the structure of the model), (b) relevance (help users to explore relevant contents of the model), (c) navigation (to enable less detailed or more detailed views on the structure of the model), (d) showing values, types and confidence of stored information (e.g. to differentiate

among data, knowledge, beliefs, goals, plans, behaviour, etc.). Finally, opening the content of user models to humans involves that their visualisations should enable also Universal Access. Enabling universal access means to cope with diversity in:

"(i) the target user population (including people with disabilities) and their individual and cultural differences; (ii) the scope and nature of tasks; and (iii) the technological platforms and the effects of their proliferation into business and social endeavours." [Stephanidis 2001]

Several challenges arise from this issue, such as (a) making virtual or augmented environments usable and universally accessible, (b) evaluate the implications on adapted content, functionalities or interactions, (c) evaluate if universal access is economically viable in the long term. For more details refer to e.g. [Stephanidis et al. 1997], [Stephanidis et al. 1998] or [Stephanidis 2001].

So far, highly relevant challenges have been identified and briefly discussed. Most of them build the basis for the design and development of the **"yo?"** modelling system, which is described in the ensuing chapter.

4 "yo?" - Modelling System

> *Basic research is what I am doing*
>
> *when I don't know what I am doing.*
>
> (Wernher von Braun)

"yo?" is a Multi-purpose Profiling and Modelling System (MPMS). This system provides services to access and administer information about individual agents in the environment of adaptation-pertinent systems. According to this last statement and within the *general context* of this book (i.e. the field of adaptation-pertinent systems), the notion of multi-purposefulness is explained as follows. The design and development of a modelling system that supports multi-purposefulness implies two main goals: (a) an MPMS should be capable of modelling not only users, but also other physical and abstract entities the adaptive system must be aware of, e.g. groups of persons, sensors, interacting devices, respectively, information or knowledge structures, activity workflows, roles; and (b) the utilisation of an MPMS should not be restricted to a certain application type or domain. The *specific context* of the "yo?" system is given by the field of Personalisation in Adaptive E-Learning, i.e. it should primarily serve as learner modelling system. The ambivalence between these general and specific contexts have one essential aspect in common: every adaptational procedure aims at fulfilling some need or requirement of a person interacting with the adaptive system. Therefore, the previous chapters have concentrated on issues regarding both learner modelling systems as well as general (or generic) user modelling systems. Based on the aforementioned statements, this chapter describes of the design and development of the "yo?" modelling system according to the following methodology. In order to define the positioning of the system in its practical and theoretical environments, the first part of this chapter provides an overview on the Application Scopes, i.e. (a) an introduction into the concrete projects in which "yo?" has evolved and is applied, as well as (b) a model that generalises the possible adaptive environments in which "yo?" shows its multi-purposefulness. The second part of this chapter presents the functional and architectural design of the system. Thereby, (a) main requirements for MPMSs are stipulated, (b) a layer-based architecture shows the main functions of the system, and (c) insights into the components and their interdependencies are described. The third part details most relevant issues concerning the system's development. The chapter concludes with a summary of main remarks showing to which extent the system fulfils the essential requirements on a MPMS.

4.1   The Application Scopes

"yo?" is the result of research and development work on distinct research projects at IICM (Institute for Information Systems and Computer Media at Graz University of Technology, Austria).

Regarding the evolution of the system, first requirements emerged for developing a User Modelling Component (UMC) for the adaptive e-learning system AdeLE (*Adaptive e-Learning with Eye-tracking*). This UMC should provide learner modelling services for the adaptive engine of the overall AdeLE system. Though, this adaptive engine was not the unique consumer and provider of user information in AdeLE. Other components should also interact with the UMC for distinct purposes, e.g. an eye-tracking system on the client-side for monitoring attention and gaze behaviour of individual learners, a content-tracking system for the analysis of learning progress, and a so-called DBL (*Dynamic Background Library*) providing domain-dependent access to additional learning materials. For another research project, called Mistral (*Measurable intelligent and secure semantic extraction and retrieval of multimedia data*), a UMC was also required to enable personalised access to semantically extracted and enriched information from multi-modal meeting recordings and topic-related documents. Within the Mistral system, also distinct components should interact with the UMC, e.g. a search engine, a visualisation framework, and again, a DBL. The DBL, which can be seen as a knowledge facilitation tool, had been tested previously and needed to be improved in a new version. Considering that on the one side, AdeLE's UMC ought to manage learner knowledge states and on the other side, that a new DBL version was needed (i.e. a new 'knowledge modeller'), the motivation arose to combine these needs in a single system.

The result for the multiple requirements stated so far was a system called **"yo?"**: a flexible service-oriented system that can be utilised e.g. as user modelling system and as knowledge modeller. Firstly, this sub-chapter gives an overview on the aforementioned projects to present briefly the practical context of the system. Secondly, this sub-chapter stipulates a general architecture for adaptation-pertinent systems in order to illustrate the positioning of the system within its theoretical multi-purposefulness.

4.1.1 Overview on Practical Context: Three Projects

The main objective of the **"yo?"** modelling system is to manage user models. For that purpose it supports the acquisition and provision of user information as well as the creation and administration of individual user models. Originally, **"yo?"** should just be developed as a pure UMC for the AdeLE system. It would interact with distinct components, such as AdeLE's adaptive engine, eye-tracking system, content-tracking system and dynamic background library. Then, for the Mistral research project, a modelling component was needed in order to interact with other types of components, e.g. a visualisation framework and a search engine.

Therefore, "yo?" emerged as a UMC and evolved in the last years into a general modelling system, for which parallel instances can be run

and configured to fulfil distinct modelling duties. At present, within the scope of the AdeLE and Mistral research projects, **"yo?"** is used as User Modelling System and as Concept-based Context Modelling System. This section introduces briefly the application scope of the underlying projects. More details about the practical use of **"yo?"** in each project are presented in the next chapter.

The *AdeLE* Research Project

The 4 year research project AdeLE (*Adaptive e-Learning with Eye-tracking*, 2003-2007) aims at developing an adaptive e-learning system that uses fine-grained user profiling techniques for the analysis of learner behaviour and learner states in real-time. One novel feature in the scope of the AdeLE research project is the application of eye-tracking technology (= flat screen monitor at left side of Figure 4.1) in combination with fine-grained content-tracking information. [Pivec et al. 2004]

Regarding the topic of personalised e-learning, one of the main goals of the AdeLE system is to be capable of adapting the presentation of learning material in real-time according to individual learning styles and to gaze-tracking inferences. For example, the AdeLE system may present the learning content shown in Figure 4.2 to those users that are considered to be 'verbalisers' (i.e. they learn better from textual content) or the content in Figure 4.3 to learners considered to be 'imagers' (i.e. they learn better from visual explanations, such as images, diagrams, tables).

Figure 4.1: Usage of AdeLE's Eye-Tracking System. [García-Barrios 2006e]

Verbalisers and Imagers represent one of the dimensions of Riding's WAVI cognitive style classification, which is explained in more detail in section 5.1.

Further, AdeLE aims at investigating adaptivity at the level of Aggregation, i.e. the visibility of some segments within single pages of the learning material might be enabled or disabled. Responsible for this action could (essentially) be the eye-tracking device. In the AdeLE system, segments of learning pages (such as text paragraphs) can be tagged with didactical goals by teachers, e.g. with 'to be read' or 'to be learned'. AdeLE's eye-tracking device detects those tagged segments and infers from the gaze behaviour of learners if these goals have been reached or not. If the learner attempts then to continue with some next page, the system delivers an interme-

diate 'warning page' composed of those elements that have not matched the goals. As the AdeLE system does not impose any instructional steps, the learner may follow the recommendation of repeating the page or just ignore it. Thus, one relevant aim of the AdeLE project is to investigate a combined application of eye-tracking and content-tracking with personalisation techniques.

From a general point of view, the AdeLE system can be seen as a Web-based Adaptive E-Learning (AEL) system, i.e. it falls into the category of AEHSs (Adaptive Educational Hypermedia Systems). The features mentioned above illustrate some of the main practical goals to be reached within the AdeLE research project. From the technological point of view, the main goals of the AdeLE project can be summarised as follows:

Figure 4.2: Partial View of AdeLE's Web User Interface; adapted Content for Verbalisers.

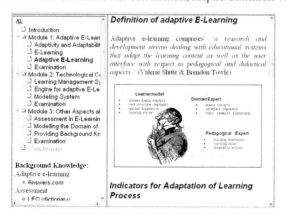

Figure 4.3: Partial View of AdeLE's Web User Interface; adapted Content for Imagers.

- implementation of an open system,

- provision of exchangeable and extensible features,

- strict modularity, i.e. development of separated and task-specific modules,

- platform independence for AdeLE's components and sub-systems,

- high scalability of the system, and

- easy integration of existing applications or external systems.

The general architecture of the AdeLE's system is illustrated in Figure 4.4. At client-side, learners interact with the Web Client as with a usual Web-based environment. At the same time, also at the client-side, real-time information about the behavioural and constitutional states is being tracked. Thus, the input for the User Modelling System consists of learner information received from the client-sided real-time systems as well as information about e.g. learners' traits, preferences and interactions gathered from the Learning Management System via the Adaptive System.

The Adaptive System personalises content, navigation and visualisation of learning material by exploiting user information from the User Modelling System. Further, course creators and teachers may use the Dynamic Background Library to define knowledge topics of material structures and their associations. The Dynamic Background Library interacts with the Search & Retrieval System to enable an access to dynamically retrieved background learning resources relying on the Web.

Following the main goals of the research project (summarised previously in this section), the AdeLE system should be developed on a service-oriented architecture. Hence, the Openwings framework specification was chosen as architectural basis for the server-side systems: User Modelling and Adaptive Systems as well as Dynamic Background Library (see e.g. sections 4.3.8 and 5.1 for more details). As indicated at the beginning of this chapter, there existed already a first prototype for the Dynamic Background Library. AdeLE's new version of this library tool is known as Concept-based Context Modelling System (CO2) and is described shortly in the ensuing section.

The DBL Research Agenda

According to [García-Barrios 2006c], the notion of Dynamic Background Library (DBL) goes back to [Dietinger et al. 1998]. Accordingly, a DBL makes it possible to access domain-dependent, topical and user-relevant knowledge resources by means of querying the services of an Information Search and Retrieval System (ISRS); these resources are meant to build a background library, because they rely beyond the internal and static learning repository of traditional e-learning settings. The DBL places these 'dynamically' retrievable resources at the disposal of a learner during the learning journey.

Figure 4.5 on the next page shows the basic architecture of a DBL (for details see e.g. [García-Barrios et al. 2002]). The knowledge Background Resources (BR) relying on the World Wide Web (WWW) can be accessed via ISRS (see BR layer on top-side of the figure). These resources are modelled within the Core System (CS) of the DBL as sets of Concepts (see CS layer). Further, DBL Concepts are assigned to Expertise levels and point to course sections of an E-Learning Repository (see arrow between CS and ELR layer). The Adaptive Content Delivery (ACD) layer adapts the contents in the E-Learning Repository (ELR) according to the interdependencies defined in CS. The ACD layer is also in charge of adapting the presentation of the deliv-

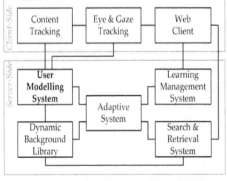

Figure 4.4: General Architecture of the AdeLE System. [García-Barrios 2006e]

ered ELR pages depending on the preferred DBL viewing mode of learners (In File, After File, After Chapter, After Content). For example, if a learner requests an ELR content page and the ACD finds concepts assigned to this page, the concepts may be attached to the page either in form of hyperlinked a 'book' icons (see Figure 4.6) or word lists (respectively, InFile viewing mode or the others). When learners follow DBL hyperlinked words or icons, a search request is sent to the ISRS using the pre-stored and concept-related queries (see arrow Activation on the left side of Figure 4.5). Thus, from the user's point of view, the final outcome of the described functionality is perceived as a 'dynamically' generated, additional Web page with search results that represent a set of accurate, relevant and topical documents from the Web, i.e. 'dynamically generated background knowledge'. The name of this tool, *Dynamic Background Library,* was derived from this point of view.

This first version of the DBL had been integrated into the eLS (eLearning Suite) of the Hyperwave Information System (see [García-Barrios 2002] for more details). For the AdeLE project, a distinct e-learning platform was given, and thus,

WWW
BR
Information Search & Retrieval System
Queries
Concepts
Expertise
CS
Chapter 1
Chapter 2
...
Chapter n
ELR
In File
After File
After Chapter
After Content
ACD
Activation
Figure 4.5: Overall Functional Architecture of a Dynamic Background Library (DBL). [García-Barrios 2006c]

the DBL had to be migrated an adapted to AdeLE's technological framework. Further, first evaluation results on the usage of the DBL were present, and improvements had to be made. In short, a new version was required.)[García-Barrios et al. 2004a] [Mödritscher et al. 2005] [Mödritscher et al. 2006c])

The most relevant goals of the AdeLE project do not include the development of mechanisms to track and model the acquired knowledge of learners. But the needs to improve the functionality of the DBL and to adapt it to be utilised within the AdeLE framework led to the idea of designing its new version as a 'knowledge-based modelling system'. The abstraction of the DBL for this idea is shown in Figure 4.7 (on the next page). The bottom part of the figure shows a mapping of the structure of a learning course (right side) into an internal representation, called Learning Context (left side).

The new DBL version should also provide teachers the possibility to define a set of 'descriptive tags' that might be related to the elements of the Context. An example: a course consists of the sections A and B, where B has two sub-sections B1 and B2. This structure can be easily mapped to an internal hierarchical model (Learning Context on bottom-left side of the figure). Further, each element in the context (i.e. each course section in the real-world) can be described in terms of e.g. 'this chapter gives an explanation of the topics X, Y and Z'. Within the scope of the new DBL version, this notion is interpreted as 'defining a set of *Concepts* to describe elements of learning *Contexts*'. Based on that, the new DBL version is called *Concept-based Context modelling system* (CO2). Further, distinct ISRSs can be regis-

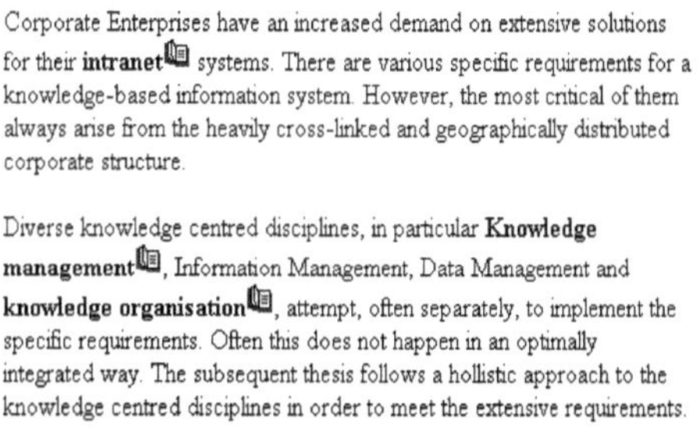

Corporate Enterprises have an increased demand on extensive solutions for their intranet systems. There are various specific requirements for a knowledge-based information system. However, the most critical of them always arise from the heavily cross-linked and geographically distributed corporate structure.
Diverse knowledge centred disciplines, in particular Knowledge management, Information Management, Data Management and knowledge organisation, attempt, often separately, to implement the specific requirements. Often this does not happen in an optimally integrated way. The subsequent thesis follows a hollistic approach to the knowledge centred disciplines in order to meet the extensive requirements.
Figure 4.6: The DBL Viewing Mode *In File* shows Web Pages adapted with hyperlinked 'Book'-Icons. [García-Barrios 2004a]

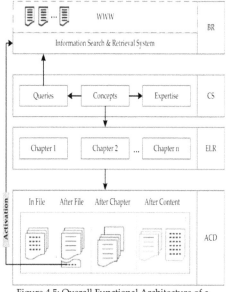

tered within the CO2 by means of their corresponding queries and requested on demand (see Q's on upper-side of Figure 4.7). In general, the way to call ISRSs comprises a stable syntax with placeholders for *keywords* that define a specific search need (e.g. 'engine.for/search?get=*keys*' can be used to call the fictive ISRS 'engine' to search for the keywords in '*keys*'). This mechanism is utilised in CO2 to give teachers the possibility to assign distinct keyword combinations (for distinct levels of knowledge expertise) to Concepts. For example, the concept 'Java' may be defined in CO2 for one ISRS as 'Java overview' for novices in the topic of Java Programming, as 'Java in a nutshell' for intermediates, and as 'J2EE design patterns' for experts. Hence, querying an ISRS with these three distinct keywords will lead to distinct sets of results.

In summary, the internal core functionality of CO2 is defined by *concept models* (graphs) which can be interconnected in two ways, first, to internal structures that model the real world (*contexts structures*), and second, to a group of attributes (*searchquery←concepts*) that can be reproduced to

reach different information seeking goals (*levels of knowledge expertise*). From the point of view of User Modelling, the idea behind CO2 is like the combination of stereotyping and overlay modelling. The following analogies can be identified. In User Modelling a specific stereotype is activated trough specific triggers; respectively, in CO2 a specific group of ISRS definitions is activated trough distinct levels of expertise. Further, in User Modelling simple overlay models may be implemented as a set of attributes in individual user models that refer to the nodes of generic structures in domain models; respectively in CO2, contexts are a set of attributes that refer to generic concept graphs. The concrete task within the scope of developing **"yo?"** is to make it possible to embed the mentioned functions in the design of the **"yo?"** system, and so, to utilise it as a CO2 system as well.

The *Mistral* Research Project

The MISTRAL research project aims at smart semi-automatic solutions for semantic annotation and enrichment of multi-modal data from meet-

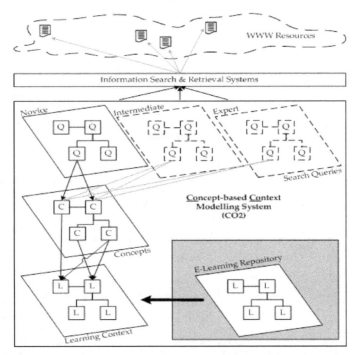

Figure 4.7: Basic Idea for New DBL: Concept-based Context Modelling System, 'CO2'.

ing recordings and meeting-related documents [Mistral 2006]. As indicated in [García-Barrios and Gütl 2006], face-to-face and virtual meetings can be recorded using different modalities. Usually, cameras and microphones are applied for this purpose, but also other specialised modalities are conceivable, such as device-usage or click-behaviour recorders. Nonetheless, all relevant information remains hidden or beyond users' reach into those distinct multi-modal streams.

Hence, to efficiently and contextually manage the knowledge addressed and generated in meetings, additional mechanisms are needed for the semantic processing, enrichment and integration of multi-modal data. To manage multi-modal data, Mistral's system consists of modules for data management, uni-modal stream processing, multi-modal merging of extracted semantics, semantic enrichment of concepts, and semantic applications (see Figure 4.8). The Data Management unit represents the main storage repository for all Mistral-relevant data. The Uni-modal unit consists of five modules for the processing of single modality streams: video, audio, speech-to-text, text and sensory modules. The Multi-modal unit merges uni-modal data and checks the confidence of comparable uni-modal extractions. The Semantic Enrichment unit detects conflicts between uni-modal annotations and infers further semantics. Finally, the Semantic Applications unit (SemAU) is the 'first consumer' of information and builds the front-end of the Mistral system for external clients. Mistral's SemAU embraces different components: Portal, Adaptation System, User Modeller, CO2 Modeller, Visualisation System and Retrieval System. The external users or client applications may access SemAU's components vía the Web-based services of the Portal. Only the Retrieval System has direct access to the Core Framework.

Mistral's adaptive behaviour is implemented in the Adaptation System. The following sample scenarios should help to clarify the practical usage of the Mistral system:

(1) Danijela is trainer in the company and in charge of instructional activities. Her current task is to prepare courses for future vocational trainings, and thus, she has decided to investigate hot topics from recent project meetings. She might use the Mistral system to search over topics addressed in meeting recordings.

(2) Fernando enrolled for a course at the company's e-learning system. Based on his preferred learning style, he wants to get illustrative examples by audio-visual media. According to the lessons' topics, the Mistral system might deliver him those segments of meeting recordings where topics had been treated under the agenda point 'discussion'.

(3) Estela is a new member of the company. To get familiar with the problem solving style and practices of the company, she must complete an experience-based training. In case of problems or questions, she can consult persons that are experts on her current task. Thus, according to her current role and to the topic of the task at hand, the Mistral system might show her those meeting participants considered to be experts.

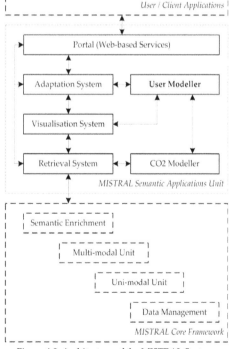

Figure 4.8: Architecture of the MISTRAL System, with Focus on its Semantic Applications Unit.

From the point of view of end-users, Mistral's SemAU provides three main functionalities: *search & retrieval, personalisation* and *visualisations*. E.g. to fulfil the information retrieval needs of scenarios 1 and 2, Mistral's Retrieval System may deliver results as shown in Figure 4.9.

Further, the outcome of Mistral's Visualisation System for an 'expert search' scenario (as required in scenario 3) is illustrated in Figure 4.10: although the topics addressed by meeting participants may diverge strongly from a general point of view (see small box on bottom-left side), an adaptation of search queries may deliver more accurate and useful information; the visualisation in the background of the figure shows

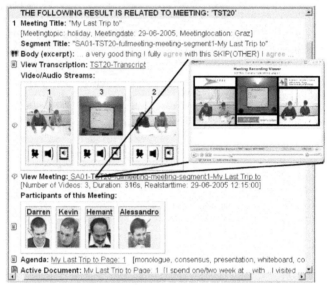

Figure 4.9: User Interface of Mistral's Retrieval System (Visualisation of Search Results).
[García-Barrios and Gütl 2006]

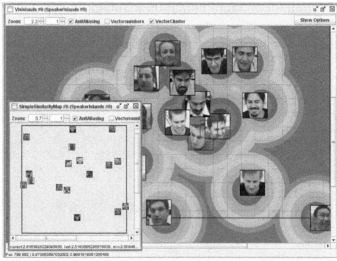

Figure 4.10: Visualisation of general vs. topic-related Similarities among distinct Meeting Participants.
[García-Barrios and Gütl 2006]

which meeting participants have addressed those topics related to the current working context of the interacting system user (expressed in similarity distances, whereby the current user is located in the middle of the window). More details and results of Mistral are shown in the next chapter (see also e.g. [García-Barrios 2006d] or [Gütl and García-Barrios 2005c]).

Hence, the **"yo?"** system (originally intended to satisfy e-learning requirements) should be able to fulfil the personalisation needs of the Mistral project as well, i.e. in a distinct application area: Retrieval of Cross-Media Semantics.

Summary

The projects introduced in this section represent the distinct application areas for the **"yo?"** modelling system. The most relevant aspects about *application- and domain-independent user modelling systems* were presented in chapter 3 (section 3.2.4). These systems are often called *Generic User Modelling Systems* (GUMS).

The term *Multi-purpose Profiling and Modelling System* (MPMS) has been introduced in this section to describe the role of **"yo?"** in the aforementioned projects: the notion of MPMS is used to underline that the development of **"yo?"** is not limited to the scope of modelling users for one application area (see also introductory section of this chapter).

This notion is similar to the aspect of 'modelling distinct agents' as coined within the scope of General User Modelling Systems, e.g. *GRUNDY* and *GUMS₁* (see section 3.2.4). Though, in order to show the functions of the **"yo?"** system as a MPMS, the focus of attention has to be brought back to the scope of application- and domain-independent modelling systems. This point of view is necessary to provide a general fundament for the proposed solution and to better convey and delimit the role of MPMSs in adaptation-pertinent systems.

The next section introduces a *general architecture for adaptation-pertinent systems*, where the generalised functional scope of **"yo?"** as a multi-purpose modelling system is defined.

4.1.2 Overview on Abstract Context: Adaptive Systems

This section is based on the findings of the investigation about adaptive systems presented in the previous chapters. It summarises most important findings that are relevant to define a *general architecture for adaptation-pertinent systems*.

From Adaptation to Personalisation

The topics of *Personalisation and Adaptation* (P&A) are neither new techniques in the field of Computer Science nor new issues of research in other fields, as can be identified e.g. in Cybernetics, Evolutionary Research, Biology or Climatology. In the last 20 years much research work took place to discuss general issues regarding P&A. In general terms, *adaptation* involves the following process: as a response to specific stimuli from its environment, an adaptive system may need to alter something in such a way that the result of the alteration corresponds to the best-known succour solution for the fulfilment of the specific need. In specific terms, these specific needs are linked to the main goals of the person interacting with a software-based adaptive system, i.e. the notion of Adaptation can be narrowed to the notion of Personalisation.

Adaptations may be differentiated according to whether they are autonomous or planned, occur in natural or socio-economic systems, are anticipatory or reactive, and thus, may take technological, institutional or behavioural forms (see e.g. [Smithers and Smit 1997], [Smit et al. 1999] or [Smit et al. 2001]). Adaptation is then a process that involves a *purpose* at its beginning and a *success-measure* at its end. On the one side, this helps to get a task-oriented view on adaptation-pertinent systems, and on the other side, distinct components are in charge of distinct duties along this process.

This brings to a systemic point of view. In concordance with [Haubelt et al. 2004], it can be stated that systems may include other systems as parts of their construction, and that a system is made of components, which interact with each other. This point of view has relevant implications since, first, it implies the possibility of constructing complex systems as integration of other simple ones (i.e. a *system of systems*), and second, it allows the introduction of the concept of modularity (a key feature in the development of service-oriented architectures). Thus, the following requirements for the design of architectures or models for general adaptation-pertinent systems can be identified:

- modularity and flexibility,
- sensitivity control, and
- purpose-oriented structuring of the system.

The second requirement is particularly relevant as it involves the internals of those components of the adaptive system, which are in direct interaction with the external environment, i.e. the external stimuli providers, or in other words, the input interfaces of the adaptive system. Hence, the sensitivity of the system is defined by means of the characteristics of its sensors. In turn, the initiation or continuation of adaptation processes is constantly 'triggered' by the stimuli over these interfaces.

Adaptation may further concern the response to a specific type of stimulus, which lasts longer than a specific period of time and may out-last the sensitivity period of the system. In order to deal with such persistent stimuli, adaptive systems should also include some integral feedback control mechanisms to enable robust adaptation. Also in this context, if a small change in an input parameter (stimulus) results in relatively large changes in the outcomes (big adaptation impact), the outcomes are said to be sensitive to that parameter.

In some cases, this may mean that the input parameter has to be determined as exactly as possible, or that the resulting outcome has to be redesigned for a lower sensitivity (see also [Heylighen 2004]). The aspect of sensitivity control is relevant for the scope of this book, because one special trait of the AdeLE system is its high sensitivity, due to the usage of an eye-tracking device. Eye-tracking systems gather big amounts of data about eye movements in very short time periods. This fact may imply that a communication bottleneck appears between eye-tracker and user modelling system. This aspect is considered in the design of **"yo?"**.

Other relevant characteristics regarding the task-oriented (i.e. purpose↔ success processes) and systemic (i.e. involved components) points of view on adaptation-pertinent systems were already identified in chapter 2 (sections 2.1.3 and 2.3), e.g. resistance, stability, feedback, adaptive capacity, adaptability, and responsiveness. Within the context of this book, the specific adaptational needs of an adaptive system are directly linked to the specific goals of the person interacting with the system. Here, a paradigm shift takes place from Adaptation to Personalisation. The most relevant details regarding this paradigm shift were considered through the investigation journey from chapter 2 (*Adaptation*) to chapter 3 (*Personalisation*).

Adaptation-Pertinent Systems

Adaptation-pertinent computer systems are able to improve the experiences of their users through personalised responses to interaction events. For that purpose, the system needs to 'know individual users' as accurate as possible, a need that could be fulfilled by explicitly asking the user for personal information constantly. As this method is too annoying for every user, a user-adapting system holds an internal representation over time and infers assumptions about the user from observation and analysis of the user's interaction history.

Further, the adaptive system may also analyse the history of all known users to find similarities and infers individual information from social behaviour. Adaptation-pertinent computer systems carry out these observations and assumptions in an iterative or incremental way to model individual users internally (i.e. it needs a user *modelling system*) and utilises this internal knowledge to decide how to personalise the outcome of the system's behaviour (i.e. it needs an *adaptive engine*).

In order to identify which are the components that are most relevant for the **"yo?"** modelling system, this section presents a first approach to a *General Architecture for Adaptation-pertinent Systems* (GAAS). Figure 4.11 shows this general architecture. It is highly relevant to point out at this point that GAAS does not represent a formal model to define generic or general adaptivity, because such a formalisation goes beyond the central scope and intentions of this book. Rather, GAAS is a systemic abstraction needed to indicate which general components build the architectural basis of computer-based interactive personalisation systems.

GAAS considers an adaptation-pertinent system consisting of two main elements: *System* and *Environment*. The System comprises all software components in computers that are needed to perform personalisation tasks. The main target of the system is a *Person* interacting with such components. To make these interactions possible, some of the System's components are defined as peripheral *Interaction Interfaces*, i.e. electronic devices that are able to interact with the system, e.g. mouse, keyboard, monitor, digital pen.

From the system's point of view, Interaction Interfaces can be *Input* or *Output* electronic devices. This differentiation is essential because some devices may be used (a) solely as sensors of the system, usually to feed the user model with personal interactions (e.g. mouse, eye-tracking device, light sensors), (b) solely as output devices, usually to express the adapted results (e.g. volume of loudspeakers, screen view on a standard display), or (c) as a combined interface (e.g. touch screen or mobile phone). Further, peripheral devices can be considered as *Other Targets* of adaptational procedures. This aspect is also rele-

vant due to the different notions of implicit and explicit personalisation, which were introduced in section 3.1.1. Based on this notion, consider the following observations and examples.

For implicit personalisation, the system needs knowledge about the specific individual interacting with it. Thus, the main target is a very specific person, and an adaptational goal might be e.g. to adapt the presentation of some document or the arrangement of some UI elements on the monitor. On the other side, consider an adaptive system aiming to adapt the brightness of the display of a mobile phone depending on the input of a light sensor. Naturally, such an adaptation goal is targeted to satisfy the user need of a clear view on the display. Nevertheless, for such an adaptation no knowledge about the individual user is needed. The goal of the system can be defined as 'if the phone is active, adapt the brightness of its display according to the illumination of its environment', and the target is 'a specific type of mobile phone'. In this case, the internal modelling system must not differentiate among individual users, it must differentiate among individual mobile phones.

In the same manner as *user modelling systems* may categorise 'learners' by means of profile attributes, stereotypes or behavioural aspects (e.g. demographic data, reading interests, learning styles, and so forth), the notion of *agent modelling systems* can be used to categorise other agents in

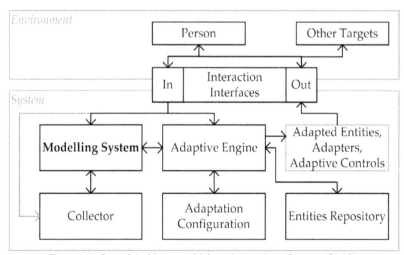

Figure 4.11: General Architecture of Adaptation-pertinent Systems (GAAS).

the environment of the system. In the example given previously, the agents 'mobile phones' can be categorise in the same way, because distinct phones have also distinctive attributes (e.g. type), common information can be stereotyped (e.g. technical description data), or they may be modelled in terms of 'behaviour' (e.g. usage, location changes, or offline time periods).

The remaining components in GAAS were derived from the requirements of modularity and separation of main duties. As indicated previously in this section, the main functional components from a task-oriented point of view are Modelling System and Adaptive Engine. The former is in charge of representing internally the knowledge about agents in the environment (in most cases, the user). The latter is essentially in charge of knowing when and what to adapt. This notion implies several important aspects.

The Modelling System is like a specialised and (mostly indispensable) assistant of the Adaptive Engine, for which it continuously acquires and analyses data from the sensors of the system. In addition, for the same purpose, the Modelling System may proactively use a *Collector* to gather information independently of the input of its sensors, e.g. through a Data Mining Tool to extract knowledge domain concepts in external documents used by a specific individual, and further infer individual topics of interest (as in personalised web usage systems or recommender systems).

Indeed, the duties of a Modelling System may not be seen as a set of actions to just gather permanently every sort of user information. Modelling the environment of adaptive systems efficiently makes only sense within the constraints of an application domain, and this makes Modelling directly dependent on the adaptational goals of the overall system. The knowledge about these goals is a main duty of the Adaptive Engine, i.e. it knows the reasons and techniques of adaptation, which can be configured through the *Adaptation Configuration* component. Thus, the Adaptive Engine is in charge of knowing and applying the general 'rules of the game'. For the successful completion of a specific personalisation process, the Adaptive Engine must

- monitor those data from Input Interaction Interfaces that are relevant to trigger a corresponding adaptational procedure, i.e. it waits for relevant stimuli;

- know which information is needed from the Modelling System in order to apply its known rules to generate, alter or select adapted results (which may be contained in an Entities Repository of adaptable objects); and

- send the adaptation result to a corresponding Output Interaction Interface.

According to the notion of Adaptors and Adapters introduced in section 2.1.1, an Adaptive Engine might be programmed on the basis of a central controller of a factory of adaptors, i.e. a manager of specialised software components that can be activated according to the adaptation task at hand. Further, the adaptable entities in the Entities Repository are represented by alterable assets (e.g. documents, data structures, sequences of actions) and Adapters, which are not alterable but used as attachments to change the appearance of adaptable entities (e.g. a personalisation result might not be a modified text in a Web page, but rather an applet for further observations of user interactions). Coming back to the example of adapting the brightness of the display of a specific mobile phone, the result of this adaptation is neither an *Adapted Entity* nor an attached *Adapter* on an adaptable entity. It is a very specific *Adaptive Control* signal that must be sent to the target device, and over there (in the external target) this signal will activate some functionality to fulfil the adaptation.

This last observation is utilised in GAAS to introduce the notion of distinct *expression types* of adaptational results, which in turn helps to distinguish between perceivable and hidden personalisation (see section 3.1.1). Therefore, the distinct expression types are related to distinct expression targets. Thus, the impact and success of personalisation can be measured in *expression places*. Consider the following simple example that combines two expression types at two expression places for two adaptation targets: within a single adaptational procedure a system may

- adapt the size of a document, i.e. compresses it due to a slow connection speed of the interacting remote target computer (expression type: hidden; place: communication channel; adaptation target: agent), and

- personalise the language of the content according to the preferred language of a specific user (expression type: perceivable; place: screen; adaptation target: person).

The necessity of a feedback mechanism has been mentioned previously, but not explained within the scope of GAAS. The task-oriented point of view has coined that a successful adaptation process is a question of 'purpose vs. success measurement'. This issue has been investigated in the previous chapters by means of John Holland's notion of *fitness of adaptive systems*. One way to explain the measurability of fitness or success of adaptational procedures is especially illustrative if taking an example from the context of Control Systems [Richter 1995]. Firstly, it must be taken into account that (in general) no uniquely valid starting point for an adaptation process can be defined, although it seems to be logical that the triggering is always an interaction with some agent of the environment. The reason is that adaptational procedures may have many purposes. [Richter 1995] stipulates three types of adaptation purposes:

- *Control*
  The aim is here to transfer the current state of an adaptable entity in a new 'desired and known' state through the experimentation of different alterations of the entity. One of the key issues in this context is the maintenance or achievement of a 'stable' state of the adapted entity. Here, the involvement of sensors as a starting point may be needed.

- *Modelling*
  Through a series of experimentation tests with different 'known' adaptable entities, an 'unknown' (but in the future perhaps needed) adapted entity has to be predicted, i.e. anticipative adaptation is needed. Here, the Modelling System could play the main role and represent a possible starting point for this type of adaptational purpose.

- *Optimisation*
  Different 'unknown' adaptable entities must be 'tried out' (i.e. generated) until the needed 'known' (but not available) adapted entity is found. This means that during the search for the best solution no modelling or stimuli recognition may be needed.

To reach these sophisticated purposes of adaptational-pertinent systems, distinct feedback loops through distinct GAAS components are needed. Therefore, GAAS does not contain any single feedback management component, rather it shows implicitly that these processes can be performed through the interaction of its components. In concrete, the duties to improve the fitness of the system can be delegated as follows.

Controlling the stability of best solutions may be a combined task for the Modelling System (because it implies observation and memory) and the Adaptive Engine (because it knows and controls the processes and goals of adaptation). The prediction of new best models could be a task for the Modelling System. This seems to break the notion of a coherent and central adaptive system, but in reality the components of user-adapting computer systems may be distributed, and in turn, this implies that the adaptable entities are part of the external environment of the Modelling System. From the point of view of the Modelling System, it is modelling just an additional agent. An automatic generation of new solutions should be carried out by the Adaptive System, which must feed itself back with fictive stimuli. These stimuli can be only interpreted by the Adaptive System, because it is the only component that knows 'the rules of the game'. An intervention of the Modelling System does not make sense, because it does know the rules to produce and check adaptable entities.

So far, the proposed General Architecture of Adaptation-pertinent Systems (GAAS) has clarified the theoretical constraints of multi-purpose modelling systems, i.e. the *abstract context* of the **"yo?"** system. Further, the concrete real-world environment of "yo?" has been introduced, i.e. the three research projects that define its practical applicability. The next sub-chapter presents the design of the internal architecture and duties of **"yo?"**.

4.2 Architectural and Functional Design

The main architectural and functional aspects concerning the internal modules of the **"yo?"** system are presented in this sub-chapter. General requirements are described in section 4.2.1, from which a general layer-based architecture (see section 4.2.2) is derived. Finally, to show the interdependencies of functional modules, some detailed insights into the overall system are given in section 4.2.3.

4.2.1 Definition of Requirements

User modelling systems are, in essence, specialised information systems. As such, they must be able to *manage classes of user information* in an efficient way. In this context, the most relevant processes in user modelling systems regarding user information management are: acquisition and provision of classified information as well as internal creation, transformation and processing of user classes.

The functionality of a Multi-purpose Profiling and Modelling System (MPMS) is not restricted to 'user' information. Rather, the main purpose of a MPMS comprises (a) to model environmental entities as agents and (b) to provide information about these entities to its clients.

Though, the services provided by an MPMS do not differ strongly from those of a GUMS, and therefore, the requirements shown in this section converge at several points. The first important aspect is to distinguish among the different clients (users) of such systems. The system must be able to provide *distinct views and access mechanisms* for the following clients:

- *System administrators*
  i.e. persons that are in charge of the configuration and maintenance of the system. This requires that the internals of the system should be designed and developed in such a way, that they can be easily transformed from a machine-readable into a human-readable form. Usually, system administrators are technicians, and thus, a shell interface is useful.

- *Users that the system models*
  i.e. persons which know that they are modelled. This issue is especially critical and a hard task in terms of scrutability of the models. Commonly, users that are modelled in the system lack of sufficient technical expertise or are not interested in technical explanations. Thus, the transformation of models from machine- into human-readable formats is not trivial anymore. Some more sophisticated tools for information visualisations are needed.

- *Other systems*
  i.e. software components that interact with the MPMS, e.g. adaptive engines, knowledge modelling components, rule engines, user interfaces, distinct sensoring devices. These components can be external providers or consumers of information. Moreover, some of them can be both, as for example an adaptive engine that supports feedback controlling or predictive feedback. Still, the communication among software components is not trivial, because also here a transformation might be needed in terms of machine-machine transformation of data structures. A system may show limited communication capabilities with other systems if is based on a specific architecture (e.g. client-server, peer-to-peer, agent-based or service-oriented) or a certain standard (e.g. Web Services, JXTA, RDF).

As indicated in the previous chapter, the list of requirements for (user) modelling systems is very long and depends also on the overall goals of their clients. Nonetheless, some general requirements can be identified and categorised as follows. These general requirements regarding *information interchange*, acquisition and provision:

- *Managing big data*
  Provide mechanisms to avoid communication bottlenecks caused by information providers which send data chunks that may need a complex or long processing, e.g. big packages of data may lead to bottlenecks at storage level, or a high frequency of real-time data delivery may block the overall communication level.

- *Rapid initialisation*
  Provide mechanisms to accelerate model ini-

tialisation and solve the cold-start problem. This issue is critical for user modelling systems, which aim to learn the users as quick and accurate as possible. When using explicit feedback, the system should include techniques to reduce the number of questions to an acceptable minimum.

- *Detection and tracking of contexts*
  In an unobtrusive way and only with the explicit permission of users, the situational context or actual task of system users can be captured (or inferred) from their connectivity context, i.e. their Personal Area Network (PAN).

- *Exchangeability of information*
  Provide open interfaces and support standards e.g. to avoid redundant data in the models or to compare/share the contents of models.

- *Human-readability and visualisation*
  Enable the scrutability of the models for humans. This requirement implies the provision of tools for the description and visualisation of data, internal representations as well as reasons for inferred assumptions.

- *Separation of unused data*
  Prioritising of information according to the relevance of the task at hand may reduce information overload. Further, unused or 'old' information can be archived in separate storages and made accessible on-demand.

- *Privacy and security*
  Provision of all needed mechanisms to ensure that private and confidential data is thus utilised as wished by the user; moreover, respecting the legality of privacy rights.

General requirements regarding *modelling and reasoning*:

- *Semantic degree*
  Provide a separation of techniques for Profiling and Modelling. Raw data (i.e. acquired and provided 'as is', e.g. static or long-term user information) and simple profiles should be accessed without the intervention of reasoning modules. Semantic data (i.e. enriched or inferred information), may imply the intervention of complex processing techniques (e.g. cognitive or behavioural traits). In other

words, if some information must not be post-processed or interpreted internally, it should find the most direct way to the storage device.

- *Model certainty*
  If the system distinguishes between facts and assumptions about the user, it should attach to each assumed trait a level of confidence, e.g. through machine learning methods or direct feedback from the user, the confidence level may be increased. This issue is related to scrutability, as coding machine-assumptions about humans implies the need to include their corresponding human-readable justifications.

- *Multiple reasoners*
  Distinct techniques are used to infer distinct characteristics of a user, e.g. to predict future behaviour, infer assumptions from other data, observe and compare behaviours, trigger stereotypes, cluster user traits, or identify goals from low-level interactions. Many of these techniques can be isolated as reusable algorithms. Thus, if feasible, the system should treat these techniques as functional black-boxes, for which some input is given and some output expected. This abstraction improves the multi-purposefulness of the system.

Of special interest are also some general requirements regarding more specialised functions of modelling systems. General requirements regarding modelling of dynamic traits:

- *Behaviour tracking*
  provision of methods to log and interpret chronologically acquired information. This requirement includes e.g. the need for techniques to store and maintain history data, observe characteristics of streams of data, or monitor the triggers of stereotypes.

- *History statistics*
  through behaviour tracking, behavioural aspects of users can be recorded, e.g. their past interactions with the system. Mechanisms should be provided to analyse or evaluate such information streams, e.g. to compare individual with group behaviours, to retrieve

segments of logs, interpret data in the logs, or to visualise logged data as diagrams.

- *Consistency and topicality*
especially if distinct clients of the modelling system share the use of some information in models, the consistency of this data must be maintained. Then, a constant evaluation of the topicality of user models should be granted. Concurrent clients operating on shared, continuously evolving information in user models may produce conflicts of access, i.e. synchronisation of access and internal permission policies are needed.

Finally, for the MPMSs it is also required to provide capabilities to manage groups of agents, such as the membership of users in given groups, the assignment of role attributes to users, or the detection of common traits among all users and cluster them into coherent groups. Within the context of this book, this aspect is called *community modelling* and leads to additional requirements:

- *Detection of common traits*
Representation of relevant common traits of agents (e.g. users) as members of a coherent community using a specific application or following a common goal (e.g. vía machine learning methods as in recommender systems).

- *Classification of agents*
Provide similarity clustering e.g. vía stereotypes or group management.

- *Fellow adoption*
Integrate individual user models into canonical community models. The more users are adopted into canonical models, the more grows the multi-purposefulness of the system. E.g. if using a canonical model as 'root' element of a stereotype hierarchy, each new relevant trait in individual models may contribute to generate more accurate substereotypes.

- *Fellow propagation*
In contrast to fellow adoption, this aspect implies the integration of typical traits of community models into individual user models. Thus, each new or modified trait in the community model leads to the same

changes in the individual models. From the point of view of the community models, the notion of fellow adoption and propagation is comparable with traditional Pull&Push techniques of information flow (i.e. adopt=pull, propagate=push).

As the development of the AdeLE and Mistral systems should follow a *service-oriented approach*, the development of this MPMS should meet the following technological requirements as well:

- *openness* of the overall system,

- *exchangeability* or *extensibility* of functionalities,

- strict *modularity*, i.e. development of separated, specialised, task-specific modules,

- *platform independence*, and

- *ease of integration* into existing applications or external systems.

Based on the requirements depicted so far, a general architecture for an MPMS can be defined as shown in the ensuing section.

4.2.2 Architecture and Workflows

This section introduces the general architecture of the **"yo?"** system. The layer-based architecture presented in this section should build the basis to identify the main logical parts of the system as well as their general functions and interactions. The goal of this section is to give an overview on those basic elements of the system, for which the highest level of abstraction is defined, i.e. the components shown here are defined as main containers for specialised modules. Thus, the next section shifts to a view on the specialised components and services of the system.

The *global tasks of user modelling systems* have been investigated in the previous chapter and can be used to define the main processes in such systems (see section 3.2.2): acquisition of user information, creation of individual user models, administration of the models, and provision of user information. According to this point of view and under consideration of the requirements specified in the previous section, a general <u>L</u>ayer-based <u>A</u>rchitecture of the "<u>yo</u>?" <u>S</u>ystem (LAYS) can be derived, as shown in Figure 4.12.

The LAYS architecture comprises seven layers: Consuming and Providing; Managing; Viewing; Assisting; Modelling; Profiling; and Persisting.

Consuming and Providing: This layer builds the environment of the system and comprises all possible external clients. These clients may be interested in feeding the system with data about agents that have to be modelled or in fetching accurate information about individually modelled agents.

Managing: In order to communicate with its external clients, the system must manage different levels of communication and access. At the upper level of Managing, message-driven communication is supported. In general terms, the system may accept requests from and create notifications for its clients. Hence, Managing involves the duties of checking the access permissions of the clients. In addition, Managing implies the automatic control of two-way dispatching. This means that on the one side, Managing includes the recognition of that layer below it that is in charge of processing the message. On the other side, if the system should provide an answer or create a notification, Managing involves also a scheduling of processes needed to send the message to the external client. Thus, the Manager steers access and communication among external clients and internal layers.

Persisting: The Persisting layer is in charge of providing the lowest level of transparency for the system. It provides an application interface to upper layers to access permanently stored data structures, such as single attributes and profiles of agents as well as permissions and data from the clients. Thus, it controls all functions regarding the permanent storage of data structures. A Persisting layer must be transparent in order to enable upper layers to handle just a single (but flexible) data structure. At the bottom of this layer there may exist e.g. distinct types of file systems or databases. These distinct storage mechanisms support distinct languages to deal with data (e.g. it is not the same to read data from a file system and from a relational database). For the upper layers, this heterogeneity is abstracted within this layer through a transparent and internal application interface.

Profiling: This layer represents the simplest way to model agents. It knows exactly which basic data structures can be persisted and provides all functions to manage them. Accordingly, it holds and can differentiate among the most usual types of profiling information, such as lists, trees, sets or graphs. Clients that use the system as a simple profiling component and do not need any complex reasoning methods, may access profile information directly through the Managing layer, i.e. without the intervention of any other layer.

Modelling: The layer above Profiling represents the reasoning level of the system. This layer provides more sophisticated functions as Profiling, such as inferring additional information from profiling data, making assumptions about logged behavioural data, creating clusters of agents from the analysis of common data in distinct profiles, or observing changes in the states of some profile contents. The Modelling layer operates on the data structures of the Profiling layer. It may access the Persisting Layer to persist data about reasoning methods or structures, but not to store and retrieve user information. This is the only way for a clean separation of Profiling and Modelling. In fact, before some complex user models can be persisted by the Managing layer, the functionality of the Profiling

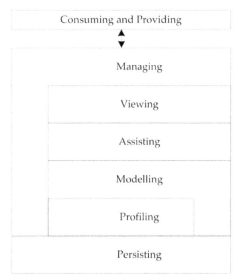

Figure 4.12: Layer-based Architecture of **"yo?"**.

layer must be extended accordingly.

Viewing: This layer embraces all mechanisms to enable a direct interaction with humans, i.e. it provides basically all needed mechanisms to create and manage user interfaces or visualisations.

Helping: All systems need some kind of utility box to ensure a high degree of openness, flexibility, exchangeability or scalability. This is enabled in LAYS through the Helping layer through e.g. tools for data encryption and decryption, import or export of data in distinct formats, or to enable the transformation of internal data into the syntax of a certain standard. Thus, where the Managing layer enables the correct transport of messages among heterogeneous systems, the Helping layer enables the correct interpretation of message contents among heterogeneous languages. Based on LAYS, the following essential functions are defined: *Configuration and Scrutability* of the system and the models, *Initialisation* of models, *Profiling* through the system, and *Modelling* through the system. Further, the representation of these functions on LAYS enables to identify and explain the main streams of information flow.

Configuration and Scrutability: The main persons that are allowed to interact directly with a user modelling system are system administrators and users that the system models (see previous section). As shown in Figure 4.13, for both types of system interactors, the Managing layer must fetch the corresponding user interface or visualisation from the Viewing layer and if necessary, activate a corresponding tool to support machine-human readability transformations. Based on that, the information of interest must be collected from the Modelling, Profiling, or Persisting layer. The output of the lower data layers must be then prepared according to the Viewing requirements and returned to the user.

Initialisation: As in most modelling systems, the appearance of a new agent that must be modelled may imply a lookup in the Persisting layer (see arrow '(1)' in Figure 4.14). If a user does not exist, the system must reply with the corresponding minimal set of initial questions that are required to initialise the user model. The most usual method to get initial information is through direct feedback, and therefore, a ma-

chine- to human-readable transformation of data must be performed. Finally, the corresponding Viewing for the transformation must be fetched and delivered to the user. In a second step (see arrow '(2)' in Figure 4.14) the feedback of the user can be post-processed (Modelling or Profiling) according to the method at hand (e.g. stereotyping). If the information from a first feedback is not enough to initialise the individual model, a

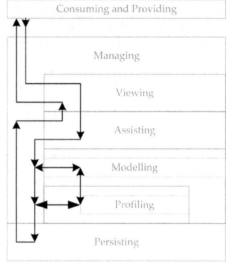

Figure 4.13: Configuration and Scrutability in LAYS.

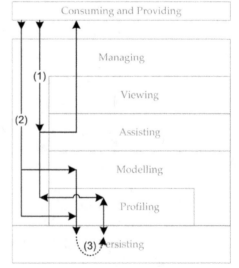

Figure 4.14: Model Initialisation in LAYS.

second feedback loop must be started (see arrow '(3)').

Profiling: The Profiling process workflow can be represented in LAYS as shown in Figure 4.15. If the system needs a transformation of internal data, the Assisting layer may be also needed. Otherwise, user information is retrieved or update directly through the Profiling layer. To give an example of the necessity of a transformation, consider a system that can provides data according to the GESTALT standard and a system that may have access to consume this information, but understand only IMS LIP.

Modelling: As for Profiling, also a transformation of model information might be needed while Modelling, and thus, the Assisting layer may be helpful (see Figure 4.16). Further, the Modelling process must access user information through the Profiling layer and in addition, may also need data about reasoning methods or complex models (see arrow '(1)', respectively arrow '(2)' in Figure 4.16).

So far, a general overview on requirements, architecture, functions and process regarding the **"yo?"** system as a multi-purpose modelling system have been presented. The internals of the system are shown in the subsequent section.

4.2.3 Services and Components

From the *Layer-based Architecture of the* **"yo?"** *System* (LAYS) a direct mapping into its main components can be performed. The result of this mapping is shown in Figure 4.17 and called CAYO (*Component Architecture of "yo?"*). Basically, the notion of actions (verbs) in the layer of LAYS is mapped to nouns in CAYO. Further, the Assisting layer is mapped to a Tools component.

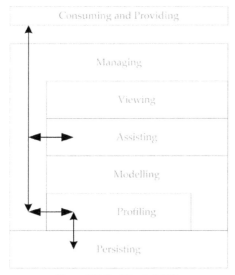

Figure 4.15: Profiling in LAYS.

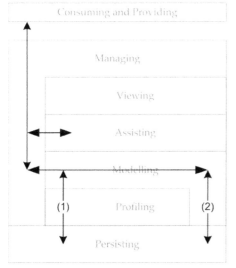

Figure 4.16: Modelling in LAYS.

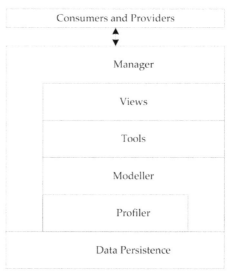

Figure 4.17: Component Architecture of **"yo?"**.

The reason for this mapping has a practical background.

The software development of **"yo?"** is based on a service-oriented framework. As will be shown in the next sub-chapter, one of the first challenges in the design of the system was given by an investigation of *service granularity*. For the context of this section it is just relevant to mention that the components in CAYO represent the main services in the **"yo?"** system. Still, it is relevant to emphasise that the components in CAYO and not the modules are implemented as services.

CAYO components are containers for specialised modules in **"yo?"**. The most relevant functionalities provided by each module are defined in the service interface of the corresponding component. Hence, this functional encapsulation of components in CAYO by means of services fosters the modularity and transparency of **"yo?"**, and consequently, it leads to a well-formed logical encapsulation of duties for the layers in LAYS.

The architecture of the Manager component of **"yo?"** is shown in Figure 4.18. Providers and Consumers are external client of the system that may access its functionalities, but solely through the Manager. The Manager represents then the *Single Point of Access* for external applications. Nonetheless, as indicated previously in this chapter, distinct instances of **"yo?"** may be distributed for distinct application purposes; such instances are called Siblings and may access directly the other components of the system. This mechanism enable to deploy e.g. proactive sensors in the Tools layer of Siblings, which may distribute information directly to trusted ('familiar') systems.

The Communication Handler in the Manager component is in charge of managing distinct types of communication and discovery alternatives, such as Java RMI vía Jini, Java RMI vía RMI-Registry, HTTP vía Servlets, HTTP vía SOAP (Web Services), or HTTP vía REST engines (for technical details see section 4.3.2). The first 'obstacle' for external clients is represented by the module Access Filtering. This is the central controller of permissions to access further components of **"yo?"**. Only external clients that are registered as trusted systems (for which Policies must be created) may communicate with internal components. By means of Triggers (rules of permission policies), Observers check constantly the messages of clients, and may allow or not a forwarding to the Dispatcher.

The Scheduler in the Dispatcher controls the flow of information within the system, i.e. it must know which components are responsible for the processing of which messages. The assignments of messages to components as well as their interpretation and conflict resolution are carried out by the Resolver. Further, expensive calls to the system may be stored temporarily in a Cache to improve the response performance of the system. If the Cache does not include an answer to a given request, the Dispatcher may continue and delegate the request to the responsible component.

The architecture of the components Tools and Views of **"yo?"** is shown in Figure 4.19. Both components are containers of functionalities that may be deployed in the system in order to make

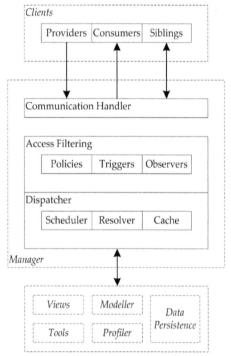

Figure 4.18: The Manager Component in **"yo?"**.

it functionally extensible and to improve its scrutability.

The system provides some default modules in these components. A Formatter enables to deploy Security methods for encryption-decryption of messages as well as Exchange converters for different standards. In addition, a container for the configuration of simple proactive collector scheduling is given: for example, (a) so called 'listeners' may be configured to open a specific interface of an internal component in specific time intervals to accept data from external systems, or (b) 'gatherers' may be deployed to interchange messages with Siblings. The Views component of the system is also a container of frameworks. Per default, the Visualisations module includes some graphical user interfaces and shells for modules in the Profiler and Modeller components as well as Web-based templates for questionnaires (e.g. used for the initialisation of stereotypes). The Universal Access module complements the Visualisations module. It ensures a strict separation of content and layout of user interfaces, e.g. for Web-based questionnaires, just distinct style-sheet definitions are allowed in the Forms module, whereas their corresponding XHTML template must be deployed in the Universal Access module.

The architecture of the Profiler component of "yo?" is shown in Figure 4.20. In essence, the Profiling Manager is the central part of the Profiler and represents a general provider for internal data structures. It is on the one hand, the unique module that has direct access to the Data Persistence component, which is denoted by the arrow on the right side of Figure 4.20. On the other hand, it places structures at the disposal of other components in order to be e.g. (a) manipulated on run-time by the Modeller, (b) visualised by the Views component, or (c) exchanged with external systems through the Tools component. The distinct data structures are hold and managed within specialised containers, such as Profiles (simple hierarchies of groups of attributes), Stereotypes & Communities (distinct types of canonical models for the construction, assignment and reactivation of individual models) and Loggers (list of chronological log entries). At the bottom of this component, a Structure Factory module checks the validity of internal structures with the interface capabilities of the Data Persistence component. This last mechanism is needed to keep data structures consistent and reusable.

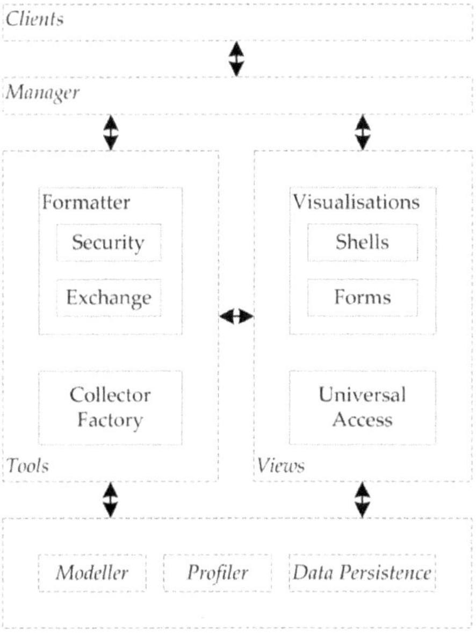

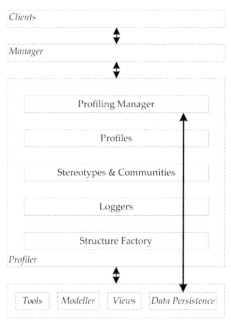

Figure 4.19: Tools and View Components in **"yo?"**.

Figure 4.20: The Profiler Component in **"yo?"**.

The architecture of the Modeller component is shown in Figure 4.21. This component comprises all reasoning capabilities of the system. Data structures are maintained in the Profiler, whereas they are mainly manipulated through the Modeller. Thus, the module Reasoners is a container of e.g. (a) machine learning algorithms that can operate upon individual profiles or groups of them, or (b) stereotyping reasoning rules that control the triggering of stereotypes.

The Knowledge Modelling module (KMM) in the Modeller component of **"yo?"** is a specialised unit that can produce and manage associations among data structures of the Profiler. Thus, the KMM may create metadata on contents of the profiles and models, or even interrelate models through descriptive linkages (details about this module are shown in section 4.3.8 and sub-chapter 5.2). The Behaviour Tracking module operates on Loggers of the Profiler and infers assumptions upon sets of log entries (details about this module are shown in section 4.3.7 and sub-chapter 5.1). The expandability of the reasoning capabilities of the Manager component is ensured through the FunctionLets module in the Modeller component. A FunctionLet is **"yo?"**-compatible software that can be deployed into the system. The system is implemented on the service-oriented Java-based framework 'Open-wings' (see next section), and therefore, FunctionLets must be written in the Java programming language.

The architecture of the Data Persistence component of **"yo?"** is shown in Figure 4.22. For the purpose of operating upon the data structures of profiles and models in the system, the Data Persistence component provides an Application Programming Interface (API) that can be used by the Manager, Profiler and Modeller. In concrete, this component is an abstraction layer that ensures the transparent use of distinct storage types. As indicated in section 4.3.4, this component is implemented in **"yo?"** on the basis of the Java-based open source persistence framework 'Torque', and hence, regardless of the underlying database type, the Profiler and Modeller components operate on Java-based data structures and utilise the Java-based API of the Data Persistence component to create, read, delete or update them permanently on a physical storage device.

So far, requirements, design, architecture and

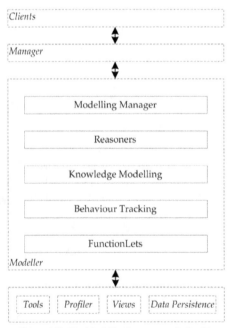

Figure 4.21: The Modeller Component in **"yo?"**.

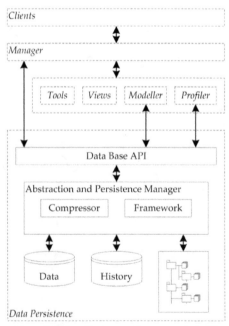

Figure 4.22: Data Persistence Component in **"yo?"**.

functions of the **"yo?"** modelling system have been presented. The next sub-chapter concludes the description of the system with an overview over concrete aspects about its development.

4.3 Development of the System

The software implementation of the "yo?" modelling system is the result of aiming at a Multi-purpose Profiling and Modelling System (MPMS) that should meet the requirements of three distinct projects (see section 4.1).

The previous sub-chapter showed the main issues concerning the design of the overall architecture of the system. Its practical application within the project prototypes is shown in chapter 5 ('Applicability and First Experiences') of this book. This sub-chapter presents relevant issues concerning some milestones in the development process of **"yo?"** as a MPMS. This sub-chapter deals with the following software development aspects:

- Service-Oriented Paradigm (see section 4.3.1)
- Deploying Services (section 4.3.2)
- Micro vs. Macro Services (section 4.3.3)
- Accessing Services (section 4.3.4)
- Security, Privacy & Scrutability (section 4.3.5)
- Profiles vs. Models (section 4.3.6)
- Observing Behaviour (section4.3.7)
- Modelling Knowledge (section 4.3.8).

4.3.1 Service-Oriented Paradigm

As shown in [Gütl et al. 2004], the applicability of many adaptive system solutions is restricted to very specific application domains. Further, the practical benefits of utilising adaptive techniques are demonstrated for specific application domains. These and other features have been evaluated and compared, for example: adaptivity degree for adaptive hypermedia systems [Cini and Valdeni 2002], comparison of techniques for recommender systems [Burke 2002], evaluation of user models [Chin 2001], Bayesian algorithms for student modelling [Millan and Perez 2002], analysis of commercial user servers on the WWW ([Fink and Kobsa 2000], [Kobsa and Fink 2003]), and evaluation of adaptive hypermedia techniques for user modelling [Hothi and Hall 1998].

Indeed, the research community gives excellent answers to complex problems. Nonetheless, the majority of implemented systems is not reusable. Also within the broad context of adaptive e-learning, solutions are necessary that enable the integration of different didactical techniques on a modular and flexible system, and therefore, these solutions should be built on reusable and scalable architectures. Systems that are based on a *service-oriented architecture* (SOA) have emerged in the last years as response to the above depicted problems. The SOA paradigm aims at developing distributed systems that ensure transparency at the levels of platform and communication heterogeneity.

In contrast to interface definitions in a component-based paradigm, services provide a higher transparency regarding transport and communication protocols through open connectors and standardised contracts, i.e. a service in SOA is defined by a software unit that must meet the following requirements:

- service interfaces are platform-independent,
- location and invocation of services can be performed dynamically, and
- services are self-contained.

Thus, implementing under the SOA premises implies the techniques of dealing with the inter-communication of 'software components' that are loosely coupled and reusable, and further, services should be invoked through platform-independent interfaces. There already exist distinct initiatives for the standardisation of the technological SOA aspects, e.g. Openwings [Bieber 2002], DINO [Schmaranz 2002], Web Services [Booth et al. 2004], CORBA [OMG 2006], JINI [SUN 2006], JXTA [JXTA 2007] and more. For further details on the SOA paradigm see e.g. [Benjamins et al. 1998], [Sollazzo 2001], [Trastour and Bartolini 2001], [Bieber 2002], [Bieber and Carpenter 2002], [Papazoglou 2003], [Endrei et al. 2004], [Wang and Fung, 2004], [Fröschl 2005], [Nickull 2005].

The **"yo?"** has been developed on a service-oriented middleware framework, which provides, on the one hand, the required reusability, openness, flexibility and transparency at the upper layers of LAYS (see section 4.2.2). On the other hand, the middleware framework should also provide independence from the underlying computer architectures and operating systems. Examples of such frameworks are RIO[36], Openwings [Bieber 2002], and OSGi[37]. Some differences between these frameworks are given in [Fröschl 2005]. In particular because Openwings met better the requirements of the AdeLE project, this framework has been selected to build the middleware framework of **"yo?"**.

4.3.2 Deploying Services

The **"yo?"** system is developed on the run-time implementation of the service-oriented Openwings specification. It provides capabilities to develop independent and reusable software components as well as to deploy them in form of services. For detailed descriptions on the Openwings specification see e.g. [Bieber 2002] [Bieber and Carpenter 2001] [Bieber and Carpenter 2002] [Bieber and Carpenter 2003] [Bieber and Crumpton 2003] [Bieber and Thrash 2003] [Bieber et al. 2003]. An insight into the technological issues of Openwings and its application in AdeLE are presented in [Fröschl 2005].

In summary, Openwings provides *service interfaces* as Java interfaces. It supports also asynchronous service interfaces. Openwings provides an easy configuration concept through so called *policy interfaces* that enable an access to the locally stored policy files. A *container service* in Openwings monitors the run-time behaviour of all processes; if a service throws an exception, it is restarted automatically. In the Openwings framework, perhaps the most relevant strength is its provision of distinct communication and discovery technologies, e.g. (a) self-developed discovery add-ons may be deployed into the system, or (b) distinct connector factories can be deployed in order to integrate communication protocols. [Fröschl 2005]

The main components of the **"yo?"** system, as described in the previous section (Manager, Views, Tools, Profiler, Modeller and Data Persistence) are deployed within Openwings as services, i.e. the modules in each **"yo?"** component are group into Openwings services that run as independent processes within the framework.

The Openwings Explorer is a comprehensive visualisation of the run-time behaviour of inter-communicating platforms and their internals (see Figure 4.23). The highlighted parts in the figure show the relationship between executable components, service interfaces and running processes for the sample "yo?" service Manager (with the

Figure 4.23: Openwings Explorer showing Part of the run-tine Structure of the AdeLE Platform.

---

[36] http://rio.jini.org
[37] http://www.osgi.org

prefix 'AMS_'; above, the executable component, in the middle the service interface, and below the running process).

4.3.3 Micro vs. Macro Services

The application of *service oriented-architectures* (SOAs) in software solutions implies using and integrating reusable, platform-independent and standardised technologies in systems of collaborative distributed components. From the point of view of software development, two alternatives arise: either single components or groups of them are deployed as single services. In addition to this assembling problem: how much functionality should be encapsulated in each component or service? The construction of modules and the distribution of services have direct implications on architecture, functionality, flexibility and scalability of the system as well as on a possible communication overhead. These aspects embrace the *problem of service granularity* (or *service atomicity*). [Gütl and García-Barrios 2005a]

As stated at the beginning of this chapter, the initial design of the **"yo?"** system was determined within the AdeLE project, and thus, a SOA solution was required. To solve the problem of service granularity, two solution approaches should be implemented and inspected for AdeLE's learner modelling system (**"yo?"** in first versions): (a) *micro-service approach*, i.e. the system consisted of small, specialised services (*atomic services*), and (b) *macro-service approach*, i.e. the number of services was reduced by grouping the atomic components of the first solution into main *service constructs*. Thus, both solutions were functionally equivalent.

The service topology of the first m<u>i</u>cro-service version of **"yo?"** as learner modelling system is shown in Figure 4.24. Services are illustrated as labelled circles, whereby the tip of an arrow denotes a service provider interface, and on its opposite, a service consumer interface. A classification of the services according to their User-Modelling-related duties leads to the following service groups: Managing, Tools, Profiling and Modelling. For the context of this section, the specific function of each service is not relevant, because the task at hand is to show the implications of service granularity by means of commu-

nication performance. For details on the specific context of Figure 4.24, please refer to e.g. [Fröschl 2005] or [García-Barrios 2006e]. Nonetheless, an example should clarify the flow of information through that system.

The communication process between external services and the Modelling System is managed by the Communicator (CO). For further processing, CO delegates the communication message to the Communication Interpreter (CI), which parses the message by means of a corresponding auxiliary service, e.g. the message of the external Adaptive System (AS) service is handled by the Adaptive System Communication Interpreter (AI). Next, CO sends the result to a Context Manager (CM), which resolves further delegation, e.g. to the Profile Modeller (PM), and this in turn, delegates messages for further processing to more specialised services. Considering that 'user behaviour information' (e.g. information about consumed learning materials or about system actions invoked by the user) is handled by the Behaviour Handler (BH), the chain of processing continues from here to the Instruction

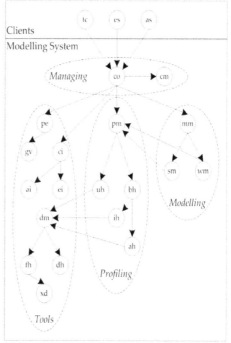

Figure 4.24: First M<u>i</u>cro-Version of **"yo?"** in AdeLE.

Handler (IH) or to the Action Handler (AH). Finally, the information may be persisted in a database by means of the Data Manager (DM) and the Database Handler (DH).

The service topology of the second solution approach, which was functionally equivalent to the previously described, is shown in Figure 4.25: first *macro-service* version of **"yo?"**. The degree of specialisation was reduced to almost one service per group, i.e. service structures in the micro-approach were integrated into their root nodes to build more general services in the macro-approach, e.g. *micro{ci(co) + cm(co)} = macro{ma}*. Accordingly, *micro{gv(pe)} = macro{ev}*, *micro{fh(dm) + dh(dm)} = macro{dm}*, *micro{uh(pm) + bh(pm)} = macro{pr}*, and *micro{sm(mm) + wm(mm)} = macro{mo}*. On the basis of this mapping from micro- to macro-approach, as can be seen when comparing Figure 4.25 with Figure 4.17, the component architecture of the **"yo?"** system was derived (CAYO; section 4.2.3):
macro{ma=manager} → *cayo{Manager}*,
macro{ev=editors&visualisation} →
cayo{Tools&Views},
macro{pr=profiler} → *cayo{Profiler}*,
macro{mo=modeller} → *cayo{Modeller}*,
macro{dm=data manager} → *cayo{Data Persistence}*.

Within the Java-based Openwings framework, communication among single services is carried out through RMI technology (see e.g. [Bieber and Carpenter 2001]). This characteristic may not be advantageous for systems of services like the mi-cro-version of AdeLE's user modelling system, because information flow within the system must be as fast as possible and thus, independent of high-level technologies.

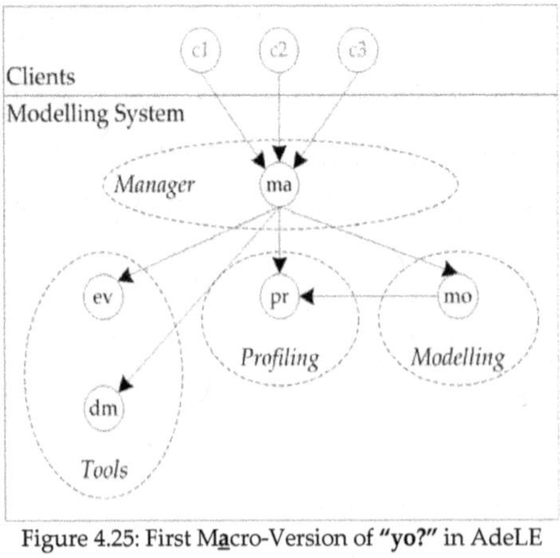

Further, an activated security option of Openwings around each service decreases performance significantly if the processing of information involves the intervention of many atomic services. An example of concrete evaluation results is shown in Table 4.1. For each approach there is little difference between the two types of requests *set(p)* and *get(m)* although distinct chains of services are affected. Nonetheless, the macro-service solution is (in average) five times as faster as the micro-approach (approx. 5.075; for more details refer to [Fröschl 2005]).

On the other side, a rapid role-based *inspection of the system* was conducted as well. The *roles* for this inspection were created and defined in the same way as for *user scenarios* within the scope of user-centred design in Human Computer Interaction (HCI) (see e.g. [Bødker 2000]). The following roles were defined for this inspection of the AdeLE's learner modelling system: *developer* (DEV), *application user* (AU), and *administrator* (ADM). The evaluators utilised the following inspection heuristics: *performance, reliability, clustering, handling* as well as *privacy & security*. The inspection method based on a questionnaire distributed to software engineering experts, where each of them assumed one of the distinct roles at distinct points in time of the development process.

For each criterion, the parameters to compare both solution approaches were defined by means of a scale with 5 grades: -2 (*strong drawback*), -1 (*weak drawback*), 0 (*no significant difference*), +1 (*weak advantage*), +2 (*strong advantage*). A sample

Figure 4.25: First Macro-Version of **"yo?"** in AdeLE

	Micro		Macro	
	set(p)	get(m)	set(p)	get(m)
50 requests	96	104	18	21
500 requests	1016	1080	199	215

Table 4.1: Comparison of SOA Granularity Approaches: *Micro* vs. *Macro* – Part 1. Example for Evaluation Results on High-Traffic Requests: 'set(p) = writing attributes in profiles' vs. 'get(m) = reading outcome of models'. *Legend: numbers in Milliseconds.*

summary for results of the inspection is given in Table 4.2. Details on the inspection are depicted in [Fröschl 2005].

According to the overall results, the macro-approach is preferable for the following reasons. Application users are mostly interested in aspects like performance, reliability and privacy&security. In the case of the inspected system, these criteria are guaranteed by the Openwings framework. Further, the macro-approach has advantages for administrators compared to the micro-approach, mainly because of they must handle less modules. For developers, there is no significant difference between the two solution approaches.

This last result is presumably given because the implications of the modularity of distributed systems (e.g. extension, migration, robustness, scalability, reuse.. of modules) are software quality aspects that depend also on designers and testers, and thus, is not perceived as a duty of developers. Usually, developers concentrate on the efficient application of software techniques for the specific task at hand and think about 'elastic users' while coding the system (see sections 3.2.1 and 3.2.2). In contrast, end-users and administrators concentrate on ease-of-use, i.e. usability of the system. In sum, the more transparency is guaranteed towards end-users of distributed systems, the more comfortable users feel while interacting with the system (see e.g. [Tanenbaum and vanSteen 2007]).

In general terms, as a result of evaluating both approaches, the most critical technical problem

was found in the application of the micro-service approach, because in fact, the high number of specialised services resulted in a *communication overload* in the traffic of messages.

In addition, regarding the micro-approach, *security problems* had been expected, because each service had to be defined with open interfaces. Though, Openwings enables to control access to services through an integrated management of 'users and roles within contexts', which can be applied to service interfaces (next section discusses this aspect more in detail).

An additional technical obstacle in the micro-approach arises when the distribution of duties to process one type of user information is too high across single services. Consider for example dynamic user states, which may vary continuously over very short periods of time, like real-time indicators captured through the eye-tracking system of AdeLE (e.g. the binocular tracking frequency of such systems is approximately 50 Hz, see sub-chapter 5.1). The work on each chunk of data from such sensors may be a long lasting task until its transformation in useful and usable information for other system clients (acquisition, revision, interpretation, compression, storage, comparison and reasoning, etc.). The management of such user states on run-time implies the application of effective (and relatively complex) algorithms for a consistent and synchronous updating of models.

Adaptive engines interacting with the modelling system are continuously monitoring these states for distinct individual user models (e.g. in order to adapt the presentation of content and navigation). Thus, the modelling system must be fast enough to simulate a real-time snapshot of user traits and states. And consequently, the micro-service solution approach creates communication bottlenecks in the system.

Finally, the management of collaborative services based on the micro-service approach is not trivial anymore, because of the increased potential complexity of the system. For example, starting and scheduling each service separately during the system's initial setup is an extra-workload. To solve this problem, Openwings provides the possibility to configure the initialisation of coordinated services through the inte-

Criteria (for micro-approach)	DEV	AU	ADM
Performance	0	-2	0
Reliability	-1	-2	-1
Clustering	0	0	-1
Handling	-1	0	-2
Privacy & Security	0	-2	-1

Table 4.2: SOA Granularity Inspection. Criteria applied on Micro-services when compared with Macro-services. For details see [Fröschl 2005].
Legend: -2 (strong drawback), -1 (weak drawback), 0 (no remarkable difference), +1 (weak advantage), +2 (strong advantage).

gration of an *Installation Service* [Fröschl 2005]. Thus, a single service can be implemented and configured to assume the life-cycle management of all services on the platform. From the programming point of view, it can be stated that the Openwings specification is very flexible and powerful, but it might be also very complex. Developers of complex service-oriented systems require enough time to get familiar with this technology.

4.3.4   Accessing Services

The previous section has shown that the design of architecture for a *service-oriented system* (i.e. based on SOA principles) implies to take care of the granularity of its interacting parts. Thus, from the point of view on intra-system communication it is relevant to keep the number of services as small as possible. The findings from the previous section were restricted to the internal architecture of the **"yo?"** modelling system, i.e. the task at hand was to improve performance in(!) the system by means of architectural principles and their implication on overall efficiency. In this section, also within the scope of performance improvement, a shift to technical aspects is undertaken. First, results on the use of *Caching* in **"yo?"** are shown, i.e. again intra-communication is the focus. Second, other results are presented with respect to improving the inter-communication of **"yo?"** with external systems.

User Modelling Components (UMCs) are usually placed at the back-end of adaptive systems. On the one side, they are constantly dealing with several sensoring components to acquire as much information about users as possible. Technically speaking, these are mostly *Write* operations on the UMC. On the other side, UMCs are constantly providing user information to their consumers. These are mostly *Read* operations on the data of the UMC. If the UMC is application-independent, the number of these operations may increase in an unpredictable degree due to the variety of types of interacting clients. This critical issue has direct implications on the internal performance of the system, because the processing of many distinct requests may concern the same internal service and thus, may lead to block external clients in their duties. The usual

solution to overcome this problem is the use of intermediate stations of storage for frequent requests, *Caching*. Several caching problems are known and must be considered, such as the optimal placement for the caching module, the number of caches along the system workflow, as well as the consistency and topicality of the data in the caches (see e.g. [Friedman 2002]).

In the case of **"yo?"** as a UMC within an adaptive system, the notion of CRUD operations (Create, Read, Update, Delete; see section 3.2.2) helps to identify which types of requests are being resolved by its Manager component. Concretely, after inspecting where the use of a caching mechanism brings more performance benefits, the Cache Module was place in the Dispatcher of the system, on the side of the Resolver (see Figure 4.18). First considerations aimed to put a cache on the lowest data processing level, i.e. in the Data Persistence layer, but this solution would not bring the expected results, because the messages would be already passed through other components and modules. Further, the intention was to improve the performance for the overall system in terms of inter-communication and not in terms of data storage, which can be performed e.g. through an optimised configuration of the data base management system.

For the scope of this section, the CRUD operations on a modelling system are generalised as *Write* and *Read* operations. These operations are encoded in those messages that external systems send, because each client defines a request according to the expected result. The Resolver module in the Manager of **"yo?"** can distinguish between a SET and a GET request (i.e. *Write* or *Read* operation). This is ensured by allowing each module in the system to register in the Resolver the signature of their interface methods. Therefore, the Resolver is not only in charge of the discovery and look-up of services for further delegation of duties, it also is able to read the syntax of client messages and steer accordingly the behaviour the cache. Using such a cache in the system takes advantage of transparency at a front-end level of the system and thus, increases the scalability of internal services. According to [NISO 2006], caching *can reduce bandwidth consumption and also decrease the latency of requests making the service appear faster and more responsive. Caching of content can also reduce load on the web service by reducing unnecessary requests. [...] Enabling [...] responses to be cached can drastically reduce the load on back-end systems such as databases.*

Naturally, the **"yo?"** system does not cache SET requests since their effect is an alteration of the states on user information. Further, for each GET request on user information, there is normally a corresponding SET request, e.g. in the AdeLE system there exist user-specific requests like SET_GAZEINFO_LAST vs. GET_GAZEINFO LAST, or SET_MODEL_STATE vs. GET_MODEL_STATE. Taking for granted that these calls are registered in the Resolver module, each SET request will lead to an update of the data in the cache (for the corresponding GET call).

Furthermore, the caching mechanism in the **"yo?"** system supports a *collaborative* improvement of performance among its internal services. UMCs support usually reasoning techniques in such an extent that the acquisition of some user trait may lead to a chain of inference steps with different outcomes, i.e. some GET requests aim at data that is generated in the system and was not supplied directly by an information provider. Such inference modules collaborate with the Cache by supplying the relationship of one SET to (inferentially) related GET requests. This mechanism is part of the registration of services in the Resolver.

The Cache of the **"yo?"** system supports two techniques: FIFO and LFU. Depending on the given memory capabilities and the requirements on the size of the cache, solely a certain number of entries are admitted in a cache to ensure its performance. Thus, if a cache if full and a new entry must be stored, the First-In-First-Out (FIFO) technique removes the first element that was stored first, i.e. the oldest element will be always removed. The Least-Frequently-Used (LFU) technique adds to each element an *access counter*, which is incremented with each *Read* operation on it. Thus, in LFU, not the oldest element as in FIFO, but the element with the smallest value in this counter would be removed. Other optimisation techniques for caching might

be possible, such as Pass by Reference for immutable types (String and primitive types), Copy by Reflection for Bean-type and array-type objects, Java serialisation for serialisable objects, SAX event sequences, and so forth (see e.g. [Takase and Tatsubori 2004]). Other caching techniques are applied at client-side to ensure offline operation e.g. for mobility services (see [Terry and Ramasubramanian 2003]. In the field of Web Services further techniques have found successful application, such as Similarity-Based Multicast [Khoi et al. 2006], table-driven XML caches [Ng 2006], etc. For more details see e.g. [Fernandez et al. 2005], [Suzumura et al. 2005].

Different simulation tests have been conducted to check the benefit of the cache in **"yo?"**. As stated before, the general request load onto a user modelling system is not predictable. For the evaluation, one *Simulation* is a cycle containing a sequence of 14 distinct AdeLE-specific requests for one specific (fictive) user:

INIT_USER_INFORMATION,
GET_USER_PARAMETER,
GET_USER_PARAMETER,
SET_GAZEINFO_LAST,
GET_GAZEINFO_LAST (2x),
SET_MODEL_WAVI, GET_MODEL_WAVI
(2x), GET_MODEL_WAVI_IS_W (2x),
GET_MODEL_WAVI_IS_A (2x),
GET_MODEL_WAVI_VI (2x).

Further, the simulation client waited for a response from the system and sent immediately the next request (i.e. no further processing steps were undertaken). The results for 100 Simulations are shown in Figure 4.26 and Figure 4.27 on the next page.

Within the constraints of this evaluation, the LFU method seems to be the right choice. As depicted in Figure 4.28 (also on the next page), the FIFO method needed 4% more time than LFU for the 100 Simulations. Without a cache, the system is slower by almost 18%, compared to LFU. The simulated client was implemented as additional service in Openwings, but was running on a remote machine (note: both machines built a two-end network, so neither other machines could reach them nor they were interacting with other systems). The aim was to investigate how the UMC would react for many adaptive engines concurring within a distributed Openwings framework. But under real conditions, **"yo?"** serves mostly as information provider for *non-Openwings* systems.

The communication performance of service-oriented systems is not restricted to the critical aspects discussed so far. Services interchange information by means of a specific underlying technology. Consider the Web Service standard for message interchange: SOAP (Simple Object Access Protocol - http://www.w3.org/TR/soap); the services as endpoints of a communication and the messages they may interchange are described in (SOAP-based) WSDL interface defini-

tions (Web Services Description Language; http://www.w3.org/TR/wsdl); the discovery of services must be done through a UDDI server (Universal Description, Discovery and Integration; http://www.uddi.org). Using a discovery server makes sense, because service-oriented solutions must count with the possibility that services are mobile, and therefore, a discovery server can be used to keep track of this locomotion.

Though, most Web Service solutions do not use a UDDI discovery server, because their development requirements take mostly for granted that they know each other. This is one of the reasons why Web Services are often compared with RESTful applications [Muehlen et al. 2005]. REST (Representational State Transfer) is an HTTP-based architectural and design style for networked solutions that follow the CRUD principle to operate on remote resources (see [Fielding 2000]).

In the SOA-based Openwings middleware utilised for **"yo?"**, services s mobile across platforms, thus a discovery service is used: Sun's sample implementation of the Jini lookup service, Reggie (http://www.jini.org). The synchronous communication technology utilised in Openwings is RMI (Remote Method Invocation; http://java.sun.com/javase/technologies/core/basi c/rmi).

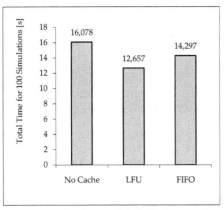

Figure 4.26: Comparison of Cache Performance in "yo?". Total Time for NoCache, LFU and FIFO.

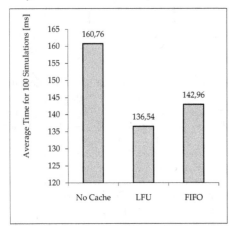

Figure 4.27: Comparison of Cache Performance in the "yo?" Modelling System. Average Time for NoCache, LFU and FIFO.

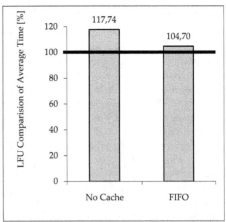

Figure 4.28: Comparison of Cache Performance in "yo?". Benefit of LFU technique compared with No-Cache and FIFO (in %).

Based on the statements introduced so far and considering that (within the AdeLE project) an eye-tracking system would be a sensor producing high traffic of information on the input side of the user modelling system, the need arose to investigate which could be the best technological way for communication. For that purpose, an external tool (relying on the outside of the Openwings framework) was implemented to simulate an arbitrary number of clients interacting for different purposes.

The configuration interface of the simulation tool *Load Simulation Client* (LSC) is shown in Figure 4.29. The overall architecture of LSC and its possible ways to interact with "yo?" are shown in Figure 4.30; on the bottom-right side the modelling system is represented by its *RMI connector* in Openwings. A *Client* of the system may communicate (from right to left in the figure) directly vía *RMI* after locating the Manager Service of "yo?" with the aid of the (a) *Jini Registry* (the lookup service per default integrated in the *Openwings* core) or (b) the *RMI Registry*. Further, Web Service communication is also supported by means of sending *SOAP* messages to the Axis engine of the *Tomcat Web Server* (which

is a stand-alone server, but in this case also part of the Openwings framework). An *HTTP* connection was also enabled through an *HttpServlet* and a *RESTfulServlet* to take advantage of the *Servlet Engine* of the Web server, i.e. HTTP and REST communication was also supported.

The LSC can be configured to run different simulations. Each simulation is defined by one communication alternative, running as a single Thread: *RMI* (RMI + RMI Registry), *JINI* (RMI and Jini Registry), *WS* (Web Service = SOAP + HTTP), *REST* (RESTful servlet) and *SERVLET* (HTTP servlet). Thus, one simulation represents the work for one user (i.e. for one interacting learner). For each simulation, sub-Threads are created, one for each type of sensor: e.g. adaptive engine, browser, CO2, eye-tracker. For each sensor the number of sent requests can be also configured. Finally, the number of simulations can be also given to define the number of iterations through each sequence of concurrent sensors.

The results of 300 simulations for one Client (and for all five communication technologies), are shown in Figure 4.31 and Figure 4.32. The diagrams show that the most effective way to

Property	Value
edu.iicm.perf.AdaptiveEngineInterval	2
edu.iicm.perf.AdaptiveEngineMessages	10
edu.iicm.perf.BatchOrder	JINI,RMI,HTTP,REST,WE
edu.iicm.perf.BrowserInterval	1
edu.iicm.perf.BrowserMessages	10
edu.iicm.perf.CO2Interval	2
edu.iicm.perf.CO2Messages	10
edu.iicm.perf.EyeTrackerInterval	2
edu.iicm.perf.EyeTrackerMessages	10
edu.iicm.perf.LogLevel	2
edu.iicm.perf.NumberOfRuns	500
edu.iicm.perf.RMILookupName	test
edu.iicm.perf.Impl.HTTP	edu.iicm.perf.connectic
edu.iicm.perf.Impl.JINI	edu.iicm.perf.connectic
edu.iicm.perf.Impl.REST	edu.iicm.perf.connectic
edu.iicm.perf.Impl.RMI	edu.iicm.perf.connectic
edu.iicm.perf.Impl.WEBSERVICE	edu.iicm.perf.connectic
edu.iicm.perf.url.HTTP	http://localhost:8880/se
edu.iicm.perf.url.JINI	localhost
edu.iicm.perf.url.REST	http://localhost:8880/se
edu.iicm.perf.url.RMI	localhost
edu.iicm.perf.url.WEBSERVICE	http://localhost:8880/as

Figure 4.29: Partial View of the Configuration Interface of the external Load Simulation Client for the Service-based Learner Modelling System of

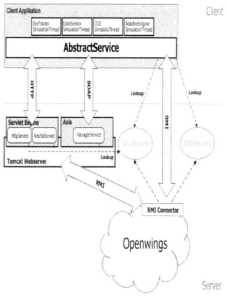

Figure 4.30: Analysed Possibilities of the Load Simulation Client to communicate with the Service-based Learner Modelling System of AdeLE.

communicate with AdeLE's modelling system is RMI and JINI, i.e. using RMI as communication technology and either the RMI Registry or Jini Registry as lookup service. Utilising the Web Server results in approximately 30% of loss in total communication performance.

A remarkable finding from this evaluation (as shown in Figure 4.31 and Figure 4.32) is that a communication over Web Services takes on average 10 times longer than through RMI/Jini. Further, the WS communication lasts almost 7 times longer than utilising REST (RESTful request to the Web Server). As previously mentioned, this last aspect represents an ongoing and well-known debate over the value of SOAP vs. REST. In real, complex, service-oriented applications, a decision to use SOAP or REST should not rely only on matters of performance; however, this evaluation shows that indeed, Web Services are slow. One of the reasons is this 'big' standardised envelop over each message (SOAP), which must be parsed each time on a specialised engine at server side (see Figure 4.30; *Manager Service* in the *Axis* engine of the *Tomcat Web server*).

The result of another evaluation with LSC was conducted for 300 simulations and 5 clients. A second very relevant finding could be gained: Web Services are not always the slowest. The graphic illustration of this additional result is shown in Figure 4.33 (see next page). Although the comparisons *WS vs. REST* and *WS vs. HTTP* kept in analogy to the previous example, the average time per simulation for RMI and JINI increased continuously and drastically. As indicated in Figure 4.30, in order to reach the Manager of the learner modelling system, all alternatives MUST communicate vía RMI in the last step. Thus, the only reason for the identified communication bottleneck could be the use of a lookup service, either the Jini Registry or the RMI Registry. In further debugging trials it could be found out that the lookup intervals represented just a minimal and insignificant proportion of the whole delay. Literature survey showed that similar behaviour could be also identified in other performance evaluations of Jini technology combined with RMI, e.g. [Lenders et al. 2001] and [Huang et al. 2002]. [Wang and Zhang 2007] show in their report 'Performance Evaluation & Optimization about Lookup Service in Jini Archi-

tecture' that the dispatching time within each RMI operation increases in linear form (but drastically), if the number of requests is 'too' high.

After a more detailed analysis, Wang and Zhang identified that the RMI server instantiates a thread to process each arriving client request. Therefore, too many concurrent threads decrease substantially the overall performance of the system (e.g. in terms of concurrent CPU time slots,

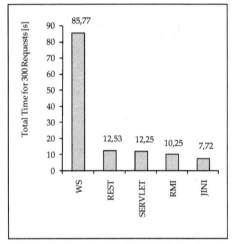

Figure 4.31: Comparison of Communication Technologies from external Components; Total Time needed for 300 Requests.

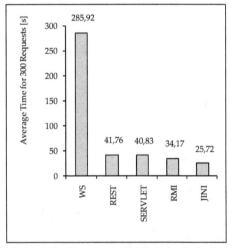

Figure 4.32: Comparison of Communication Technologies from external Components; Average Time needed for 300 Requests.

limited CPU speed, limited free space in memory).

4.3.5 Security, Privacy and Scrutability

The problems of managing information in a modelling system may have distinct causes from the development point of view. From the point of view of software design, in the case of **"yo?"** as a learner modelling system in AdeLE, general and detailed use cases were defined in order to delimit the flow and control of information through the system, e.g. the general Use Case 'Profiler' shown in Figure 4.34 (next page). Such a use case is initiated by the Adaptive System (AS) of AdeLE. In this general example, the main actions of the AS are *Check if a User is Available, Send User Data, Query User Data, Create a New User* and *Construct a new Profiler*. The latter one makes it possible to add profiles of individual learners by any trusted external system. But to which extent is a system a trusted system? Moreover, is it sufficient to develop secure systems to ensure e.g. privacy?

The Openwings framework provides security concepts, which enable to build secure services. This is a highly relevant aspect in the field of multi-purpose (user) modelling systems, because the User Modelling Component in an application-independent adaptive environment should avoid access to private information for unauthorised information consumers.

Thus, a trusted and secure environment is a *must have feature* of such systems. On the other side, a *trusted local environment* is an additional pre-requisite of the system, because e.g. the security policies of the overall Openwings environment as well as those of the services deployed in the system are persisted on local plain text files. This might be a vulnerable point for the security policies and attackers might gain a way to alter these policies. Within the Openwings framework, services must be deployed within *context services*, which are secured containers of the system avoiding an external untrusted access. Each context service comprises a set of *secure connectors* and *security roles* (including passwords). This internal Openwings mechanism can be seen as code elements that may be only access with valid signatures. [Fröschl 2005]

Further, from the point of view of software development, sequence diagrams were also helpful in the implementation of **"yo?"** to get insights into the main functionalities of these Openwings issues, as e.g. illustrated in Figure 4.35. This figure shows how the **"yo?"** Manager (despite of its inherent functionalities, as indicated in the pre-

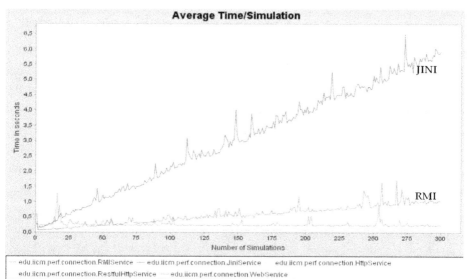

Figure 4.33: Comparison of Communication Technologies from external Components; Graphical Illustration of Behaviour for 300 Requests and 5 Users.

vious section) delegates security and validation issues regarding the general access of external services to specialised Openwings services (Security Manager and Context Manager).

These Openwings-internal sequence steps lead either to the authorisation or refusing of client requests. In concrete, the Security Manager of Openwings is part of the Access Filtering module of **"yo?"**, and the Context Manager of Openwings is invoked within the Dispatcher module to support the lookup and resolving of **"yo?"** components.

Regarding the aspect of *Privacy*, it must be

emphasised again that this issue is not a main aspect of research within this book. Nonetheless, as stated in [Westin 1970],

> *"Privacy is the claim of individuals, groups, or institutions to determine for themselves when, how, and to what extent information about them is communicated to others."*

This definition is highly important, because, within the context of User Modelling Systems (UMSs), it should be absolutely required that that modelled users have the possibility to define which *level of privacy* they want. Further, the system must provide the mechanism to enable users to select which information is allowed to be shared and with whom. Jörg Schreck claims in [Schreck 2001] that the *confidentiality* of individual information in user models may be implemented by means of anonymity (without revealing any private data) or pseudonymity (usually, choosing a self-defined combination of pseudonym and password). Between anonymous and pseudonymous identification, according to [Schreck 2001], the best choice is pseudonymity, because it is the best intermediate solution between privacy demands of users and requirements of UMSs. Pseudonyms make it possible to link user model and corresponding user without revealing the identity to information consumers

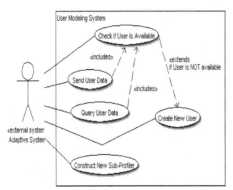

Figure 4.34: Sample Use Case in "yo?": Profiling.

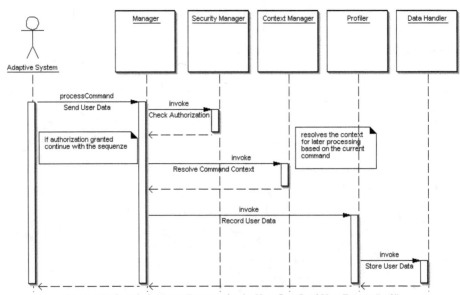

Figure 4.35: Sample Sequence Diagram for the User Case Send User Data in "yo?".

of UMSs.

Summa sum arum, if the modelling system offers a mechanism to recognise and choose among distinct levels of privacy, the acceptance of personalisation systems may increase. Thus, this issue brings inevitably to the context of *Scrutability* of user models. In essence, the aim of developing scrutable modelling systems is a very hard task, because

- the information in the user models is mostly not proactively acquired, it is delivered by information providers; the modelling system is like a data (base) management system in the background of other applications and is mostly not used to communicate with the user. Thus, either it has no possibility to access the front-end of the adaptive environment, or it has not the possibility to translate its data into human-readable information.

- a user modelling system is a reasoning machine, which is able to infer information from other information. Usually, the reasoning mechanisms lead to the creation of complex models (i.e. machine-readable data structures) that provide machine-readable information for its consumers (also machines).

- the involvement of user modelling system in adaptive environments is mostly considered a back-end task. Indeed, users are confronted with personalised information and may want to know the reasons for the adaptation. Further, knowing that the system is just delivering part of its internals may lead to a curiosity chain of side-effects. Hence, users want to the system to reveal (a) which are the reasons for adaptation, and (b) which things are hidden for that particular user, but visible for others. The crucial point in this context is that the user modelling system cannot provide this information, because it is in charge of knowing the user, but not of deciding what to adapt and why to adapt. These tasks are the duties of the adaptors (e.g. adaptive engines), which are exclusively implemented to meet the adaptational requirements of the application at hand.

The **"yo?"** system includes functions to support scrutability as far as described in the previ-

ous section. How the AdeLE system and the CO2 application make use of these functions, which are embedded in the components Views and Tools, is shown in the next chapter when presenting the main applicability aspects of **"yo?"** in that projects.

4.3.6 Profiles vs. Models

This section presents critical issues concerning the separation of duties in service-oriented User Modelling. The CAYO architecture of the **"yo?"** modelling system (see section 4.2.3), comprises two dimensions of deployment: *Information Flow* and *Human-Machine Transparency* (see, respectively, top and bottom part of Figure 4.36.

The *Information Flow* dimension enables to separate the duties of the system regarding (a) *Access* to information, (b) *Processing* of information, and (c) permanent *Storage* of data. Also in this order, the following benefits can be identified: (a) on the one hand, considering a general point of view on adaptive systems, the upper level of this dimension contributes to determine the sensitivity and vulnerability of the system, and on the other hand, considering a user modelling component as independent system, it provides a mechanism to configure the security and privacy level of user information as well as represents a single point of access to a distributed system of tasks; (b) the middle level provides an encapsulation of user profiling and modelling functionalities; and (c) the bottom level enables an internal separation and transparent access to distinct types of storages techniques.

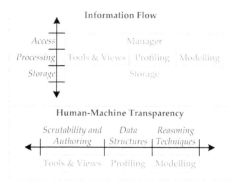

Figure 4.36: Development Dimensions of "yo?".

The *Human-Machine Transparency* (HMT) dimension of CAYO permits a separation of duties for User Modelling Techniques, as follows. Each technique applied in CAYO must be separated and distributed along the HMT dimension. This aspect is shown in Figure 4.37 and Figure 4.38 respectively, for the modelling techniques of Stereotyping and Behaviour Tracking.

Stereotyping is a technique for User Modelling that is deployed in the **"yo?"** along the following line (see Figure 4.37 from left to right side). The HMT level of Authoring and Scrutability comprises all software parts for stereotyping that must be delivered to specialised clients. This includes e.g. authoring tools for user model developers, and human-readable descriptions on each created stereotype for modelled users. The HMT level of Data Structures comprises the functionalities for creating and managing stereotype profiles. Stereotype Profiles can be deployed in **"yo?"** using Profiler modules (see also Figure 4.20), e.g. registering attributes, lists and trees from the 'Structure Factory', configuring them into hierarchies of profiles in 'Profiles', and implementing activation for individual user in 'Stereotypes and Communities'. Finally, the HMT level of Reasoning Techniques comprises

the definition and implementation of e.g. rules for triggering for Stereotype Reasoning from the 'Reasoners' module in the Modeller component (see Figure 4.21).

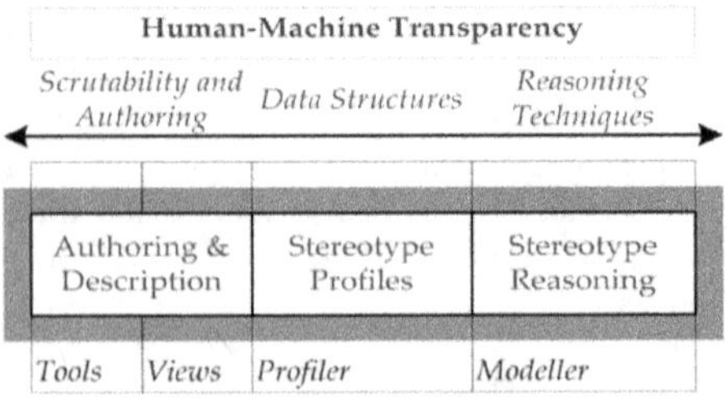

Behaviour Tracking is a technique in the **"yo?"** system to work on chronological data about user actions. In the AdeLE system (see Figure 4.4) this technique is utilised to support the adaptive system with functions of logging and analysing gaze-tracking and learning progress data. In essence, the Behaviour Tracking (BT) technique is a *Logging* technique, where data streams can be collected into profiles in a chronological order of delivery. Thus, a BT profile provides information consumers basic functionalities for storing and retrieving data in a time-dependent way, i.e. request on BT data may have the form (a) 'provide the last five entries in the profile of user A', (b) 'provide those entries regarding user A in the time interval [x,y]', or (c) 'provide those entries in the profiles of user A and B for which the text *xyz* is contained in delivered profile entries since time point X', and the like. The capability of responding to such requests is ensured in **"yo?"** through a communication protocol (*BT-Protocol*). The general principle of this technique is defined along the HMT dimension:

- The HMT level of Authoring and Scrutability provides visualisation possibilities for user model developers, e.g. authoring of BT profiles, or comparing entries of distinct users in BT profiles and visualising them in diagrams.

- The level Data Structures comprises the basic functionalities to operate on sequences of data (BT profiles) to respond to simple requests, such as the above given examples (a) and (b). For this purpose, the module Loggers in the Profiler component is used (see Figure 4.20).

- The level Reasoning Techniques comprises the utilisation of the Behaviour Tracking module in the Modeller component of **"yo?"** (see Figure 4.21). This module comprises more sophisticated methods and is able to interpret and analyse the BT-Protocol. Thus, inferences on sets of BT profile entries can be done, e.g. to respond to the request in example (c), given the case of stored according to the BT-Protocol.

Human-Machine Transparency

Scrutability and Authoring	Data Structures	Reasoning Techniques
Authoring & Description	Stereotype Profiles	Stereotype Reasoning

Tools	Views	Profiler	Modeller

Figure 4.37: Stereotyping with HMT.

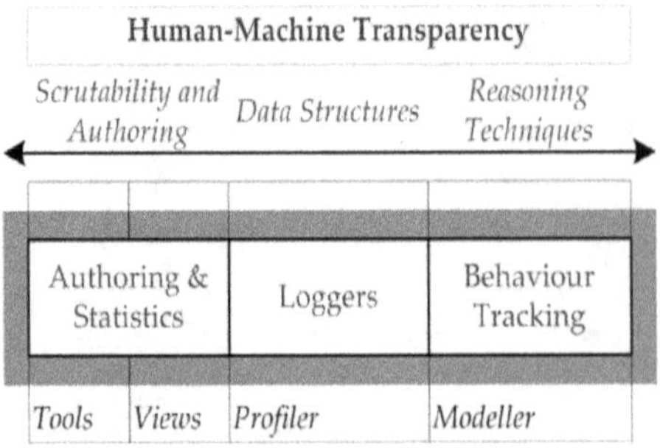

Human-Machine Transparency

Scrutability and Authoring	Data Structures	Reasoning Techniques
Authoring & Statistics	Loggers	Behaviour Tracking

Tools	Views	Profiler	Modeller

Figure 4.38: Behaviour Tracking with HMT.

These two examples indicate the distinct levels of sophistication between low-level management of user profile information (through *Profiling*) and higher-level reasoning operations on user profiles (through *Modelling*). This separation of user modelling duties is very helpful for the application of distinct personalisation methods on same profile structures for distinct purposes.

4.3.7 Observing Behaviour

Within the scope of this book, *Behaviour Tracking* denotes a technique of monitoring traits of users of interactive systems in real time by means of 'analysing log entries before they become history'. For that purpose, data must be collected through the sensors of the system, such user interfaces or observation devices. To express an evidence on user behaviour, the system analyses the collected data as they flow into the system and provides useful and topical inferences, e.g. the personalisation engine of an adaptive e-learning system could infer that a user 'behaves well' (in terms of effective learning progress) based on the analysis of 'just-in-time tracked actions', such as consumed learning materials and assessment results. Further, such tracked actions are persisted and might be utilised by the user modelling component in order to e.g. infer learning styles or interest from previously consumed materials.

The **"yo?"** modelling system provides to its software clients (e.g. adaptive engines) a way to *observe user behaviour* through the cooperation of three modules in the system (see Figure 4.39): *Logger, State Modeller* and *Behaviour Tracker*. In correspondence to the overall architecture of the system, the Logger is a module in the Profiler component, State and Behaviour Modellers build the Behaviour Tracker in the Modeller component (see Figure 4.20 and Figure 4.21). The *Sensors* of the system send the *Logger* streams of tracked user actions, e.g. clicks on browser buttons, nodes in navigation paths through menu items, visited hyperlinks, visual focuses of attention during reading (from gaze-tracking devices, as within the AdeLE system), head movements or gestures (from cameras), locomotion data (from GPS sensors), etc. The streams of tracked

data are accessible to the *Behaviour Modeller* (BM) and to the *State Modeller* (SM).

The BM comprises a set of rules that can be applied on stored Logger data in order to retrieve specific chunks of information, i.e. the BM provides stored information on-demand. The application of rules on Logger data is possible due to the utilisation of a communication protocol (*BT-Protocol*), which must be respected by Sensors. The BT-Protocol defines the syntax of correct log entries in the Loggers. As shown in Figure 4.40 (on the next page), each log entry must contain:

- A unique identifier (log_ID): assigned by the modelling system to ensure the uniqueness of logged entries;

- a time information (timestamp): in order to build the chronology of log entries;

- information about the logging sensor or system (log_level, log_system): where log_system is the identificator (name) of the system, and log_level provides a system-dependent mechanism to e.g. filter log entries during retrieval or to show/hide them in visualisations;

- information about the agent being modelled

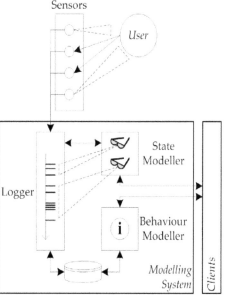

Figure 4.39: Observing User Behaviour in "yo?".

(log_profile, user): to build the relationship between entries and individual users and user profiles (this profiles are called Behaviour Tracking Profiles, BT-Profiles);

- a sequence of message triples identifiable through the prefix 'msg_' (e.g. {msg_1_datatype, msg_1_key, msg_1_value}, {msg_2_datatype, msg_2_key, msg_2_value}, etc.): each message has to be defined in terms of these triples to ensure that BM and SM can interpret and combine them by means of their data types ("yo?" distinguishes among String, Number and Boolean); and

- an attachment (additional_data): this last part of the protocol provides sensors to add binary data to each entry (e.g. video or audio files).

Hence, the BM operates on the basis of this protocol and provides methods to retrieve information chunks by means of operators that depend e.g. on the *datatype, keys* and *values* of messages within log entries: *And, Or, consistsOf, GreaterThan, LessThan*.

Clients of the system must be aware of the syntax of the protocol as well as of the possible combinations of operators. Therefore, it is relevant to state that this mechanism does not interpret logged information in terms of 'reasoning about data to provide meaningful responses to Clients'. Rather, it represents a mechanism to retrieve chronologically stored information with the aid of an agreed protocol and agreed operations. Thus, the goals that can be reached through this mechanism depend on the cooperative work of Sensors and Clients.

Further, in contrast to the BM, which retrieves information chunks from the stored data, the SM works on the data streams of the Logger by means of operational rules on the messages of log entries (see Figure 4.39). Clients of the system must firstly configure which rules must be applied on the data they need, and subsequently, ask the Behaviour Tracking module to create one instance of the SM to work on this data. Thus, one instance of the SM *observes* continuously the data streams of the Logger, fetches each entry of interest, checks each entry according to the rules that have been previously configured, and collects the results of checked data until the next request of a Client.

The practical application as well as the benefits of this mechanism is explained in sub-chapter 5.1 within the scope of the AdeLE research project. First of all, segments in the learning material of the system have to be previously marked by teachers in terms of '*text passage X must be read*' or '*text passage Y must be learned*'. This information taken for granted, on the one side, AdeLE's eye-tracking system monitors and interprets continuously the gaze movements of learners, and sends this tracked data (respecting the BT-Protocol) to the Logger of the modelling system in terms of e.g.

'*text passage X, which is to be read, was read by user A*', or '*text passage X, which is to be read, was read by user B*', or '*text passage Y, which is to be learned, was read by user A*', or '*text passage Y, which is to be learned, was learned by user B*', and so on.

On the other side, the adaptive system of AdeLE is aware of an existing State Modeller instance that continuously observes this gaze-tracking data stream and checks which learners '*had read what had to be read*' and '*had learned what had to be learned*'. According to this example, the next request from the adaptive system to the State Modeller regarding the learner A will result in a '*OK*' response, whereas the response for learner B will lead to a '*NOT OK*' status including logged entry that has broken this 'rule of learning'. All aspects regarding this sample use

Uniqueness	*log_ID*
Time	*timestamp*
System	*log_level*
	log_system
Agent	*log_profile*
	user
Messages	*msg_X_datatype*
	msg_X_key
	msg_X_value
Attachment	*additional_data*

Figure 4.40: Behaviour Tracking Protocol (left side: description; right side: protocol).

of Behaviour Tracking are presented in sub-chapter 5.1.

4.3.8 Modelling Knowledge

As stated in section 3.2.3, the knowledge acquired by users of e-learning systems during their learning journeys, can be represented in learner models with the aid of the overlay technique. Thereby, individual learner models are subsets of an external expert or knowledge model. Such external models are usually represented as networks of concepts. From the technological point of view, the key idea of deriving individual knowledge from generic models is to separate the duties of representing general knowledge (in the external model) and of assigning it to individuals (in the learner model). Further, in adaptive hypermedia systems, the main source for the creation and description of knowledge structures, is hypermedia content (see e.g. sections 2.2.4 and 3.1.2). Thus, a knowledge modelling system represents also an intermediate level between the modelled information about users and the sources from which knowledge is extracted. Details on the topic of this section can be found in [Gütl and García-Barrios 2005b], [García-Barrios 2006a] and [García-Barrios 2006c].

In the introductory part of this chapter, the requirement has been mentioned regarding the attempt to embed in the "yo?" modelling system a mechanism to represent knowledge structures in order to utilise it also as 'knowledge modeller'. It is worth emphasising at this point that the aim was neither to convert a user modelling system into a general knowledge modelling system, as those introduced previously, nor to investigate the application of modern semantic technologies within user modelling systems. The main idea is to integrate in the "yo?" system a *data structure* that allows to build attributes (concepts) and moreover, to interrelate them in order to build associative networks. This idea provided the motivation as well as the technological basis to show its multi-purposefulness serving as underlying framework to design and implement a Concept-based Context (CO2) modeller. This section introduces the technical issues about the data structure utilised for that purpose in the Knowl-

edge Modelling module of the Modeller component of "yo?" (see Figure 4.21). The practical application and benefits of this structure regarding the CO2 system is described in sub-chapter 5.2.

Biological and artificial life-forms - in the general terms of adaptation - tend to build own internal representations of their environment by means of the inputs of their sensory components. This *internal view*, on the one hand, is well modelled by neuronal networks in human-beings and on the other hand, it can be modelled in software systems by artificial intelligence methods, such as logical agents, connectionist and logic-based knowledge bases. Taking a closer look at human-beings, they tend to simplify, unify and cluster traits in phenomena (e.g. observations from environment, or complex systems and processes) as well as to interrelate them into semantic structures for their thoughts and notions. Considering the latter aspect and analysing research work in Cognition Science and Social Science, the idea of *concepts* can be identified. For the term *concepts* there exist several perceptions in different research areas, e.g. in Semantic Web [OWL 2007], in Knowledge Management [Rollett et al. 2001], and in Learning Science ([Smith and Zeng 2004], [Beasley 2002]). In this section, the focus is set on the description of the technical realisation of concepts through so-called *Concept Molecules*. From a geometrical point to view, the notion for a Concept Molecule (CM) follows the definition of concepts in the field of *Concept Modelling* [Gärdenfors 2001]. Further, the idea of symbolic and associationist representations of concepts (as first developed in the 1960s by J. D. Novak) is considered in CMs: accordingly, a concept may be defined in a non-textual form (e.g. images or sound) and the connections among concepts define their semantic proximity and pragmatic interpretation.

The general idea of Concept Molecules is depicted in Figure 4.41 on the next page. The upper box depicts the *Principle* of CMs: spheres contain the definition of concepts and arrows represent the capability of connecting to other molecules (thus, to build a compound). In addition, connector capabilities might be single or multiple. The upper-right illustration is an example of a compound of 4 CMs.

In contrast to e.g. Concept Maps, Topic Maps or RDF structures (see [Gütl and García-Barrios 2005b]), where a distinction is made between *the concepts* and *the associations among them*, in the Principle of CMs an association is also a CM, because *'everything is a concept'*. Nevertheless, some concepts have an implicit compounding nature and therefore, a representation may be reduced to one *association*, called then *connector concept*. Another key aspect of utilising CMs is that they provide the possibility to distinguish between *semantics and pragmatics*, i.e. between *the meaning of something and its practical intention*:

- Semantics

  The middle box in Figure 4.41 shows some examples of semantic views on CMs following the associationist notion: accordingly, (a) hypernyms and hyponyms describe super- and subordinate concepts as found in taxonomies or classifications in semantic-oriented systems (e.g. an orange is a hyponym of fruit, and fruit is a hypernym of orange), (b) holonyms and meronyms describe an is-part-of relationship, enabling a constructivist point of view (e.g. a planet is part of a solar system), and (c) synonyms and antonyms are used for analogous respectively contradictory or opposing meanings.

- Pragmatics

  This point of view on concepts was derived from the field of Concept Modelling. Consequently, the dimension in a CM defined by extensions and intensions represents its connotation and denotation, i.e. extensions are instances of a concept and intensions are attributes to describe the concept (see lower box Figure 4.41 under Pragmatic view). For example, one extension of the textual concept 'summer' may be an image of a beach in a sunny day, whereas one intention may be the textual attribute temperature. Each of these endpoints is also a concept, as temperature itself is a concept e.g. within the domain of physics. The main characteristic to distinguish between semantics and pragmatics is given by a special type of concept, a Context, and is explained as follows.

The notion of contexts (a) within the scope of this book and (b) regarding its usage in relation with CMs, is depicted in Figure 4.41 (Pragmatic view). Contexts are applicable 'forces' that separate or integrate concepts. Hence, different explanations to a concept may be represented with the aid of contexts (*cx1* and *cx2* in the figure).

For example: consider the common case when trying to explain some topic to persons of differ-

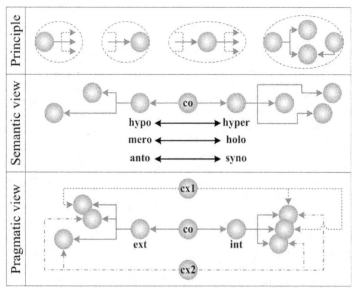

Figure 4.41: The Idea behind Concept Molecules. [García-Barrios 2006a]

ent ages and with distinct notions of the topic. Usually, only those extensions and intensions will be used as vocabulary, which fulfil the suitable intentional need (*pragmatics before semantics*), e.g. consider a person explaining to a child and to an adult the notion of exponentiation of numbers. Let's say, for the child, the person might use apples or other symbols, and for the adult, a mathematical equation could be enough.

Further, the situational aspect of contexts (i.e. its time dimension, as e.g. explained in [Kaenampornpan and O'Neill 2004]) is omitted here, because, if necessary, it can be integrated as an attribute (intension) for each concept, e.g. for logging and managing behavioural data. In summary and from an ontological point of view, dimensions of CMs can be used to find the semantic and pragmatic value of conceptual spaces.

The technical realisation of Concept Molecules is shown in Figure 4.42 by means of a data structure model expressed as an XML Schema Definition. The upper part of the figure shows a concept as a structure with the following elements:

- *uid*: a uniqueness identifier for each stored concept;

- *type*: distinguishing among (a) primitive, (b) text, (c) resource and (c) attribute, in order to describe, respectively, (a) a new isolated concept, i.e. added and not yet classified or described, (b) concepts expressed as text, (c) non-textual concepts, e.g. images or audio data, and (d) key-value pairs,

- *name*: the textual formulation, label or description of a concept, and

- *relatedConcept*: to represent a wrapper for connectors and concepts, such as holonym, synonym, etc.

The bottom schema in Figure 4.42 depicts the general *model* for concepts, connectors and contexts. *Connectors* are defined by their type and linked to a context. The type of a connector might be an association (e.g. for undirected relations), a pointer (e.g. to represent directed graphs), or a

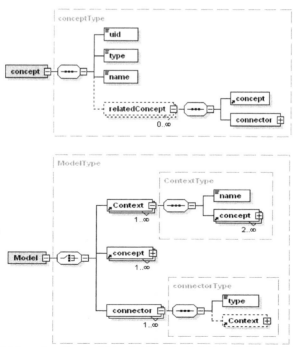

Figure 4.42: XML Schema for a Concept Molecule (above the Definition of a Concept; below, the General Model for Concepts, Connectors and Contexts). [García-Barrios 2006c]

specific type of concept (hyper-, holonym, etc). A *Context* is a container of concepts with a specific *name*. Thus, the schema is used in the **"yo?"** system as basis to implement the notion of Concept Molecules as data structures representing a set of graphs, i.e. nodes (*concepts*), collections of edges that connect pairs of nodes (*connectors*), and collections of nodes (*contexts*).

So far, this chapter described the main issues regarding requirements, design, architecture, functions and implementation of the **"yo?"** modelling system. Based on that, the next sub-chapter concludes the description of the system with a brief summary.

4.4 Concluding Remarks

This chapter has presented the overall design and most relevant implementation aspects of the **"yo?"** multi-purpose profiling & modelling system.

Firstly, the chapter gave a general description of those research projects that had influence on its design and implementation: AdeLE (Adaptive e-Learning with Eye-tracking), DBL (Dynamic Background Library) and Mistral (Measurable intelligent and secure semantic extraction and retrieval of multimedia data). This description has been given in order to show the implications on the design and implementation of the system. In particular, the (apparent) technological constraint of needing a service-oriented solution (given implicitly by the projects), has contributed to allow the development of a highly modular, flexible and scalable solution. The applicability of the solution (and consequently, the practical results of applying the system within the projects) is demonstrated in the next chapter.

The comprehensive investigation work presented in the chapters 2 and 3 as well as the derivation of most relevant challenges and requirements on user modelling systems (as shown in sub-chapter 3.3 respectively section 4.2.1), had direct effect on the results presented in this chapter.

Firstly, having chosen a service-oriented solution approach has shown several advantages, in particular, the following:

- The loosely coupled components and modules of **"yo?"** following the SOA solution approach leads to a high flexibility in terms of *compositional agility*, i.e. processes can be easily modified in response to changing requirements. Thus, a loosely coupled architecture *increases the adaptability and scalability of the system* as well as *makes components reusable*. In turn, the SOA-based solution of **"yo?"** *supports the multi-purposefulness* of the system.

- The modularity of the SOA-based **"yo?"** system *makes software development, deployment, and maintenance easier*, because it reduces the complexity of problems. Thus, its modularity and service granularity enable the optimisation of solutions under the premise of 'Divide and Conquer', i.e. solve a big problem by breaking it down into sub-problems and provide optimal solutions for them. From the point of view on a long-term evolution of software systems, *modularity and separation of duties accelerates the deployment of additional functionalities*: new complex functionalities can be resolved taking advantage of the composition of existing specialised services.

- The *platform-independence* of **"yo?"** makes it independent of software and hardware choices, i.e. it *makes integrated multi-source solutions possible*, and thus, *increases the scalability of solutions across heterogeneous distributed systems*.

Nonetheless, critical problems may arise while developing SOA-based solutions, e.g. performance is a problem that should not be ignored (see section 4.3.4). In general, increasing the communication performance of a service-oriented system is not only a matter of technological decisions, such as the choice of (a) standardised protocols or data representations, (b) applicable communication technologies, (c) utilisation of lookup services, or (d) a combination of (a) (b) and (c). Rather, *developing SOA-based solutions is a matter of overall design* and implies distinct aspects that must be considered according to the requirements of the solution, e.g. service granularity, minimising the number of requests, several tests at distinct degrees of transmission load utilising distinct standards for the task and application domain at hand, or considering usability and transparency to have higher priority than performance, and so forth. For example, within the AdeLE system, **"yo?"** communicates with the eye-tracking and the adaptive engine through RMI/Jini, because the lookup is undertaken once, which leads to an optimal performance tendency (see Figure 4.31 and Figure 4.32).

Regarding the concrete topic of User Modelling, the advantages of **"yo?"** as a service-oriented system (e.g. openness, platform-independence) as well as its architectural modularity and the flexibility of its services meet many requirements of user modelling components in adaptive systems, such as:

- Ease of *integration of different modelling techniques*, e.g. stereotyping vs. overlay models vs. group modelling.

- Reaching a better *balance for task distributions*, e.g. persisting vs. processing vs. visualising information.

- Clear definition of the functional boundaries between user modelling and external components, i.e. a *clear separation of duties* has been feasible *at the level of external interactions and general roles*. E.g. calculating the cognitive style of a user is a duty of the user modelling system, but not the selection (or adaptation) of the corresponding learning material (which is a duty of the adaptive engine).

- Stipulation of the internal functional boundaries among the modules in the user modelling system. E.g. *specialised services strictly separate the internal duties* of profiling vs. modelling.

Further, the integration of **"yo?"** in a smart framework that *meets the main goals of distributed systems* (i.e. integrating the services of **"yo?"** in the Openwings framework) makes it capable of fulfilling special requirements like *mobility of code, support of heterogeneous platforms, openness of interfaces, secured service accessibility*, as well as *ensuring failure handling, concurrency, and distinct levels of transparency*.

Finally, some of the techniques implemented in the **"yo?"** system support directly the fulfilment of 'typical' user modelling needs, e.g.:

- It enables the creation and usage of static *hierarchical profiles* and dynamic *hierarchical stereotypes* (in Profiler component).

- It *records behavioural aspects and interaction histories* (Behaviour Tracking module).

- It *supports distinct levels of scrutability and visualisations* (through HMT dimension).

- It is *capable of showing the topicality of user data in real-time* (State Modeller module).

- It *supports basic issues of privacy and security* (by means of security policies and run-time behaviour of the Openwings framework).

Still, it is not completely clear how the **"yo?"** system meets the requirements of the projects AdeLE, CO2 and Mistral. These open issues represent the main focus of the next chapter.

5 Applicability and First Experiences

In theory there is no difference between theory and practice.

In practice there is.

(Yogi Berra)

The development of the **"yo?"** multi-purpose profiling and modelling system was mainly influenced by the requirements and evolution of distinct research projects (and their corresponding research areas): AdeLE in the field of Adaptive E-Learning, DBL in Knowledge Exploration, and Mistral in Cross-Media Semantics (see also sub-chapter 4.1). The **"yo?"** system meets the distinct personalisation requirements stipulated within these projects. Thus, on the one side, the general application area of the proposed and developed solution is defined by the field of adaptive systems; more precisely, the multi-purposefulness of **"yo?"** enables its integration in personalisation-pertinent systems. On the other hand, due to the fact that AdeLE has been the research project with most implications on the system, the specific application area of adaptive e-learning has determined its specialisation as a learner modelling system. As an innovative solution within the context of user modelling systems, the combination of two very specific aspects makes of **"yo?"** a modelling system with remarkable capabilities: it is service-oriented and multi-purpose. The former aspect has mainly a technological background and its main issues have been presented and analysed in the previous chapter. Consequently, the service-oriented architecture and implementation of the system provides the basis for being multi-purpose. This chapter shows this capability through the description of role and contributions of the **"yo?"** system within the aforementioned projects and application areas.

5.1 "yo?" in AdeLE: Supporting Adaptive E-Learning

The research project AdeLE (*Adaptive e-Learning with Eye-tracking*), as already described in section 4.1.1, aims at developing an adaptive e-learning system using fine-grained user profiling an modelling techniques. The system observes learner behaviour and interactions in real-time by applying eye-tracking technology (see Figure 5.1) and fine-grained content-tracking information. This sub-chapter extends section 4.1.1 and describes in more detail the main aspects of the AdeLE project.

5.1.1 Project Description

As stated in [García-Barrios et al. 2002] (see also section 2.2.1), the knowledge transfer process in technology-based teaching and learning systems

is composed of two streams: teaching and learning. Further, e-learning is a large and complex research area comprising several learning and teaching paradigms, e.g. symmetric and asymmetric, face-to-face, managed learning.

Figure 5.1: Eye-Tracking System ('Tobii 1750') of the AdeLE Research Project . From '*Keep an Eye On*', AdeLE public presentation folder (online available: http://adele.fh-joanneum.at).

At present, it is possible to identify, analyse, track and monitor relevant aspects of instruction, e.g. different velocities, paths or strategies of learning. Moreover, social problems from conventional face-to-face learning, such as censorship of information or racism, can be regulated or partially solved through mechanisms of e-learning. Thus, the combination of e-learning with face-to-face meetings allows a regulation of symmetrical and asymmetrical learning. Among others, these aspects motivated the AdeLE research team to develop a system capable of binding advanced technologies (e.g. eye-tracking) and software development approaches (e.g. service-oriented approaches) to improve the knowledge transfer streams, and in turn, to increase learning performance. Literature in the field of cognitive psychology indicates that people show significant individual differences in how they learn (see e.g. [Glaser 1984], [Honey 1986], [Schmeck 1988], [Bransford et al. 2000]).

According to [Pivec et al. 2005], individuals with stronger visual than verbal skills find text-based material harder to learn. In face-to-face instruction the teacher has the opportunity to immediately adapt or explain learning material according to individual needs. In e-learning environments, firstly, the teacher is frequently absent, and second, didactical material is mostly delivered to every student with the same presentation. Here, adaptive e-learning comes into play, and AdeLE. Information about the learner's gaze behaviour could provide an opportunity to personalise learning material to individual needs: e.g. if a learner shows a preference to consume text, ignoring images, the number of images might be reduced in subsequently delivered material. Further, AdeLE's eye-tracking system could help to identify the topics of most relevant areas of attention and provide further information in that topic (e.g. because a learner is highly interested in the topic or has problems to understand the area).

Mainly, the research efforts within the AdeLE project are focused on two aspects: (a) development and investigation of methods to extract individual learning styles from learner's gaze behaviour and based on that, adapt the provision of learning assets; and (b) development and investigation of methods for real-time analysis of consumed information assets (e.g. words, text passages, areas in images or tables) in order to provide additional topic-specific information. To show how these goals might be reached with the AdeLE system, its technological realisation is presented in the ensuing section.

5.1.2 AdeLE System

The general architecture of the AdeLE system is based on the technical requirements originally stipulated within the project development process, and can be summarised as follows: easy extensibility and open interfaces; strict modularity and high scalability; encapsulation of different scopes of functionality as well as specialisation of functional ranges; ability to (easily) integrate networking functionality; exchangeability and replaceability of software components; utilisation and integration of well-established standards; interchangeability with external system modules. [García-Barrios et al. 2004b] [Pivec et al. 2004] [Gütl et al. 2005b]

The general architecture of the AdeLE system is depicted in Figure 5.2. At the Client-Side of the AdeLE system, the Web Client renders the e-learning content provided by the Learning Man-

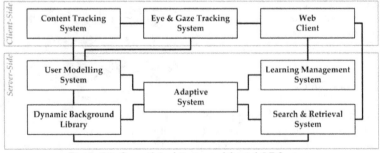

Figure 5.2: General Architecture of the AdeLE System.

agement System (LMS) at Server-Side. Simultaneously, real-time information about gaze movements as well as constitutional states of the interacting user is collected and compressed at client-side; this is done, respectively, by the Eye & Gaze Tracking System (EGTS) and the Content Tracking System (CTS). Thus, at server-side, the input for the User Modelling System (UMS = "yo?" as learner modelling system in AdeLE) consists of (a) observed individual user information received from AGTS and CTS, as well as (b) information about individual user interactions gathered from the LMS vía the Adaptive System (AS). Further, the AS consumes continuously information about individual learners from the UMS. To close this first cycle of description, the LMS is steered by AS and delivers the user-adapted e-learning content and user interface to the Web client.

A second cycle of processes runs in parallel to the aforementioned. Hereby, at client-side, the Dynamic Background Library (DBL) may interact with an internal, specialised Search & Retrieval System (SRS) and with external search services; this SRS builds an index of resources from selected information spaces of the Web and keeps track of their relevance and topicality. The DBL allows teachers to define relevant course-dependent concepts (i.e. topics expressed as keywords) and corresponding search queries for the SRS. With each interaction from a user (vía Web client) needing new learning material, the AS requests the DBL to deliver the set of concepts that are defined for the scope of the instruction at hand. Depending on the operating mode of the DBL or on the type of request form AS, the UMS refines the queries or personalises those concepts. The DBL, after this possible intervention of the UMS, delivers the set of concepts ('ready to render on the Web') to the AMS, which attaches this information to the learning content delivered to the learner. Mainly these functions define the adaptive behaviour of the AdeLE system.

Details about the DBL (its functionality, architecture and role in AdeLE) will be given in the next sub-chapter. AdeLE's UMS (represented by the "yo?" system) has been extensively described in chapter 3 of this book; though, short overview is given on its role in AdeLE. The reminder of the section focuses on insights into AS, LMS and the eye-tracking device.

Knowing the Learner: User Modelling System

In AdeLE, the "yo?" system manages learner profiles and models in order to support the AMS to personalise the learning paths (i.e. at the level of learning instructions) and learning material (i.e. at the level of learning content) that is delivered through the LMS.

The general architecture of the UMS corresponds to Figure 4.17, the architectures of its internal components are equivalent to the illustration presented in Figure 4.18 to Figure 4.22. In summary, the components of the "yo?" modelling system are in charge of the main duties described in the following paragraphs.

The Manager is responsible for the communication with the sensors at Client-side (EGTS and CTS) as well as with internal systems AS and DBL. Thus, its serves as single point of access to and central dispatcher of the UMS. The Views and Tools components contain user interface optimised forms (layouts and contents) e.g. for the initialisation of individual models (see Figure 5.3). Further, it provides a set of Graphical User Interfaces (GUIs) to create and edit user model information.

The Profiler operates on the learner profiles,

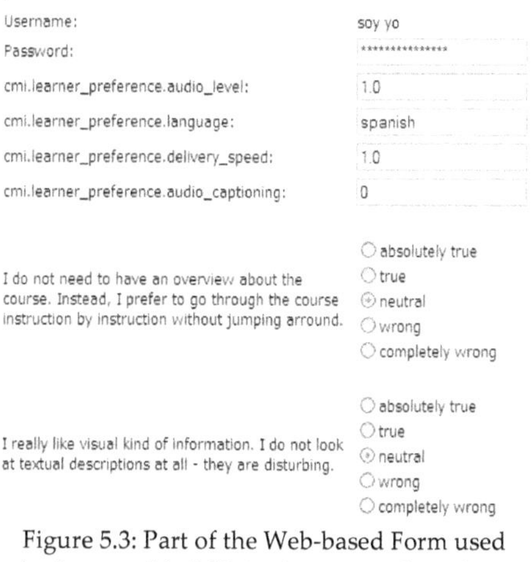

Username:	soy yo
Password:	**************
cmi.learner_preference.audio_level:	1.0
cmi.learner_preference.language:	spanish
cmi.learner_preference.delivery_speed:	1.0
cmi.learner_preference.audio_captioning:	0

I do not need to have an overview about the course. Instead, I prefer to go through the course instruction by instruction without jumping arround.

- ○ absolutely true
- ○ true
- ⊙ neutral
- ○ wrong
- ○ completely wrong

I really like visual kind of information. I do not look at textual descriptions at all - they are disturbing.

- ○ absolutely true
- ○ true
- ⊙ neutral
- ○ wrong
- ○ completely wrong

Figure 5.3: Part of the Web-based Form used for Learner Model Initialisation in the AdeLE System.

e.g. sets of learner attributes and simple (personal data for internal user, such as user name, layout preferences, etc.), interaction sequences and logging data (tracked and managed by the Loggers of "yo?"), an basic information about the cognitive style of learners (simple stereotyping over WAVI-model values; see next section).

The Modeller extracts profile information and enriches by modelling functions, mainly regarding the observation and interpretation of behavioural aspects (arriving from EGTS and CTS through the Profiles) and providing topical states concerning the current cognitive style of learners.

The Data Persistence component comprises all functions to storage and retrieval of data. It supports the management of user information in a file system or a data base. For example, Figure 5.4 and Figure 5.5 show two distinct GUIs for managing user information. Figure 5.4 shows the "yo?"-GUI for AdeLE, utilised to manage learner information (left side) and observed data (right side); the target users for this visualisation are user model developers. Figure 5.5 shows the

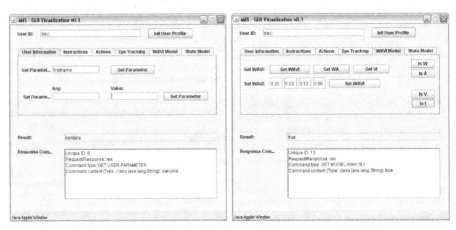

Figure 5.4: Two Views of the Main User Interface for the Management of AdeLE's Learners.

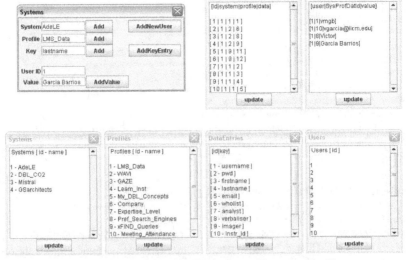

Figure 5.5: Multiple-window GUI for System Managers of **"yo?"**.

general "yo?"-GUI provided by the View component for the management of overall database entries (user target: system administrators).

How the "yo?" system contributes to provide personalisation is described in section 5.1.3.

Adaptability: Learning Management System

The Learning Management System (LMS), which is controlled by the Adaptive System (AS) of AdeLE exposes the personalisation outcome of the adaptive e-learning environment to the user. One of the critical aspects in the AdeLE project was the possibility to extend the functionality and domain area of any LMS.

Therefore, the coexistence between AS and LMS has to be highly scalable. The AS has to be able to handle the following three scenarios. First of all, the LMS may already implement adaptive behaviour, so in this case, the AS may serve as information provider. Secondly, the LMS may provide a set of commands to be controlled by AS. Finally, the AS has to implement the real functionality of the LMS, such as sampling the

content or defining the navigational elements. In the latter case the AS has to control the LMS based on the content or other e-learning features. Beside these scenarios, the AS must also be able to understand the contents that may follow a standard in the field of e-learning.

Hence, the AS could provide a service for each specification of the underlying standard. Considering all these aspects, one of the first decisions to made within the AdeLE project team has been if an own and full-featured LMS should be developed or not. To be compliant with state-of-the-art standardisation efforts, the ADL's SCORM Runtime Environment was selected as front-end LMS for AdeLE's prototype solution (see [ADL 2007]). [Mödritscher et al. 2006a]

The front-end Web interface of AdeLE's LMS is shown in Figure 5.6. In general, the layout of the interface is similar the original to Sample Runtime Environment. In concrete, the navigation frame en the left side is affected by AdeLE adaptivity, e.g. (a) automatically generated DBL concepts for the current instruction (see 'Background Knowledge' in the figure), (b) remaining

Figure 5.6: User Interface of ADL's SCORM Runtime Environment adapted through AdeLE's Personalisation Functionality (left side frame: slightly modified Tree-View Navigation and added areas 'Background Knowledge', 'Instructional Alternatives' and 'Profile'. [Mödritscher et al. 2006a]

alternatives to the current adaptive, i.e. AdeLE supports scrutability of the adaptation and controlled personalisation (see section 3.1.1), and (c) information about the learner model of the currently interacting user, i.e. AdeLE supports scrutability of user models.

Adaptivity: Adaptive System

As shown in Figure 5.2, the Adaptive System (AS) assumes a central role in the overall AdeLE system. In general, the AS interchanges learner information with the UMS and steers the LMS to express adapted learning material.

The AS is divided in two logical modules: *Adaptation Provider* and *Configuration Provider* (see Figure 5.7). These modules are implemented as services in the Openwings framework. The adaptation process itself takes place in the Adaptation Provider considering three techniques: *Sequencing, Presentation* and *Aggregation*. Adaptive Sequencing of instructions aims at tailoring the path through the course based on learner observation and didactical rules. Adaptive Aggregation selects and assembles the resources to build the content of one instruction. Adaptive Presentation visualises e.g. navigational elements. These adaptation techniques are implemented in

```
┌─────────────────────────────────────┐
│        LMS / UMS / DBL/ SRS          │
└─────────────────────────────────────┘
┌─────────────────────────────────────┐
│  ┌──────────────────────┐            │
│  │ Adaptor Factory      │            │
│  │  ┌────────────────┐  │            │
│  │  │ Sequencing     │  │            │
│  │  └────────────────┘  │ ┌────────┐ │
│  │  ┌────────────────┐  │ │Manager │ │
│  │  │ Aggregation    │  │ │        │ │
│  │  └────────────────┘  │ └────────┘ │
│  │  ┌────────────────┐  │            │
│  │  │ Presentation   │  │            │
│  │  └────────────────┘  │            │
│  └──────────────────────┘            │
│  Adaptation Provider                 │
└─────────────────────────────────────┘
┌─────────────────────────────────────┐
│  ┌──────────────────┐ ┌───────────┐  │
│  │ Data Handler     │ │ Settings  │  │
│  └──────────────────┘ └───────────┘  │
│  Configuration Provider              │
└─────────────────────────────────────┘
       Adaptive System (AS)
```

Figure 5.7: Overall Architecture of the Adaptive System of AdeLE. Based on [Mödritscher et al. 2006a].

the AdeLE system as adaptors, which are derived from an *Adaptor Factory*. The adaptation procedures in the AdeLE system are executed through these three techniques. Though, the basis for each procedure in AdeLE is an individual user model, i.e. the WAVI model for cognitive styles. [Mödritscher et al. 2006a]

The cognitive style of a person is a consistent individual method of thinking and perceiving, which in turn affects how people respond to e.g. learning, tasks, or ideas (see e.g. [Riding 1991], [Green et al. 1996] and [Rayner and Riding 1998]). There exist several methods to measure a cognitive style.

According to the survey evaluation on 30 distinct methods described in [Riding and Cheema 1991], it can be concluded that most of that methods can be divided in two basic independent dimensions: the *Wholist-Analytic* (WA) dimension and the *Verbal-Imager* (VI) dimension, both of them building an orthogonal system of coordinates (WA on horizontal axis, VI vertically) and representing the so-called *WAVI model for cognitive styles* (see Figure 5.8).

Within the AdeLE system, an individual parameter on the WA dimension is a numeric value between -1.0 and +1.0 assigned to one learner. Individual WA and VI values are calculated in the UMS of AdeLE according to each user's individual gaze and navigational behaviour. An individual value along the WA dimension indicates if a person (a) understands situations 'as a whole' and needs an overview over the course materials (wholist) or (b) understands situations in parts and does not need a general overview (analyst; such persons are also called 'sequencers', because they learn step-by-step along the course without leaving the stipulated sequence

Figure 5.8: WA and VI Dimensions of Riding's WAVI Model for Cognitive Styles in AdeLE.

of instruction). An individual value along the VI dimension indicates the degree in which a person models information while thinking, either (a) textually (verbaliser, i.e. 'in words') or (b) as mental pictures (imager, i.e. 'pictorial'). Thus, the *WA dimension* depicts how learners *process information* and is similar to holist-serialist dimension in [Pask 1976]. The *VI dimension* depicts how learners *represent information during recall*. For more details on cognitive style models and its implications on learning see e.g. [Douglas and Riding 1993], [Riding and Douglas 1993], [Riding and Watts 1997] [Rasmussen 1998], [Riding and Rayner 1998], [McLoughlin 1999], [Sadler-Smith and Riding 1999].

AdeLE's WAVI Tracking Method (AWTM) is simple and is explained as follows. If a user prefers to navigate by pressing the 'previous' and 'next' button and avoids the 'tree-view' (respectively, see upper frame and left frame of the screenshot in Figure 5.6), the WA factor is increased towards the positive A axis, otherwise the value is decreased (towards W axis). As a consequence, for positive values, the 'tree-view' is suppressed. VI values (also between -1.0 and +1.0 in the vertical system of coordinates) are modified according to learners' feedback regarding the type of resource that is preferred.

For example, Figure 5.6 shows the content of a learning Web page that contains only textual information (it was delivered according to the current model of the learner); the AdeLE system shows this learner that there is a second type of resource for the same instruction (see 'Why this way?'->'Alternative content' on the bottom-left side of the figure); if the learner clicks on the hyperlink 'visual information about...', then the corresponding individual model will be changed towards the positive VI axis (i.e. the VI value increases towards I=imager). For the initialisation of the individual user model, the CSA-Test is required (VICS v2.2b; [Peterson et al. 2003]), which is a computerised psychometric test to measure the WA and VI dimensions of Riding's model.

Observing Users: AdeLE's Eye-Tracking System

Eye movements, scanning patterns and pupil diameter are indicators of thought and mental processing involved during visual information extraction ([Rayner 1998], [Kahneman 1966]).

Thus, real-time information of the precise gaze position and of pupil diameter could be used for supporting and guiding learners through their learning journey. In general terms, eye movements can be divided into two components: (a) *fixations*, i.e. periods of time with relatively stable eye movements where visual information is processed, and (b) *saccades*, which are defined as rapid eye movements that bring a new part of the visual scene into focus. Though, more relevant indicators can be gained by analysing both components together with additionally derived parameters. [Pivec et al. 2005]

The situation in this research field is very complex; consider the following examples of problems and prospects in context. Gaze duration (i.e. time spent on an object) and fixations are not indicative of attention per se, because one can also pay attention to objects that do not lie in the centre of the focused region. But considering other indicators, e.g. eye-lid's degree of openness, saccadic velocity as well as blink velocity and blink rate, better approximations can be achieved. [Fritz et al. 1992] assumes that saccadic velocity decreases with increasing tiredness and increases with increasing task difficulty. Further, blink rate, decreasing blink velocity and decreasing degree of openness may be indicators for increasing tiredness [Galley 2001]. In this context, the AdeLE research team saw an opportunity: if tiredness is identified, it should be possible through adaptive e-learning mechanisms to suggest optimised strategies e.g. the best time to take a break. From an extensive literature survey it was concluded that the intended methodology to infer tiredness from eye-tracking technology is not accurate enough, not because of the device, rather because inferring tiredness from eye characteristics is not reliable enough. [Pivec et al. 2005]

For the purposes of the AdeLE project the outside-in *Tobii 1750* eye-tracking system was chosen. In Figure 5.1 the utilisation of Tobii 1750 within the AdeLE project is illustrated. This eye-tracking device can be used for many forms of eye-tracking studies with stimuli like Web sites, slide shows, videos and text documents, because it is integrated into a 17" TFT monitor. It does not show any problems with its functional reacquisition from extreme head-motions. [Gütl et al. 2005b]

Another advantage is given by its high tracking quality, i.e. it can be used by young or old people, by persons with dark or bright eyes, by users with different ethnically dependent anatomic eye types, by people with glasses or contact lenses. Some other technical advantages of the system are [Gütl et al. 2005b] [Pivec et al. 2005]: (a) high accuracy (0.5 degrees accuracy, bias error), (b) compensation of unparalleled quality of head-motion and drift reduction, and (c) binocular tracking with a frequency of 50 Hz. Further, the system provides a well-designed programming interface with which its automatic functionality can be configured, enabling no additional manual adjustments of parameters on the device. This interface is utilised to integrate *Eye & Gaze Tracking System* (EGTS) and *Content Tracking System* (CTS) into the AdeLE framework, as shown in Figure 5.2.

5.1.3   Personalisation: Behaviour Tracking

This section describes the processes involved in the technique of *Behaviour Tracking* within the AdeLE system. An introduction into this topic has been already given in section 4.3.7. Within the AdeLE research project, *Behaviour Tracking* is limited to the scope of *tracking the learner's gaze-behaviour while consuming online lessons*. And what for? Before answering this question, it is relevant to present two observations within the context at hand.

Firstly: *'do not teach me, let me learn'*. The best method to measure knowledge acquisition seems to be assessment. For example, by means of partial exams, teachers may identify early enough if learners have problems with the learning context and are thus able to react accordingly. Within the scope of traditional e-learning, the steps involved in an instructional cycle might take place as depicted in Figure 5.9. The cycle begins with a teacher preparing the learning materials ('Authoring') and subsequently, integrating them into an e-learning platform ('Storage'). Thus, these materials are placed at the disposal of learners ('Delivery'). When e-learners access the system,

their learning journey begins ('Learning'). Then, during the course or after it, they might be asked to make exams ('Assessment'), and in turn, the system or the teacher (or both) will evaluate the results and provide marks for each exam ('Grading'). Much time has passed: not till after this evaluation the teacher might react to problems regarding effective knowledge acquisition.

Second: *'four eyes see more than two'*. In connection with the aspect previously mentioned, one disadvantage of traditional e-learning systems (either adaptive or not) relies on the fact that teachers rarely are able to observe the learning progress on real-time. They might be provided of synchronous collaborative tools, such as chats, video streams or e-conferencing, however, in order to monitor the learning behaviour of all system users at the same time requires more than these tools. Nonetheless, this problem also exists in classroom instruction, but here, the physical nearness gives teachers the possibility to better observe the actions of learners, and consequently, to react immediately to problems of individuals or groups. They can react that fast, because they are always striving for symmetric learning. In the case of e-learning, therefore, each technological effort to achieve symmetric instruction and to foster personal contact (such as the tools mentioned) will contribute to improve the knowledge transfer process.

Although the latter aspect can support the former one, it is not enough. Consider how big each teacher's monitor should be or how many monitors one teacher might need to observe a big number of 'e-learners' and monitor them optimally at the same time.

Figure 5.9: Simplified Traditional E-Learning.

This is the answer to the question stated at the beginning of this section: AdeLE aims at providing an automatic way to simultaneously (a) observe the learners at real-time and (b) come to the conclusion if they have learned the content. For that purpose, the cycle of traditional learning (as shown in Figure 5.9) may be improved by means of AdeLE's real-time Behaviour Tracking technique (see Figure 5.10).

Beginning again with the Authoring step, teachers are provided with a tool to embed didactical goals into their (Web) materials. At this step, from the perspective of the teachers, this tool helps them (right at the outset) to define in a Web page 'which segments of the content are (a) didactically relevant, i.e. 'must be learned', (b) could be considered, 'should be read', or (c) might be ignored; see step (1) in Figure 5.10.

AdeLE's tool for teachers providing the possibility to annotate such 'didactical goals' is an extension for the Firefox Web browser and originally called STAGE, *Semantic Tagging Editor*. The screenshot shown in Figure 5.11 illustrates version 2 of STAGE (called *Stagezilla*), which is based on Annozilla (annotations tools for Web

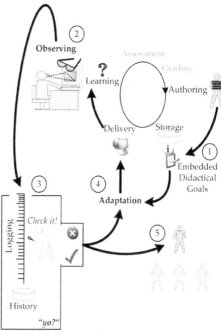

Figure 5.10: Improved Knowledge Transfer through AdeLE's Real-time Behaviour Tracking.

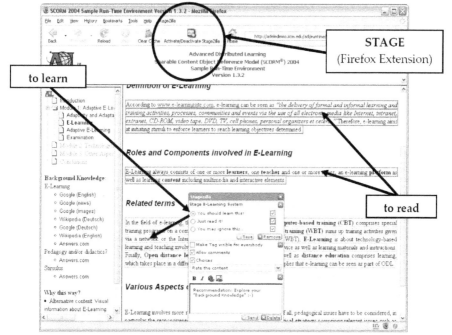

Figure 5.11: AdeLE's STAGE Tool for Teachers.

pages, as defined by the W3C Annotea project; see http://annozilla.mozdev.org).

In accordance to the architecture of the AdeLE system (see Figure 5.2), the annotated learning material is stored on the LMS, which is steered (adapted) by the AS. Thus, the steps 'Adaptation', 'Storage' and 'Delivery' in Figure 5.10 are ensured. At client-side, when learners interact with the AdeLE system through its Web Client, they get a 'normal' view on a page as e.g. already depicted in Figure 5.6. During interaction, AdeLE's eye-tracking system analyses the gaze movements of a learner: the client-sided *Eye & Gaze Tracking System* (EGTS) and *Content Tracking System* (CTS) come into play.

An evaluation study has been conducted with AdeLE's eye-tracking system at the University of Applied Sciences FH Joanneum (Graz) in order to find eye movement parameters that provide a distinction among the three different gaze behaviours *skimming*, *reading* and *learning*. For this purpose, the complexity (i.e. randomness) of a gaze path was measured in terms of its entropy (as defined in the field of Information Theory; see e.g. [Borst and Theunissen 1999]). Examples for some gaze paths derived from this study are shown in Figure 5.12. This study identified distinctions among the entropy values of the three different gaze behaviours (for more details refer to [Pripfl 2006]). Though, under consideration of the overlapping ranges as depicted in Figure 5.13 (see next page),

> *"linear contrasts [...] reached significance for learning vs. reading [...] but not for reading vs. skimming."* [Pripfl 2006]

These *Very Complex Duties of EGTS and CTS* (VCDEC) define mainly the 'Observing' process at client-side of the AdeLE system (see step (2) in Figure 5.10). In sum, the cycle of learner observation comprises the following four main tasks: (a) extraction of the annotations on the learning pages, (b) VCDEC, (c) comparison of VCDEC's results with annotations, and (d) delivery of comparisons to the **"yo?"** user modelling system at AdeLE's server side according to the Behaviour Tracking communication protocol (BT-Protocol, as explained in section 4.3.7).

Thus, having arrived at step (3) of the overall Behaviour Tracking technique (see Figure 5.10),

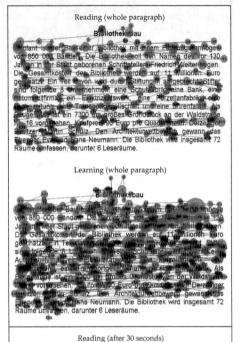

Reading (whole paragraph)

Learning (whole paragraph)

Reading (after 30 seconds)

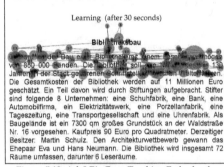

Learning (after 30 seconds)

Figure 5.12: AdeLE's Gaze-Tracking Paths for the Behaviours 'Reading' and 'Learning'.
(Original images on presentation poster [Pripfl 2006]; with permission from Dr. Jürgen Pripfl)

the **"yo?"** modelling system provides to its main AdeLE client, the Adaptive System, the possibility to *check individual user gaze behaviours* through the cooperation of three modules in its system *Logger, State Modeller* and *Behaviour Tracker*. Each entry that must be logged includes the results of the comparisons from EGTS and CTS within the three *Message Fields* of its protocol in terms of the examples given in section 4.3.7, such as '(segmentX)>(learn)>(read)' = 'segment X was marked by the teacher as to be learned, but has been only read by the learner'. Within the scope of AdeLE and its overall Behaviour Tracking technique, the State Modeller of **"yo?"** concentrates on this information.

At this point, it is worth mentioning that the clear separation of duties in adaptive systems, as claimed by the author of this book along the previous chapters, shows again its benefits and practical applicability. In particular, the duty of the modelling system is not to know what is going to happen with the observed data in terms of 'what will be adapted?'.

In AdeLE, the user modelling system focuses on 'knowing the user', and therefore, responds just to specific requests from the Adaptive System in terms of 'within the time interval [x,y], did the learner achieve the didactical goals stipulated by the teacher?', and if not 'which were the problematic areas?'. This means that the user modelling system is not aware of the extension of this time interval, and consequently, must provide an interface that is general enough to maintain time intervals and interpretations of streamed data as flexible as possible. This is the main reason for having stipulated the following two functionalities of **"yo?"**:

- The BT-protocol establishes a robust functional convention along the overall technique, i.e. it holds the technique together among the distinct components that participate, starting with the annotation tool, through the client-sided components, until the application of inferences by the adaptive engine.

- The interface for the Adaptive System is general, because it provides method calls to combine the inferences extracted from the logging entries. For example, the Adaptive System may ask the State Modeller either to collect and analyse just those entries were a specific learner did not learn the segments of text marked by the teacher as to 'learn' (and nothing more), or to collect and analyse those entries of a group of learners that did ignore some segments that have been marked as to be read, and so forth.

These issues bring to step (4) of the overall Behaviour Tracking technique (see Figure 5.10). In order to provide to the Adaptive System a way to track the behaviour of users in self-specified time intervals, the interface of the **"yo?"** modelling system includes a "CLEAR" method, which can be called at any time by the Adaptive System to clear the collection buffer of the State Modeller, i.e. 'stop tracking now and begin working again'. By means of this CLEAR method the Adaptive System may apply the intervals of time as required by its adaptational purposes. In the case of the AdeLE system this is relevant, because each interaction of a user is perceived by the Adaptive System, and therefore, it may decide to track behaviour between two page transitions, two chapter transitions, three clicks, etc. In general terms, each call from the Adaptive System leads to two possible responses of the modelling system:

- ['OK'], meani

- g 'the learner behaves as desired'. Concretely, 'since the last CLEAR no conflicts have been identified, because everything that was to be

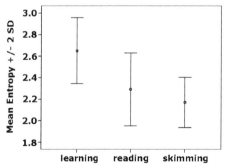

Figure 5.13: AdeLE's Results of Study on Gaze-Tracking Paths for the Behaviours 'Learning', 'Reading' and 'Skimming'. Mean Entropy values +/- 2 standard-deviations of each condition. (Original images on presentation poster [Pripfl 2006]; with permission from Dr. Jürgen Pripfl)

read has been at least read and everything that has to be learned has been read.

- ['NOT OK']+ [{set of conflictive entries}], meaning 'the learner did not behave as desired, and the problems are here..'. Concretely, the since the last CLEAR there existed segments in the content, which had to be read but have been ignored or/and had to be learned are have been ignored or just read.

So far, the most relevant details of AdeLE's Behaviour Tracking technique have been depicted, from the technological as well as didactical point of view. Though, the benefits for the learner are not yet clear. Step (5) in Figure 5.10 represents the way in which teachers are supported by means of automatic assessment (within the context of this technique), i.e. the AdeLE system responds automatically to the learner and gives feedback about the 'hidden' didactical goals stipulated by the teacher.

From the point of view of learners, the AdeLE system presents a *scrutability* corner in order to inform them about the adaptive behaviour of the system as well as about the state of their individual user models. This scrutability corner is visualised as 'Why this way?' navigational area (see bottom-left side of Figure 5.14). The third entry in this scrutability corner shows the learning status of learners: 'My learning status', whereby the word 'status' is hyperlinked and change among the colours yellow, red and green. A 'yellow' status is always shown when a learner visits the current page for the first time.

The example of Figure 5.14 shows the same content as Figure 5.11. In order to demonstrate the Behaviour Tracking technique of AdeLE without an eye-tracker device, a client-sided simulator has been developed. It can be activated through the navigation frame ('Simulate Eye-Tracker') and displays information about the tagged segments of the current page (see additional window 'AdeLE / STAGE – Runtime Eye-Tracker Simulator' in Figure 5.14). This Simulator shows a summary of the 'tagged segments' for this page, as defined by the teacher in Figure 5.11. On the right side of the Simulator, the gaze behaviour of learner can be selected and submitted to the AdeLE's server.

For this concrete example, the first three entries are supposed to be correct (read-read, read-

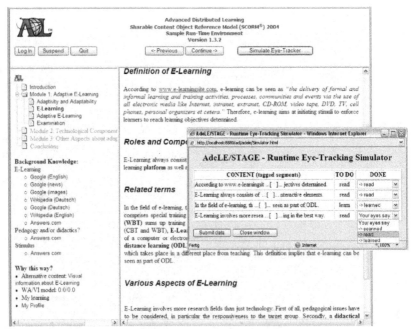

Figure 5.14: AdeLE's Behaviour Tracking Technique; Learner's View.

read, learn-learned), the last entry is being selected read-read. Thus, the submission of this status simulates a learner that has fulfilled the didactical goals of the page. The submission also simulates that the learner has requested the next page in the course. In this specific case, the result will be (certainly) the next page with a 'green' learning status in the scrutability corner. Though, if the status 'scanned' is selected for the last entry in the simulator (i.e. 'ignored'), the submission of the data will lead to a reloading of the same page with a 'red' learning status. Naturally taking for granted that the learner already knows how to interpret this functionality of the AdeLE system, the learner may notice immediately after consuming a learning until that something has been 'overlooked'.

As AdeLE supports scrutability and controlled personalisation, learners may click on the 'status' link to get information about the 'conflictive' segments of the page, but must not stick on the page until the goals are reached. Rather, they may continue with their learning journey.

At the point in time of submitting this book, the AdeLE system implements the behaviours described so far. Though, the AdeLE research project has not finalised and thus, distinct applications of the Behaviour Tracking technique may be implemented additionally. For example, it might be more useful and intuitive in the case of a 'red' status to work on the aggregation level of adaptation, i.e. the AdeLE system might reload the same instruction but showing just those segments in the content frame for which the didactical goals have not been reached. In addition, they might be highlighted and annotated accordingly, in terms of 'Sorry, dear student, AdeLE assumes that you have not learned this paragraph. Please try again...'

The next section addresses some relevant concluding remarks on this sub-chapter, i.e. about the overall AdeLE system and its adaptive techniques.

5.1.4 Conclusions

Within the context of Adaptive Educational Hypermedia Systems, as stated in [Karampiperis and Sampson 2005], the design of such systems is a hard duty, because the didactical dependencies between learner traits and learning materials are too complex, and therefore, in practical terms, all possible combinations cannot be treated. Further, this section and the previous chapters of this book have shown that the success of personalisation does not only depend on the implemented techniques in the adaptors of the system.

[Karampiperis and Sampson 2005] claims also that the complexity of didactical dependencies leads to several problems regarding the definition of adaptive rules (see also [Wu et al. 2001] and [Wu and De Bra 2001]), among them: inconsistency (i.e. semantic conflicts among rules), confluence (i.e. various equivalent rules exist), and insufficiency (i.e. required rules are not implemented). By developing systems like AdeLE, which support a clear sub-division and distribution of complex tasks, an alternative is given to move towards a solution of the previously mentioned problem.

With real-time tracking of user behaviour, relevant information can be gained about what the users is doing 'now', and thus, mechanisms can be developed and provided e.g. to improve knowledge acquisition almost at the time of consumption, or 'just' to alert teachers immediately after an 'overlooked' problem has occurred. The reasons for occurred problems might be evaluated timely, e.g. AdeLE's visual scan paths of learners might be provided to teachers as video recordings in order to inspect the usability of learning content. If learners have problems with learning content must not rely on their incapability of assimilating the topic, rather they may have got the short end of the stick. Fact is, eye-tracking technology has found a fix place in usability engineering at present. And as shown in this sub-chapter, eye-tracking devices are not 'that intrusive' anymore.

In summary, the AdeLE research project aims at enabling innovative e-learning solutions as well as improved and more profound understanding in the following areas:

- tracking of learners' behaviour in the field of Human-Computer Interaction in general as well as related to the personalisation of learning contents,

- real-time course-progress tracking and timely identification of problematic areas in the learning journey in order to reduce the response time of teachers,

- identification of most suitable media and content presentation during knowledge transfer processes,

- provision of reusable and extensible service-oriented frameworks to easily integrate or exchange e-learning tools according to the specific needs and didactical goals of instructors,

- strict integration of scrutability methods for adaptivity and user model contents, and last but not least,

- integration and careful evaluation of specialised client-side tracking sensors, but solely under the premise that system users at least 'know that they are being observed, how, why and how long' as well as giving them the possibility to disable the observation at any time.

Yet, one speciality of the AdeLE system has not been described, its Dynamic Background Library. The next sub-chapter gives insights into this system as well as explains the role of **"yo?"** in the context of 'fostering individual knowledge discovery'.

5.2   "yo?" as CO2: Fostering Individual Knowledge Exploration

The research area of Concept Modelling has been briefly introduced in section 4.1.1 within the scope of dynamic background libraries. This sub-chapter is structured as follows:

- Section 5.2.1 - Using Concepts for Teaching and Learning.

- Section 5.2.2 - Dynamic Knowledge Provisioning: an Evaluation.

- Section 5.2.3 - Improved Knowledge Exploration through Concept Modelling.

Hence, this sub-chapter begins with a general overview on the application of concepts in the topic of e-learning, presents the results of the evaluation of the first prototype version of a *Dy-*namic Background Library (DBL), and based on that, describes how the overall "yo?" architecture and functions were used to develop an improved version of the DBL, called *Concept-based Context Modelling System*.

5.2.1   Using Concepts for Teaching and Learning

The denotation of the term *Concept* within the scope of this book has been presented in section 4.3.8. In that section, also its associations with related disciplines, such as Concept Mapping or Concepts Modelling, was introduced. In this section, the focus is set on applying concepts in the field of e-learning systems (see [Gütl and García-Barrios 2005b] for more details).

Concepts might have relations among each other, which allow building knowledge structures. Knowledge in a domain, or in general knowledge, can be described, modelled and codified by a set of concepts and their relations, which are statements and assertions that describe a particular situation or a knowledge domain. Further, relations among concepts and terms may also form ontologies, classification systems, thesauri and lexica. In literature, the process of identifying concepts and relations is defined as Concept Mapping and is technologically supported by concept map tools that allow capturing, organising, communicating, transforming and assessing knowledge. [Gütl and García-Barrios 2005b]

While learning, new concepts and relations are built and integrated into prior knowledge, i.e. existing concepts and relations are adapted. This might be induced by stimuli and information in everyday life as well as initiated by an active knowledge transfer process applying different styles, such as declarative and constructivist approaches ([Gärdenfors 2001], [Chieu et al. 2004]).

From the teaching perspective, the use of concept modelling supports the identification and description of syllabi, topics and main subjects within courses and lectures. In addition, course content itself can be modelled through concepts and relations, which in turn might also help to identify the logical order to be taught. Finally, concept structures and their linked extensions can be used to transform knowledge from teachers to learners. In concrete, a conceptual space denotes a representational framework for information structures ([Ferry et al. 1997], [Smith and Zeng 2004]); though, within the scope of this book the terms concept structures and conceptual spaces are used as synonyms.

Concepts represent a powerful tool in a wide range of application domains. For example, they allow humans to symbolise simple and complex ideas, they enable concept instances to be linked (original objects or multimedia representations), and they provide a framework for modelling by attributes, language descriptions and codified rules. *Concept Modelling* enables applications to refer from concepts to illustrative examples or descriptions, and to build (automatically) further concepts and their attribute representation from training instances.

Further, concept relations have the power to structure perceptions and insights, allow expressing knowledge about a specific situation, and make possible to detect and explain dependences to and influences by other concepts. Consequently, concept relations provide means to develop applications e.g. to codify preknowledge and to modify or adapt knowledge through concept map tools, and to discover or generate knowledge from unknown or hidden relations by using co-occurrence analysis. [Gütl and García-Barrios 2005b] propose a *Classification for the Application of concepts in the field of Learning and Teaching* (CALT), which can be summarised as follows:

- *Knowledge Expression and Organisation*
  This category comprises applications that support the management of (a) concepts and their relations, (b) links to examples of concept instances, and (c) concept representations through attributes, language description and codified rules. For example, the 'ADEPT Digital Learning Environment' implements a digital knowledge base of information about concepts and their interrelations, used to organise learning material ([Smith and Zeng 2004]). Further examples in this category are the 'Concept Indexing Tool' [Voß et al. 1999], IMKA system [Benitez et al. 2001], 'CmapTools' [Ambel et al. 2004], 'NavEx' [Yudelson et al. 2004], 'Two-Phase

Concept Map Construction (TP-CMC)' [Sue et al. 2004].

- *Knowledge Retrieval and Discovery*
  Knowledge retrieval supports students and teachers in accessing and exploring concepts and their relations as well as examples of concept instances (e.g. multi-media objects). Knowledge discovery enables students and teachers to detect or reassemble a-priori unknown or unfamiliar knowledge, which was previously already codified in concepts and relations as well as in example instances or descriptions. For example, the 'Concept Indexing Tool' provides concept occurrence detection for textual documents based on manually predefined concept descriptions. Further, it statistically detects concurrence of concepts, enabling students and teachers to discover candidates of new concept relations for further exploration and validation (see [Voß et al. 1999] for details). Other examples for this category are 'NavEx' 'TP-CMC'.

- *Knowledge Exchange and Transfer*
  This category comprises supporting the exchange of personalised views on knowledge either at the level of concept structures or at the level of concept representations. The 'ADEPT Digital Learning Environment' supports knowledge transfer in face-to-face lectures by simultaneously projecting concept structures, lecture note slides and example instances ([Smith and Zeng 2004]). [Ambel et al. 2004] state that in 'Quorum', students can collaboratively create their concept maps in a knowledge domain, whereby the underlying propositions are extracted and stored in a central database. Thus, students can explore a common conceptual space and learn from each other, but they can only see those others' propositions that are linked to their own contributions.

- *Knowledge Assessment*
  Assessment of knowledge provides a powerful mechanism for assessing pre-knowledge and knowledge acquired either at the level of concepts and their structure as well as at the level of concepts and representations. The 'Automated Essay Grading' tool (AEG) extracts concepts from model answers, com-

pares them to the concepts given by student answers, and automatically processes grades (for details see [Williams and Dreher 2004]).

Being aware of the power of concepts in learning and teaching applications, research and development work in the field of e-learning at IICM (Graz University of Technology) builds on the 'concept' idea [Gütl and García-Barrios 2005b]. Some interesting examples of developed tools are: *Dynamic Background Library* and its successor the *Concept-based Context Modelling System*, which are described in the next section, as well as *E-Tester* [Gütl et al. 2005a] and *Virtual Tutor* [Gütl and Pivec 2003].

The *E-Tester* automatically extracts the main concepts from learning content in natural language, and based on these concepts, it compiles simple questions, e.g. 'What is <concept 1>?' or 'Explain <concept 2>'. In a second step, the system assesses natural language answers from students against the learning content, and finally provides users with results on a conceptual level. First evaluations about the quality of concept extraction and automatic grading compared to teachers' grading are promising, field tests and user evaluations are ongoing work. The application of E-Tester in game-based e-learning environments uses concepts for quiz-based games. In terms of the CALT classification, E-Tester contributes to the class 'Knowledge Assessment' and the E-Tester in game-based Learning to the classes 'Knowledge Exchange and Transfer' and 'Knowledge Assessment'.

Virtual Tutor, as described in [Gütl and Pivec 2003], complements traditional learning activities through dialog-based learning and experiments based on concepts and their relations to course materials. These main concepts (and their attributes or sub-concepts or related concepts) are modelled in the knowledge base of an expert system. Following the dialog initiated by the expert system, students have to answer questions about the attributes of the main concepts. Based on the answers, the expert system infers concept candidates and their probabilities. In such dialog sessions, students can learn about concepts' attributes and are encouraged to experiment, e.g. to answer questions about terms to address a specific concept. A user evaluation has shown that

in order to solve concrete tasks or questions, students found the usage of the Virtual Tutor and knowledge access by concepts significantly more helpful than using the lecture notes. The Virtual Tutor covers the sub-class 'Knowledge Retrieval' and the class 'Knowledge Exchange and Transfer' of the CALT classification.

The next section focuses on the other two developed tools mentioned previously: the Dynamic Background Library, and its successor, the Concept-based Context Modelling System.

5.2.2 Dynamic Knowledge Provisioning: an Evaluation

Before entering into the topic of *dynamic background libraries*, some clarifications must be given regarding terminology. The overall scope of knowledge provisioning is viewed from a specific point of view: courseware ('learning materials', 'learning content'). Further, e-learning systems provide the capability to store and administer courseware in 'learning repositories', an aspect which in turn brings to the scope of Learning Content Management Systems (LCMSs) or Learning Management Systems (LMSs). These terms have been clarified along the previous chapters. Usually, courseware is stored in a *static* manner, i.e. it must be uploaded into the repository and will remain there as long-term static data.

These repositories enable teachers to structure the learning materials of their courses, e.g. in chapters, sub-chapters and other sub-divisions. Some systems provide the possibility to store additional materials in *background* repositories, such as libraries or glossaries. From the point of view of learners and teacher, such additional repositories might be seen as *static background libraries*, because the resources relying on it do not change over a learning journey, moreover, sometimes they do not change over several years.

Thus, it is up to teachers and course authors to ensure the topicality and domain-relevance of the information stored in these repositories. Both, reflection and introspection of learning activities expose widely the same situation, namely that just studying one particular learning source will not satisfy a deep understanding of new knowledge and not enhance the knowledge acquisition process in terms of overall learning efficiency. In this sense and with respect to the increasing amount of knowledge sources on the Web, additional background sources are needed (or at least would be helpful) for many learning situations. Moreover, the selection of relevant background information represents an additional effort and (sometimes serious) challenge for teachers and course authors.

Among others, the issues depicted so far represented the motivation to suggest and implement the first prototype of the *Dynamic Background Library* (DBL) approach, as first described in [Dietinger et al. 1999]. In the field of static learning resources there exist various initiatives, e.g. listed in [NLII 2004]. Moreover, it is known that knowledge acquisition and learning processes can be supported by Information Retrieval (IR) activities, as stated in [Baeza and Ribeiro 1999] and [Liaw et al. 2003]. Further, some problems related with the 'freedom' in and 'chaotic growth' of the Web, e.g. censorship of information, or reliability, accuracy and topicality of resources, can be solved through techniques, such as *white lists* of servers [Lennon and Maurer 2003] and the *Quality Metadata Schema* approach discussed in [Gütl et al. 2002].

The main aspects of a Dynamic Background Library (DBL) have been presented in section 4.1.1. The aim of this section is to show some relevant evaluation results of the first prototype version of the DBL in order to expose the gaps that will be filled through by the next version. This next version is called *Concept-based Context Modelling System* (CO2) and has been implemented as a sibling of the **"yo?"** system. First experiences on the usage of a DBL have shown that learning performance can be increased. Furthermore, to offer a transparent mechanism to access additional background knowledge resources seems to promote further interest in detailed information and related concepts. An extended evaluation of a DBL implies the examination of at least the following aspects:

- functionality from the authoring and teaching points of view,

- usability of the system from the learners' point of view,

- knowledge acquisition improvement, and

- comprehension improvement regarding distinct types of learners (e.g. scholars, university students, company workers, etc.).

The main functions of this first version of the DBL are summarised as follows.

To access the background knowledge resources relying on the Web the DBL provides a communication layer to the *Information Search and Retrieval Systems* xFIND [Gütl 2002] and Google (http://www.google.com). For learners, the DBL provides the possibility of choosing the own level of expertise and one of four different viewing modi for displaying the DBL entries while navigating through the course contents. Thus, the front-end representation of a DBL entry is represented by a highlighted hyperlink (see e.g. the word 'intranet' in Figure 5.15 on the next page). The four Viewing Modi of the DBL (see also Figure 4.5 and Figure 4.6 in section 4.1.1) are:

- In File
  the content of a requested learning page is parsed and adapted (e.g. the entry 'intranet' in Figure 5.15 with its hyperlinked 'book'-icon). Each match is highlighted and linked to one Information Search and Retrieval System.

- After File
  a list of the matching DBL entries is appended at the end of the current page.

- After Chapter
  single pages are not modified. At the end of each chapter a dynamically generated Web page is provided and contains a list of those DBL entries that are specific for the current chapter and expertise level.

- After Content
  a generated page with a list of all level specific items is attached at the end of the course.

The evaluation of the DBL was conducted as an observed online lesson within the lecture 'Information Retrieval' at Graz University of Technology (summer 2004). Attendees of this lecture represented the test subject group. Details about this evaluation can be found in [García-Barrios et al. 2004a]. Participants of the experiment, in the following for short *subjects*, were requested to consume the online lessons using their own notebooks and Web browsers (the Information Search and Retrieval System 'Google' was chosen within the DBL system).

In sum, 28 different search concepts (i.e. DBL entries) were prepared, whereby each chapter of the lesson was assigned a certain number of these concepts. The experiment included the following steps (the time period for each step, in minutes, is given in parenthesis): introduction into the learning platform and DBL (15), answering pre-questionnaire before online lesson (15), consumption of online lesson (120), answering post-questionnaire after online lesson (15), and short written exam of knowledge acquisition (15). The subject group consisted of 14 male persons.

The reminder of this section presents the most relevant results of the DBL evaluation. The consumption of the main services e-mail, WWW and search was for nearly 100% 'repeatedly daily'. A majority (57%) of the subject uses digital libraries 'occasionally', while 21% at least 'daily'. A comparison between the consumption of different media types on paper and on screen showed that more than 90% of the subjects prefer to read news from screen, and more than 70% prefer to read encyclopaedias on screen. On the other side, books, newspapers and lecture notes are preferably read on paper.

Thus, it may be concludes that subjects like to use the Web for consultation and in order to acquire recent information. The learning behaviour of the subjects from the point of view of different sources for knowledge acquisition shows that the majority of subjects used lecture notes 'very often' for learning. The subjects do not really consult journals for learning, but they tend to consult or include colleagues and friends in their learning sessions. The common and regular usage of books has proved itself again. In addition to books and lecture notes for their learning activities, subjects like to gather background knowledge from the following resources: timely online information, Web sites, Internet recherché, abstracts of lecture notes, test examples, literature references, and lecture notes from other universities.

So far, it can be stated that the subjects preferred to learn from lecture notes and books on paper, but are very interested in acquiring additional (relevant and up-to-date) background knowledge and use these additional resources at their learning activities. Issues about the usability and efficiency of the DBL could also be extracted from the analysis of the post-questionnaire. The DBL was rated with an average of 1.9 points, where 43% of the subjects gave

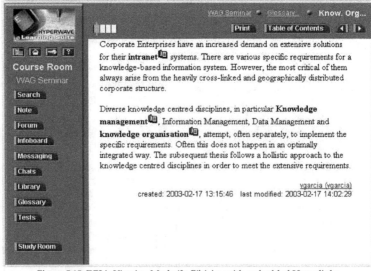

Figure 5.15: DBL's Viewing Mode '*In File*', i.e. with embedded Hyperlinks.

DBL the note '1' (1: best ... 6: worst).

A positive feedback could be gained while asking for the efficiency of the information retrieval process through the DBL. Yet, the most important findings were identified through the free-text questionnaire. Referring to the subject usability, our test persons stated - among others - the following:

- Particularly I liked:
  "Cross references directly embedded in text"; "Idea of graded information behind DBL items according to user's expertise"; "Ease of use".

- I did not like:
  "The hyperlinks in the text are too intrusive"; "I had to modify queries manually in order to find more useful information".

- Recommendations for improvement / Helpful suggestions:
  "Do not mark the embedded hyperlinks so obtrusively, it is confusing"; "Modifications in the index of Google may represent a risk, because you may get wrong documents in the future"; "Deliver Google results in the user's language"; "Build-in a direct linking to cited literature resources (if available)"; "Provide other relevant entries from online-lexica (e.g. Wikipedia or dict.leo.org) and separate them from the search results of Google".

The last part of the evaluation concerned the benefit of the DBL and the user interaction preferences. A positive tendency may be concluded from the requested personal estimation about the improvement of learning activities through a DBL; the average rate was equal to 2.6 from a 1 to 6 scale (good to bad, respectively).

Regarding user interaction preferences, the evaluation showed that it is highly important for DBL users to keep control over the individual settings. Users want to select their preferences themselves and modify the settings at their own will and convenience. In the special case of the BDL, users want explicitly to choose and change the settings for their pre-knowledge (expertise level) and DBL viewing modi by themselves.

Furthermore, when requested for the personal benefits of using a DBL, subjects annotated

(among others) the following: "user-tailored and up-to-date tips about relevant and additional literature!"; "up-to-date information and other perspectives on the subject from different authors"; "correction of individual knowledge gaps (e.g. expertise knowledge, foreign words)"; "support of autonomous investigation"; "rapid retrieval of specific terminology"; "refresh the forgotten; no problem, consult terms!".

Finally, referring to the overall functionality of the DBL, the subjects stated (among others):

- Particularly I liked:
  "Explanations appear on the place, where they are needed"; "Providing the possibility to change personal settings"; "Well structured and clear"; "The DBL works as an activator, deviating from the reading monotony".

- I did not like:
  "Design of DBL links in text"; "Not-relevant documents were also found within the Google search results"; "Partially, useless results were delivered, or results, which implied a long further search within a web site"; "Google search is too global"; "Sometimes a manual modification of the search query was needed".

- Recommendations for improvement / Helpful suggestions:
  "Enhance design of DBL links in text "; "Use tool tips"; "Term definitions, which do not have an influence in topicality should be moved to a static glossary"; "Provide a built-in library in order to find topic-specific explanations"; "Improve the load speed of parsed pages"; "If there are too many DBL links defined (e.g. in the End of Chapter viewing mode), then structure them better"; "Present word explanations in an additional encyclopaedia (e.g. Wikipedia)"

Concluding, the implementation of a DBL contributes to improve the knowledge transfer process in e-learning environments. Nonetheless, for this first prototype, the integration of adaptive techniques within the utilised framework implied complex development requirements and functional dependencies. Thus, a new stand-alone solution should be the best choice.

5.2.3 Improved Knowledge Exploration through Concept Modelling

The last section presented the results from an evaluation of the first DBL prototype version. This section focuses on the improvement of that first prototype (as shown in [García-Barrios 2006a] and [García-Barrios 2006c]), and therefore, only worst problems and most relevant recommendations are taken into account, as follows:

Problems:

- (P1) hyperlinks in the text are too intrusive;

- (P2) viewing mode In File is annoying;

- (P3) manual query refinement is required in order to find more useful information;

- (P4) using one global search engine leads to many high-ranked irrelevant resources;

- (P5) a global search service is not trustworthy (results are doubtful in little while);

- (P6) personalised search is missing.

Recommendations:

- (R1) language-dependent results would be fine;

- (R2) a separated list of DBL concepts is the best viewing mode;

- (R3) more search services for each concept would be better;

- (R4) let users re-define own queries;

- (R5) let users re-define search services;

- (R6) let users re-define own concepts.

Hence, P1 is a pure usability problem, because overwriting content with added-on-style leads always to subjective acceptance troubles and thus, was evaluated as 'too intrusive'. The problem is the noisy redundancy of DBL concepts if appearing often in a page. P2 fits only for large content, which has to be completely parsed searching for matching concepts (this may be time-consuming). P3 to P5 arose from using a global search service for the evaluation. Therefore, P6 can be seen as 'need' if using global search services.

In concrete, the efficiency of utilising DBL concepts is dependent on the didactical goal of a teacher and on the learner's expectation at the moment of detecting the 'recommended' concept. Thus, the pragmatic value of concepts is much higher than the semantic value (i.e. intention before meaning), which is confirmed from learners by R3 to R6. An example: from the teacher's point of view, a concept is defined, because it is considered as relevant within specific course segments and it will be accordingly assigned to some search results considered as useful: learners should only follow the hyperlinked behind that concept, if there is a semantic gap, i.e. they need an explanation. And exactly at this point the problem appears, because the *momentary intention* of learners is distinct from the delivered semantic need (see e.g. R1 and R3 to R6), i.e. they expected distinct explanations behind the hyperlinks: translations, definitions, graphics, etc. Finally, R2 underlines the relevance of P1 and P2. Though, it must be said that the log-files demonstrated that the majority of test subjects preferred the viewing mode with embedded concepts (In File).

The improved version of the DBL, called Concept-based Context Modelling System (CO2), is applied within the AdeLE research project. The *CO2 modelling system* aims also at *providing accurate background knowledge* and thus, it *assists users in their teaching or learning activities*. Within the AdeLE system, in contrast to the first implementation prototype of our DBL, the concepts of CO2 are provided through a unique viewing mode: as additional navigational elements.

As depicted in Figure 5.16 on the next page, the CO2 entries are visualised under the area 'Background Knowledge' (bottom left side of the figure); there exist three entries for the current learning page, one of them visualised as 'Service-Oriented?' (not visible, the concept behind this entry is 'SOA'); for this concept, three Information Search and Retrieval Systems (ISRSs) are registered, one of them being 'Wikipedia (Castellano)'; clicking on the hyperlink leads to the additional window in the figure, for which the keyword 'SOA' was packed into the search query of the Spanish version of 'Wikipedia'.

Because AdeLE's User Modelling System (AUMS) is implemented as a multi-purpose modelling system (i.e. **"yo?"**), it builds the technological basis for CO2. From this point of view and according to **"yo?"**-terminology, both systems are *siblings*. Thus, their service-oriented implementation ensures modularity, flexibility, extensibility and reusability. The main differences between them are identifiable from the logical point of view: (a) user profiles in AUMS are called contexts in CO2, and (b) the main data structure in CO2 is a Concept Molecule, whereas in AUMS mainly stereotypes. Thus, where AUMS is used to model users profiles (in essence a set of attributes for different users), the CO2 system is used to model information spaces (in essence a set of concepts for different contexts). The major improvements in the context of this section are explained as follows.

The first DBL version delivered each concept as a hyperlink to the ISRS. There was only one global ISRS and the system connected each concept to one search query. As already stated at the beginning of this chapter and following the results from the evaluation of the first version of the DBL, *distinct learners expected distinct results 'behind' the concepts* (e.g. translations, explanations, images, etc.); it is highly relevant to mention at this point that this problem cannot be accurately predicted and thus, can hardly be solved through adaptivity. To overcome this problem, in CO2 a teacher may register several ISRSs, and in

turn, define several specialised queries for each concept. This helps in solving the problem of identifying *distinct semantic needs*.

With distinct semantic needs, it is meant that learners' necessities to fill a knowledge gap vary, which consequently, leads to varied expectations on the results of a DBL entry. Therefore, the problem goes beyond semantic representations and enters the field of *unpredictable pragmatics*, i.e. the intention(!) of an individual learner to follow a hyperlink appears just-in-time and is dependent on the current state of the learner's mental model, and thus, is meant to be 'unpredictable'. Now, this unpredictable pragmatic problem behind the intention of following one DBL concept is solved by the CO2 delivering for each concept *a set of hyperlinks*, i.e. it breaks adaptivity and leaves the decision to the learner.

Instead of trying to adapt automatically to something unpredictable, AdeLE's adaptive engine (a) retrieves all associations for a concept from the CO2 system, (b) does not try to adapt navigational elements to an unpredictable need but (c) transfers the decision to the learner, and therefore (d) visualises the different known or predefined solution possibilities as additional context-dependent navigational elements.

Thus, according to the example shown in

Figure 5.16: Visualisation of CO2 Concepts in AdeLE.

Figure 5.16, (a) if the learner needs an explanation in Spanish, 'Wikipedia (Castellano)' will lead to the corresponding result, (b) if the learner expects results from a specific Web space (i.e. needs a smart search service that works on a white list of relevant servers), 'xFIND (smart search)' will help, or (c) if the learner needs more graphical hints related to the concept, the hyperlink 'Google (images)' will lead to accurate results from the Google Images search service.

The search queries behind a concept depend on the expertise knowledge level of the learner and are configurable. Concepts, contexts, search queries, and e.g. synonyms, etc., are configured through a Java GUI (see Figure 5.17) or by editing the corresponding XML files of the CO2 system. [Safran et al. 2006]

As shown in [García-Barrios 2006a] and [García-Barrios 2006c], the use of *Concept Molecules* within the CO2 modelling system enhances the previous version of the DBL, e.g. through the following features:

- the use of various search services is made via extensions,

- synonym- concept-molecules are used to match different appearances of concepts in the courseware (e.g. phrases vs. abbreviations) and to enable multi-lingual definitions,

- a taxonomical ordering of concepts is given through hyper- and hyponyms,

- all key=value pairs for the description of users, such as expertise_level=expert or name=xyz, are defined with the concept-molecules-elements type=name,

- assignment of concepts to course contexts is enabled by connectors (in this case, pointers), and

- a semantic description of concepts is defined through intensions.

Within the scope of the AdeLE system, both siblings CO2 and AUMS run as distinct instances of **"yo?"**, i.e. following the idea of a strict separation of duties the CO2 system should not model any user traits.

Though, in order to test the **"yo?"** system as integrated solution for CO2 and AUMS, the CO2 system (within AdeLE) models the individual expertise level of learners. This test has been made in order to gain first experiences for the utilisation of **"yo?"** within the Mistral system, as shown in the next sub-chapter.

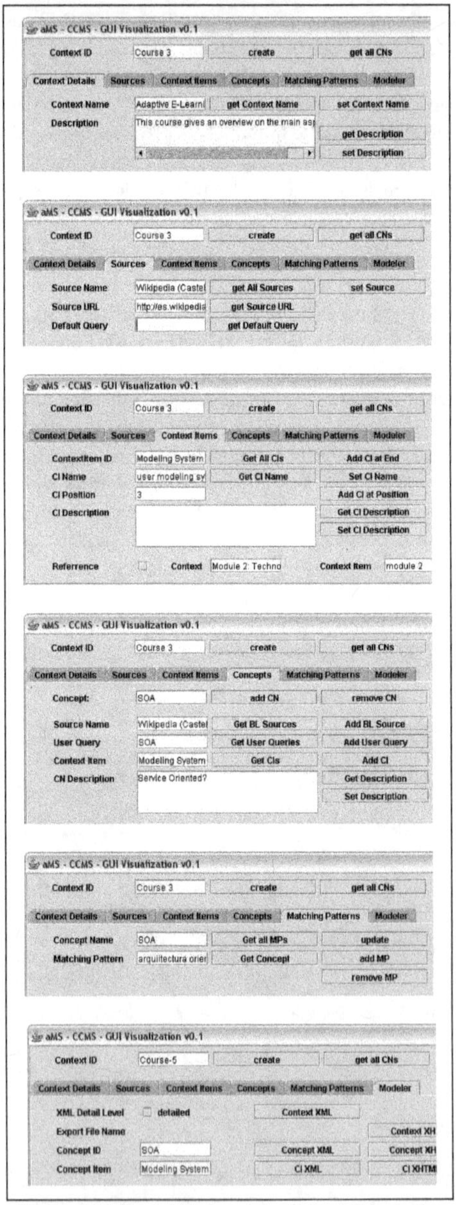

Figure 5.17: Partial Views of the distinct Tabs in the Java-based GUI for experienced CO2 Editors and Administrators in AdeLE.

5.3 "yo?" in Mistral: Facilitation of Cross-Media Semantics

The research project Mistral[38] (*Measurable intelligent and secure semantic extraction and retrieval of multimedia data*) has been briefly introduced in section 4.1.1. Accordingly, it focuses on the analysis of multi-modal data from meeting recordings as well as on its semantic enrichment, retrieval and visualisation. The Mistral project has been financed by the Austrian Research Promotion Agency[39] within the strategic objective FIT-IT (project contract nr. 809264/9338). This sub-chapter is structured as follows: General features and Requirements (Section 5.3.1), and Semantic Applications Unit (Section 5.3.2).

General descriptions of the system units can be found through the project-related publications list (see footnote 35 on this page), as for example in [Tochtermann et. al. 2005l] or [García-Barrios and Gütl 2006]. This sub-chapter extends section 4.1.1 and focuses on Mistral's Semantic Applications Unit, in which the **"yo?"** system has been integrated. Firstly, the next section focuses on the overall requirements on the Mistral system regarding personalisation needs. Based on that, the integration and role of the **"yo?"** system is described in section 5.3.2 by presenting the some results of those sample Semantic Applications that make use of it.

5.3.1 General Features and Requirements

This section summarises the descriptions given in the publications [Gütl and García-Barrios 2005b] and [Gütl and García-Barrios 2005d] regarding features of and requirements on the Semantic Applications Unit (SemAU) of the Mistral system. From the point of view on the functionalities of SemAU, the focus is set on the integration of meeting information into knowledge transfer processes in companies. Some of the requirements on the system were stipulated as follows:

- Provision of personalised support for meeting attendees and absentees to enhance the development and integration of knowledge,

e.g. user-tailored access to meeting information based on semantic annotations;

- Improvement of knowledge transfer to other members of the company, e.g. depending on position and current job tasks;

- Processing and archiving of knowledge created and transferred in meetings as integral part of the corporate memory;

- Visualisation and retrieval of multi-modal semantics;

- Linkage of knowledge captured in meetings to learning and training activities, e.g. its application in experiential-based learning.

In accordance to the duties of Mistral *units* (as introduced in section 4.1.1) and to its general architecture (depicted in Figure 4.8), relevant meeting information processing *feature groups* can be identified, e.g. (a) localisation and recognition of meeting participants, (b) speech-to-text and text processing, (c) object and sound recognition, (d) processing of sensor data streams (e.g. tracked click streams), and (e) semantic enrichment. These feature groups are explained as follows.

- Meeting Participant Localisation & Recognition: This feature group allows recognising and localising meeting participants. The uni-modal audio and video sub-units of Mistral process these features independently. The audio unit supplies the system with voice characteristic features, e.g. gender and age of participants. The video sub-unit tracks e.g. locomotion information or facial expressions.

- Speech-to-Text and Text Processing: Speech-to-text is the extraction of as much as possible textual information from oral talks based on a corresponding phoneme dictionary. The uni-modal text sub-unit of Mistral processes speech-to-text information as well as text-based meeting-related documents, e.g. agenda, project documents, etc. The outcome of these processes are called 'high-level features', i.e. additional contextual information extracted from the 'raw' data recordings, e.g. extracted concepts, summaries, text classifications, and semantic content clusters. In this context, the multi-modal merging unit tries to detect and to correct extraction errors.

---

[38] http://www.mistral-project.at
[39] http://www.ffg.at

- Object and Sound Recognition: In this context, the uni-modal sound sub-unit of Mistral delivers information about sound events, such as phone ring, laughing, clapping hands, etc. The video sub-unit performs the recognition of trained objects, e.g. mobile phone, briefcases. The multi-modal merging unit e.g. combines sound and video events to perform spatio-temporal synchronisation.

- Sensor Data Streams: Within the Mistral project, sensor data is restricted to the interactions with a presentation device; therefore, this additional modality is also called Laptop Usage. Thus, laptop usage data in combination with the aforementioned feature groups provides further useful information that can be e.g. processed by the text sub-unit or synchronised by the multi-modal merging unit.

- Semantic Enrichment: Based on the domains of meeting scenarios, the semantic enrichment unit creates further high-level features and semantic annotations. On the one hand, the semantic enrichment unit is in charge of conflict detection and resolution within the results of the aforementioned groups, e.g. a person cannot sit in the foreground and stand in the background at the same time (which could be an assumption of the video sub-unit after detecting a picture of one meeting participant on the wall). On the other hand, this unit performs inferences, e.g. the person who opens the meeting, introduces the other participants, asks many questions and closes the meeting, 'is' the 'meeting moderator'.

Mistral's *Semantic Applications Unit* (SemAU) builds on the feature groups discussed so far and on the functionality of its components (e.g. Retrieval System, Visualisation System, User Modeller, etc.; see section 4.1.1) in order to support e.g. knowledge management activities as well as learning or training on the job. The front-end to the users of the system and to other external services is called Web-based Portal component. Thus, the main requirements for SemAU can be identified from three distinct points of view, *user roles*, *information needs* and *functions*, and can be summarised as in the following paragraphs (see [Gütl and García-Barrios 2005b] for details).

Requirements on User Roles

- Personalised support for meeting attendees and absentees to enhance knowledge development, retrieval and integration.

- Improvement of knowledge transfer to other members of the company.

- Support for trainers, tutors and staff members responsible for vocational training, e.g. identifying interesting and emerging topics or recurred problems.

- Personalised access for learners to knowledge in recorded courses.

Requirements on Information Needs

- Retrieval and combination of basic meeting information, for example:
  - Organisational information: meeting data (date, place, duration), project information (project descriptions), additional meeting-related documents (agenda, description of products, organizational chart).
  - Participant information: invited persons, attendees and absentees, roles of participants.
  - Traced object and sound occurrences: spatial and temporal information.
  - Meeting topics and time-specific information: speech-to-text and related documents, topics & content abstracts, click data streams.

- Retrieval of meta-knowledge inferred from the basic meeting information, e.g.:
  - Discovering knowledge assets addressed in meetings: either following the topics in the agenda or representing 'off-subject' assets for further reuse.
  - Clustering similar topics.
  - Access to the background information, e.g. through the integration of a meeting-specific dynamic background library.
  - Provision of learning material according to given concepts as well as providing model answers and automatically generated tests in order to assess knowledge acquisition.

- Linking meeting information with related information, for example:
  - Identification of similar meeting information based on specified documents or a meeting scenario (i.e. 'search by example').
  - Linking relevant meeting topics with related learning material.
  - Storage and management of knowledge cre-

ated and transferred in meetings as integral part of the corporate memory.

- Personalised and contextualized information retrieval, for example:
  - Provide views on the meeting corpus depending on user profiles.
  - Filtering of information assets with respect to role, group membership or access rights.
  - Modelling of personal tasks, needs and traits, as well as the combination of user-related and company-related profiles.
  - Support a modelling of different concept-based information spaces.

Requirements from the Functional Viewpoint

- Retrieval of information implies the definition of functions to identify, pre-process, access and recall relevant meeting information from Mistral's Data Management Unit.

- Management of distinct sorts of modelling information, such as user and user group information, context information, tasks, etc.

- Adaptation and personalisation in order to provide user-tailored information.

- Information visualisation to provide useful services for external systems. Based on the aforementioned information needs, SemAU should provide distinct search result presentations, e.g. linear lists in table form, structure-dependent visualisations through metaphors for hierarchies, graphs, similarities.

- Access to SemAU has to be provided to users

as well as to 'internal applications', such as to a corporate knowledge management system, the intranet system, or an e-learning system.

Based on the features and requirements stated so far, the next section gives an overview on the results of the SemAU prototype.

5.3.2 Semantic Applications Unit

In accordance to the description of the Mistral system given in section 4.1.1 and to its general architecture, depicted in Figure 4.8, the following general workflow of information processing can be identified (see Figure 5.18).

Mistral's duties begin with the recording of meetings, whereby the distinct multi-modal data streams as well as further relevant meeting documents are stored in a central repository (see top left side in the figure). All data collected for one meeting is called a *meeting corpus*. The analysis and processing of these recordings takes place through Mistral's units in its Core Framework. Thereby, descriptive topic-based and interactions-related characteristics are extracted from the uni-modal video and audio sub-units, such as the identity of the speaker, the time points at which utterances have been made, the location and gender of participants, locomotion of participants, topics in the agenda document, most relevant topics addressed by speakers, etc. These extracted features represent the input for the multi-modal merging and semantic enrichment units, which will resolve possible conflicts, infer

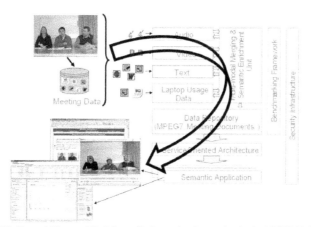

Figure 5.18: General View on Workflow of Information Processing in the MISTRAL System.

further descriptions and assumptions. The outcome of units is placed at the disposal of SemAU in a central data repository. This semantically enriched meeting corpus is the main information source for the components of SemAU.

It is worth emphasising at this point that the outcome of the Mistral system, in terms providing a front-end to its users, is defined by means of its Portal. This portal is a Web site from which users may access the results of distinct Semantic Applications (see bottom left side of Figure 4.8). Therefore, distinct views on the Mistral system are possible; they depend on the collaborative work of Mistral units as well as of the components of SemAU. The reminder of this subchapter focuses on those SemAU results that deal with the integration, retrieval and visualisation of meeting information. As stated in the previous section, the main requirements for SemAU are divided in three categories: user roles, information needs and system functionality. The introduction of these points of view is relevant to the context of this book, because (as just stated) SemAU is not meant to represent a single application. Rather, it embraces several components that provide services, which can be orchestrated in order to fulfil different requirements.

With respect to the design and development of the "yo?" system, SemAU and the overall Mistral system also aim at providing a modular, flexible and extensible architecture in order to enable the orchestration of distinct applications. For that purpose, a service-oriented architecture was chosen. Among others, three software applications of SemAU resulted from development work: *Semantic Meeting Information Application* (*SMIA*) and *Explorative Visualisations Framework* (*EVF*; to be more precise, a *Multi-Dimensional Metadata Visualisation System* has been implemented, MD²VS) and *Personalised Concept-based Information Access* (PCIA). The **"yo?"** system has been integrated into the Mistral system in order to provide personalisation functions for these three sample semantic applications. First experiments have been conducted within this context by integrating both already presented siblings of **"yo?"**: the user modelling and CO2 systems (see Figure 4.8).

Semantic Meeting Information Application

A detailed description of SMIA can be found e.g. in [Gütl and García-Barrios 2005b] or [Gütl and García-Barrios 2006]. The main functionalities of SMIA are contained within SemAU's Retrieval System, for which the open source search system xFIND represents its core module (see [xFIND 2007] for details). The screenshot of SMIA depicted in Figure 5.19 shows one search result for a *SMIA Simple Search* (SSS). Also, Real Player's streaming output is shown (after click-

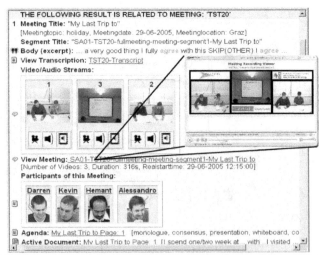

Figure 5.19: Web-based User Interface of SMIA, a Sample Application of the MISTRAL System.

ing on the corresponding hyperlinked elements).

Unlike common search engines, where one result corresponds to one document, each result of SSS visualises the most relevant information about an entire meeting corpus regarding the submitted search keywords. Thus, SSS is not meant to find single documents or modalities, rather it matches search queries in order to find relevant meetings. To identify the document where the keywords were found, SMIA highlights them in excerpts of indexed content (see *agree* in *Body (excerpt)* on Figure 4.8, left side).

In addition, visual information about meeting recordings is provided through thumbnails. Hyperlinked titles and icons (for audio, video or audio&video) enable users to access these specific video recordings via SemAU's *Streaming Server* (which forms part of the Retrieval System). This enhanced linear visualisation of search results is made possible, because of the MPEG-7 descriptions from MCF. Thus, an MPEG-7 file represents the semantic envelope for each meeting, enabling an easy and compact exploration of the semantically enriched multi-modal data.

Further, SMIA's *Extended Search* user interface (UI) supports users to specify special needs by simply typing keywords and selecting options in a search form. Yet, as the usage of extended features in search engines is hard to learn for novice users, SMIA places two special interfaces at the disposal of users: *Special Search: Speaker* (see Figure 5.20, left side) and *Special Search: Agenda*. These two special UIs are examples for special instances of SMIA's Extended Search, where specific time segments of meeting recordings can be retrieved, depending on the given *granularity of meetings*. These granularities of meetings are explained as follows: meeting recordings can be explored from distinct perspectives, which describe distinct divisions of the meeting time duration (see Figure 5.20, right side). This means that one meeting may be not only segmented in periodic time periods (e.g. 5 minutes, 35 seconds, etc.), rather it can be divided in segments for which their duration depend on the output of multi-modal analysis, i.e. on event-driven MPEG-7 annotations, as for example the following granularities:

- Speaker
  Visible or audible modalities (i.e. the output of Mistral video and audio units) provide non-periodic segments that may define those time periods at which meeting participants give audible utterances (audio) or are visible. This is relevant because it is assumed that Mistral users are not interested in revising e.g. the whole video&audio recording of a meeting looking for somebody that have addressed some relevant topic, rather they need just those short segments of the recordings where their 'search keywords' match.

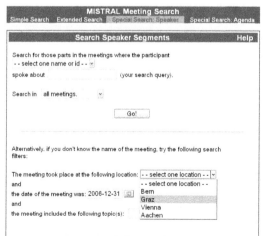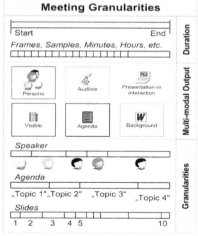

Figure 5.20: Speaker Search UI in SMIA (left side) and Meeting Granularities (right side).

- Agenda topics

  The agenda topics are delivered by the text unit of Mistral (extracted from text documents). As the aforementioned speaker segments, agenda topics represent also a possible division for the duration of a meeting. Thus, SMIA's 'Special Search: Agenda' provides users a possibility to retrieve relevant information within each time period during which an agenda topic has been addressed.

- Presentation slides

  Mistral's uni-modal Laptop Usage sub-unit provides a division of meeting duration with respect to click-to-click time segments. Therefore, this granularity might e.g. help users finding a 'topic X that had been addressed by speaker Y while the slide Z was displayed'.

Explorative Visualisations Framework

The *Explorative Visualisations Framework* (EVF) of SemAU aims at placing distinct visualisation possibilities at the disposal of users to extend the UI-possibilities of SMIA and to enable multi-dimensional perspectives on the semantic spaces of meeting corpora. In EVF, the core module is called *Multi-Dimensional Metadata Visualisation System* (MD²VS) and is a Java Applet at the client-side of Mistral's system (see Figure 5.21).

MD²VS is a smart framework that is able to control and show different visualisations, such as a multi-dimensional explorer or a data clustering tool (see [Mader 2007] for details). The dimensions in MD²VS correspond to set of metadata attributes in search results of SMIA. In order to make metadata values comparable, the dimensions in MD²VS have to be normalised to proper representations.

These normalised representations are of distinct types: (a) *numbers* and *dates*, whereby a comparison is implicitly possible by the numeric distance between the different values, or (b) *enumerators*, representing equidistant textual values between 0 and 1, e.g. for the dimension

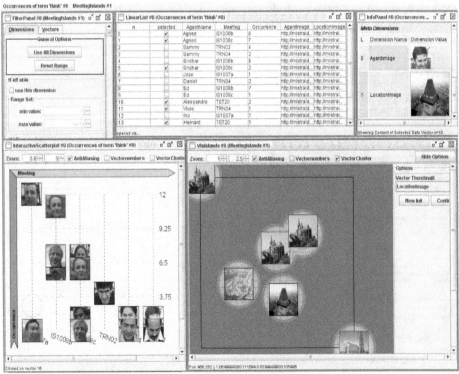

Figure 5.21: Multi-Dimensional Metadata Visualisation System in Mistral.

'*Speaker*' the metadata set ['*Angel*', '*Charly*', '*Eros*', '*Fido*'] is normalised to [(0,'*Angel*'), (0.5,'*Charly*'), (0.75,'*Eros*'), (1,'*Fido*')]. Thus, the data input for MD²VS is a set of search results with comparable metadata values.

The first prototype implementation of MD²VS supports the following tools (see Figure 5.21): *Filter Panel* (upper left box), *Linear List* (upper middle box), *Info Panel* (upper right box), *Interactive Scatterplot* (bottom left box), and *VisIslands* (bottom right box). With MD²VS, users may explore several semantic coherences among metadata defined in specific search queries, e.g. users may find the *hot topics* addressed in several meetings by identifying concepts that increasingly co-occur in meetings (vía VisIslands), or a user may also infer from the Interactive Scatterplot which meeting participants with a certain role have given most utterances related to specific topics of presentation slides. This can be useful to identify topic-dependent dialogs and discussions.

One innovative value of MD²VS is given by its integrative implementation of distinct visualisations for metadata in multi-modal corpora. Another relevant added-value of MD²VS relies on its synchronised update of data within the distinct visible visualisations, i.e. user interactions or modifications in one visualisation propagate simultaneously to the others. On the other side, a critical aspect of this system is its usability, as novice users need time to get familiarised with this types of visualisations, whereby the problem is not the visualisation itself, rather how to interpret the visualised data.

The role of the „yo?" system within SMIA and EVF is restricted to a user profiling system in which the distinct data about meeting participants is stored and from which it can be retrieved. For each participant, data is stored in specific profiles, e.g. *Business Card, Company Role, Meeting Attendance* or *Addressed Topics*. There profiles are mainly used by the Retrieval System of SemAU in order to filter search results or visualise their content.

As the overall Mistral architecture is based on the service-oriented approach, the integration of „yo?" in Mistral as user profiling system implied the implementation of the interface of its Manager service in SMIA. Further, this integration of the modelling system into Mistral's SemAU was possible, because both the „yo?" system and SemAU are Java-based implementations.

Personalised Concept-based Information Access

As exposed so far, the role of the **"yo?"** modelling system in the Mistral project has an experimental character and represented an attempt to test its applicability in other applications areas outside the scope of adaptive e-learning. Its integration as user profiling system for the previously introduced sample semantic applications (SMIA and EVF) could be performed seamlessly.

Thus, the idea appeared to incorporate also its sibling, the CO2 system, into the context of Mistral in order to place background information from meeting corpora at the disposal of company workers e.g. during their learning-on-the-job or work-flow activities. Note that in this case, the *information consumer system* (ICS) of **"yo?"** is not an adaptive engine as in AdeLE, rather it is represented by the number of possible information systems outside the Mistral environment that implement the service interface of **"yo"**, e.g. a content management system (CMS), workflow system (WFS), knowledge management system (KMS), learning management system (LMS).

To give an example, the focus is brought back to e-learning in terms of learning-on-the-job. According to [Rosenberg 2001], e-learning includes three different focuses: *training, knowledge management* and *performance support*. Within the scope of this section, the last issue is underlined. Rosenberg argues that performance support may, indirectly, assist humans to enhance their work by means of speed, efficiency and cost reduction. It is provided in different forms, such as books, mentoring, software tools, or checklists. The integrative usage of software solutions for this purpose combines hypermedia systems, expert systems, adaptive systems, real-time support systems and help of real experts.

For the software application presented in the ensuing paragraphs, the focus is set on context-based solutions. A critical aspect of personalisation systems within the scope of learning-on-the-job activities is given by the fact that these activities may change continuously over time. Therefore, the modelling system(s) in such environments must be continuously aware of situational changes and require advanced and long-term tracking mechanisms to ensure or boost the confidence of inferences. Regarding the aforementioned issues, it is relevant to state at this point that Mistral's SemAU should not be seen solely as information retrieval system, e-learning system or knowledge management system. Rather, its innovative value relies on the integrative character of its components (including the **"yo?"** system), and therefore represents a set of goal-oriented and configurable tools (software modules) which are able to support company member in their learning and working activities.

As shown within the context of SMIA and EVF, in particular, the services provided by the components of SemAU enable 'knowledge retrieval and visualisation of semantically enriched multi-modal data in meeting corpora'. The application introduced in the ensuing paragraphs is based on the descriptions given in [Gütl and Safran 2006], in which the utilisation of **"yo?"** as User Modelling System (UMS) and CO2 system is showed through an example of the Mistral system interacting with ICSs. With respect to the Mistral system, this application represents a further sample semantic application within the scope of SeMAU and is referred to as *Personalised Concept-based Information Access* (PCIA).

According to [Gütl and Safran 2006], the solution approach and first prototype of PCIA concentrates on an integrated access to relevant meeting information in an employee's working environment. The corresponding architectural design for this solution is shown in Figure 5.22. The figure shows both siblings of **"yo?"** (see 'User Modeling System' and 'Concept Modeling System') outside the Mistral system in order to separate their collaborative duty regarding the provision of personalised access to context-relevant meeting information from the general Mistral-specific duty of delivering semantically enriched meeting recordings.

Within PCIA, an 'Information System' plays the central role regarding interactivity with a user, i.e. it represents one possible instance of ICSs, as the examples given previously (WFS, KMS, LMS). This central container of ICSs is an abstraction in order to show how flexible the applicability of service-oriented modelling systems can be. The information about individual users in the User Modeling System of PCIA (i.e. **"yo?"** as user modeller) is managed in profiles and models, e.g. user record data, working field expertise, user roles within projects, data usage behaviour. On the one side, all information about users can be accessed directly by an Information System through direct interaction. On the other hand, certain type of information (e.g. profiles or subprofiles) is used to personalise the output of the Concept Modeling System (i.e. **"yo?"** as CO2).

Regarding the application of CO2 in PCIA, it allows a semantic and personalised access to meeting recording information that is modelled internally by means of context items, linked concepts and proper query templates. Thus, in essence, the CO2 provides to PCIA the similar functions as to the AdeLE system, but for more general purposes. This *multi-purpose capability* of the CO2 system is given because its functionality follows the main architecture of **"yo?"**, where a strict separation is granted between information processing and provision, i.e. the components *Profiler and Modeller* manage and compose CO2 data, whereas *Tools* provides distinct exchange formats and *Views* distinct administration GUIs (see architecture figures in section 4.2.3 and CO2 explanations in previous sub-chapter).

These aspects lead to one of the most remarkable advantages of the **"yo?"** modelling system as CO2, that is, external systems may request data in distinct formats according to their own needs regarding data processing and visualisa-

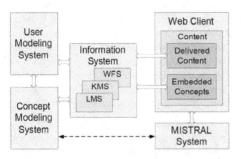

Figure 5.22: Overall Architecture for Personalised Access to Multi-modal Meeting Recordings, according to [Gütl and Safran 2006].

tion. The Adaptive System of AdeLE as well as the Information System in PCIA may visualise CO_2 data as they want. Therefore, the UI output of the Information System in PCIA, shown in Figure 5.23, looks different as the output of the Adaptive System in AdeLE (delivered to the user client through its LMS), shown in Figure 5.6, Figure 5.11, Figure 5.14 or Figure 5.16. In general, a trusted ICS may get personalised information from the CO_2 by combining user data with context and concept information in each request. In addition, each request may contain information about the desired representation format of the response. Thus, CO_2 results can be either delivered e.g. as serialised Java objects (for Java clients), in XML (e.g. for Web Service-based applications) or XHTML (for systems with own Web-based rendering engines). This third technique is the one utilised by AdeLE's Adaptive System and the ICSs of PCIA.

Regarding the first implementation prototype of PCIA, the practical context of the snapshots shown in Figure 5.23 can be described through the following example [Gütl and Safran 2006]:

- The new project manager in the company, Claudia, is in the planning stage of a project and needs information about the particular technology topic 'speech recognition'. Ac-

cording to her current task, assigned and accessible vía a WFS, she has to 'get information about most relevant tools as well as identify advantages and disadvantages of the tools'. The company's WFS is connected to an LMS that contains general information about that topic (see main content frame of the front window illustrated in Figure 5.23). Before the LMS delivers the Web page, it firstly interacts with WFS and UMS, and then requests XHTML-based output from CO_2 for the user 'Claudia', context 'project planning', context item 'methods and problems' and representation 'XHTML'. The LMS attaches the CO_2 response on the navigation frame as shown in Figure 5.23 (see navigation area 'Your Background Knowledge'). In order to know who has addressed the topic at hand in previous meetings, Claudia may follow the CO_2 hyperlink 'find speaker segments', which leads to those search results of Mistral's Retrieval System that are partially shown in the left window of Figure 5.23. Thus, the search query corresponding to these results has been contextualised by CO_2 and personalised through the information in the UMS. Further, the navigation area 'Your Background Knowledge' shows all CO_2 entries valid for the current user, topic and context, e.g. in as-

Figure 5.23: Personalised Concept-based Information Access (PCIA). [Gütl and Safran 2006]

sociation with meeting attendance of absence, meeting participants that are experts in the topic, etc.

According to [Gütl and Safran 2006], first experiences with this approach have shown that the CO2 approach

"can hide complexity of the Mistral Retrieval System and enables novice users to perform powerful information requests on a semantic level";

further, its configurability regarding search services allows predefining

"queries, which can be initialized and rewritten easily and efficiently for each concept";

and the combination of UMS with CO2 seems to be a promising approach for efficient

"personalized access to background knowledge".

This sub-chapter has presented three experiments in which the **"yo?"** modelling system has been integrated into the Mistral system. The main goals of the Mistral research project are defined within the scope of 'semantic-based cross-media exploration and retrieval employing concepts at different semantic levels'. Thus, on the one side, the *technical key issues* addressed in Mistral comprise essentially the extraction, merging and enrichment of semantic information from cross-modal data. These duties are addressed by the units in Mistral's Core Framework (MCF, see e.g. section 4.1.1). On the other side, the *practical key issues* are connected to the (distinct) possible needs of users, and therefore, the Semantic Applications Unit (SemAU) had to be designed in a flexible way as to provide distinct examples for the efficient use of the results of the units in MCF. This sub-chapter has introduced some *sample solutions of SemAU*, which concentrate primarily on the user-oriented topics of exploration, retrieval and visualisation. Within SemAU, the **"yo?"** modelling system represents an additional service that shows how to integrate personalisation features into the Mistral system.

The next section concludes this chapter summarising relevant findings and conclusions regarding the application of the **"yo?"** modelling system within the projects described so far.

5.4 Summary

This chapter has presented the application of the **"yo?"** modelling system into three research projects: AdeLE, CO2 and Mistral. The integration of the system into these projects has shown several benefits. Though, also some expected and unexpected open issues have been clearly presented. Accordingly, this section summarises the most relevant findings.

The system's capability of *supporting multi-purposefulness has been demonstrated in a practical context* by fulfilling the distinct personalisation goals of these projects within distinct application areas, i.e. adaptive e-learning (AdeLE), knowledge exploration (CO2) and cross-media retrieval (Mistral).

The original goals and development issues of the system have been defined within the AdeLE research project. Therefore, its functional scope was limited at the beginning of its design process to the application area of *learner modelling*. Within the service-oriented environment of the AdeLE system, this multi-purpose profiling and modelling system successfully assumes its *main role as central collector, manager and provider of individual learner information*.

The strict *separation of processing and reasoning duties* (Profiler and Modeller services) as well as the *provision of a central communication and dispatching service* (Manager service) have allowed to efficiently meet the specific requirements of distinct components of the AdeLE system (e.g. analyse gaze movements from eye-tracker, provide individual user information to the adaptive system, infer just-in-time learning performance from visual reading/learning behaviour, enhance the functionality of the dynamic background library).

AdeLE's highly relevant *Behaviour Tracking technique* (i.e. accurate modelling and adaptively responding to gaze-tracking changes) has been solved through a *simplification, distribution and synchronisation of specialised tasks within the system* (e.g. among the modules Logger, State Modeller and Behaviour Tracker).

Riding's WAVI model for cognitive styles has been integrated into the system as observable system of coordinates for individual learners. Though, focusing on the corresponding duties of the user modelling system and from the technical point of view, the solution is not a big challenge e.g. because there exist psychological tests that can be utilised to initialise and store the individual values of the WAVI model. Further, regarding the adaptive behaviour of the overall e-learning system and from the practical point of view, the efficiency and benefit of the utilisation of such a model depends on the collaborative work between user modelling system and adaptive engine; yet, within the AdeLE research project this issue has not been evaluated and represents thus an open issue.

The utilisation of the system in distinct projects required the design and development of application-independent, reusable and flexible components (as shown in the previous chapter). For that purpose, the service-oriented implementation of the "yo?" system has proven its *value regarding technical reusability*. In this context, reusing profile structures and adapting the behaviour tracking technique for the Mistral purposes (i.e. personalised information retrieval) represented just a simple task of reconfiguration. In addition, the *conceptual design to support multi-modelling* in Mistral (PCIA semantic application described in the previous section) shows e.g. the theoretical applicability of the "yo?" system as task or workflow modelling system (see also [García-Barrios 2006d] and [García-Barrios and Gütl 2006]).

The decision to deploy the "yo?" modelling system into the service-oriented Openwings *runtime framework provided the means to meet the security and flexibility requirements* of the projects at hand.

The evaluations regarding performance (see section 4.3.3 and 4.3.4 of the previous chapter) led to an *efficient optimisation of the system by means of service granularity, caching and openness of connectors*.

From the point of view on 'low-level' technical aspects, the utilisation of the *abstracted persis-tence layer* (Torque) as well as the *flexibility of the database schema* (see e.g. Figure 5.5) have also contributed to prove the system's value as multi-purpose solution. No changes had to be made for the application within the three projects.

The consideration of *scrutability* within the design and development of the system (i.e. the strict separation of processing and visualisation duties) has shown *useful practical advantages in terms of standardised information interchange* and *highly flexible visualisation of model contents*. The "yo" system operates on internal data structures that can be expressed easily in XML format. Two concrete examples underline this achievement: both "yo?" siblings in AdeLE, the CO2 and User Modelling (UM) systems, provide human-readable information in XHTML notation for free visualisation possibilities through CSS layout definitions. The CO2 system provides an XHTML snippet describing concept-based structures. This CO2 snippet is delivered vía AdeLE's Adaptive System to the Learning Management System, which can thus apply any CSS layout to the snippet; the result is represented by the area 'Background Knowledge' of the navigation frame of AdeLE's Web UI, as shown in Figure 5.6. In analogy, the UM system provides an XHTML snippet describing relevant user model contents; the result is represented by the area 'Why this way?' in Figure 5.6. Thus, the *delivery of information in XHTML notation* has shown *remarkable advantages regarding the support of open standards and usability*, i.e. this simple mechanism gives Web-oriented information consumers the possibility to visualise the information as desired.

In sum, the service-oriented multi-purpose profiling and modelling system "yo?" showed several advantages of its design and implementation as well as demonstrated (within the scope of the given projects) that it can be easily reused for distinct application areas.

6 Summary and Outlook

The outcome of any serious research can only be
to make two questions grow where only one grew before.

(Thorstein Veblen)

This chapter aims at summarising most relevant conclusions and findings on the research presented in this book. Firstly, a brief summery concerning the topics addressed along the structure of the book is given. Thereafter, a final review on reached vs. expected research goals is presented. Then, the proposed and implemented solution is compared with existing related works. This chapter finalises with a discussion of open issues as well as giving an outlook for future work.

6.1 Summary

This book has presented a solution approach for a service-oriented multi-purpose profiling and modelling system, called **"yo?"**.

The research goals stipulated in chapter 1 are discussed in the subsequent sub-chapter. Chapter 2 provided an investigation on the main general issues regarding the concepts of adaptation, of adaptive systems as well as of adaptive e-learning. The chapter gave a general view on those topics, whereby terminological aspects were also treated.

In concrete, it was shown that the concept of adaptive e-learning is difficult to define clearly. This difficulty has several causes, e.g. the diversity of applicable didactical paradigms, the increasing amount of specialised computer-based tools, or the big number of perpetual beta solutions, but also the influences of and implications on other fields, such as politics, psychology, and social behaviour. Therefore, distinct denotations and connotations of the term can be identified for distinct theoretical and practical scenarios. In particular, to reach a common understanding of the term Adaptation is a hard task, as cited in [Reeve and Sherman 1993]:

"The difficulty of the concept adaptation is best documented by the incessant efforts of authors to analyse it, describe it and define it." (E. Mayr, 1983)

In accordance to these observations and as also shown in chapter 3 of this book, the variety of existing and still emerging models that explain and define adaptive behaviour reflects the complexity of the topic.

These models were analysed based on two points of view. Firstly, a point of view from outside the user modelling system helped to identify which are the most relevant services placed at the disposal of the main 'clients' (i.e. information providers and information consumers) of the modelling system. This point of view allowed focusing on the aspect of information interchange, which in turn led to propose a central communication manager in **"yo?"** as single point of access. Secondly, a point of view on the internals of the modelling system was required to focus on an internal separation of duties, which decouples the tasks of the system from the communication level, i.e. the focus was set on the issues regarding the processing of the modelled information.

The overall design and development of the proposed system was presented in chapter 4. Internally, five distinct service-oriented components have been proposed: two of them (Profiler and Modeller) are in charge of information processing; one enables a transparent access to heterogeneous storage environments (Data Persistence); another is responsible for visualising information (Views); and a last one is an extensible container for additional functions (Tools).

The resulting modelling system ("yo?") and its underlying service-oriented framework (Openwings) make it possible to create various instances of the component services for distinct purposes as well as to move them across platforms (also at run-time). These characteristics of the implemented solution allowed easily integrating the "yo?" system and reusing it in the projects *AdeLE, CO2* and *Mistral* (see chapter 5).

Due to the fact that the implementation of the system is the result of more than 20 persons programming during several months, a description of all technical details would run out of the scope of this book. Therefore, considering that the two most relevant characteristics of the "yo?" system are its underlying *service-oriented* approach and its *multi-purpose* capabilities, the forth chapter has focused on the former aspect and the fifth chapter on the latter one.

Hereby, the majority of focused issues were chosen to meet the main contributions of the author of this book, e.g.:

- the overall architectural and functional design of the system,

- the optimisation on service granularity and openness of interfaces,

- the stipulation of clear separated duties (regarding e.g. profiling vs. modelling, processing vs. visualising information; providing vs. composing services),

- the utilised solution for real-time behaviour tracking based on gaze movements,

- the notion of concept molecules to enhance the modelling of information and knowledge structures,

- the realisation of multi-purposefulness for modelling components within adaptive systems.

6.2   Research Results

The research in this book has concentrated on the issue of designing and implementing a service-oriented multi-purpose profiling and modelling

system that should meet the requirements of distinct domain and application areas. Thereby, special focus has been set on general and specific research goals, which are discussed below.

6.2.1   General Research Goals

How to reach a high level of abstraction in the design and implementation of a modelling system that ensures multi-purposefulness

On the one hand, a comprehensive theoretical research has covered terminological and functional distinctions. This theoretical research has been presented in the chapters 2 and 3, whereby the investigation has covered most relevant topics from the general concept of *Adaptation*, through the roots of *Adaptive E-Learning*, to state-of-the-art solutions in the context of *User Modelling* (including principles, techniques and systems).

Based on this research, a high-level of abstraction could be reached through the identified *main global tasks* and *main components* concerning application- and domain-independent user modelling systems. In accordance to the required service-oriented solution approach, these issues enabled the evaluation and optimisation of service granularity and composition. The resulting architecture is shown in section 4.2.

How to identify and establish a clear separation of duties between a modelling component and the other parts of an adaptive system

Regarding the separation of duties between modelling tasks and the other tasks of adjacent components in an overall adaptive system, a *general architecture for adaptation-pertinent systems* (GAAS) has been presented in section 4.1.2. This general architecture explains not only which are the main components in an adaptive system, it discusses further their main roles and tasks.

Thus, it clarifies not only the specific duties of a modelling components, it comprises an explanation of the duties of the interacting components. From this point of view and as result of the theoretical investigation, global tasks as well as main components in the user modelling system have been derived and implemented.

In addition, from a point of view on the internal tasks of the proposed solution, the service-oriented architecture of the modelling system enables also an *internal separation of profiling and modelling duties*. This is an essential requirement on multi-purpose user modelling systems in order to clearly distinguish between the processing of 'raw data structures' and 'inferred information', i.e. the profiling component operates on fixed data structures, whereas the modelling component has the freedom to create new structures as consequence of its knowledge reasoning methods.

How to find a flexible and reusable solution for a modelling system that ensures its domain- and application-independence

The chosen service-oriented solution approach as well as the proposed architectural framework has enabled distinct levels of transparency.

On the technical level, for example the hierarchical profiles and individual user attributes of **"yo?"** as learner modelling system in AdeLE could be successfully reused in the CO2 system to represent hierarchical information spaces and their specific concept-based descriptions.

On the abstract level, section 4.3.6 shows how to deploy two distinct techniques into the system reutilising its service granularity, i.e. the traditional user modelling technique of Stereotyping and the self-developed technique of Behaviour Tracking could be deployed along the Human-Machine Transparency development dimension.

This is explained as follows. In order to ensure distinct ways of serving to user information consumers (humans as well as software systems), the architecture of the **"yo?"** modelling system establishes two main dimensions of development (i.e. two main levels of transparency, see section 4.3.6):

(a) The *Information Flow* dimension represents a single and abstract point of access for distinct applications, i.e. a central and generalised interface is placed at the disposal of external systems. Thus, external systems need solely to be aware of 'what' to CRUD (create, retrieve, update and delete) in/from the modelling system, but need no

knowledge about 'how' these operations are processed.

(b) The *Human-Machine Transparency* dimension is a general deployment rule that separates the operation on data and information structures from the possibilities of exchange and scrutability, i.e. it follows strictly the rule of separating content (data values) from presentation (data visualisations).

6.2.2 Specific Research Questions

How is it possible to support gaze-tracking in real-time in order to enhance the learning performance of users of the AdeLE system

Within the research scope of the AdeLE project, a so-called 'real-time *Behaviour Tracking* technique' has been created and developed in order to enclose the whole cycle of adaptation regarding the topic of personalisation based on gaze movements. This gaze-based technique takes advantage of a differentiation between 'reading' and 'learning' text passages. Thus, independent processing steps have been distributed among the components of the AdeLE system.

As far as the user modelling system is concerned, an integrative and efficient technique has been created and implemented: at the profiling side a *Logger* has been deployed, and at the modelling side a *Behaviour Tracker* and a *State Modeller* have been integrated into the Behaviour Tracking module. This technique has been presented and discussed in the sections 4.3.7 and 5.1.3.

Moreover, the multi-purpose capability of this technique is given by its open interface, the simple logging protocol and the flexibility of configuring the State Modeller. Hereby, the technique makes it feasible to support real-time tracking of individual user behaviour from distinct real-time sensors for distinct purposes (e.g. sensors for environmental light intensity at client side, GPS-based user locomotion trackers, or body temperature sensors). In practical terms, either (a) these devices implement the BT-protocol of **"yo?"** or (b) they send their raw data to the modelling system whereby this data must be interpreted in the Formatter module of the Tools component of **"yo?"**.

How can knowledge of users be well organised for or in a user modelling system

Based on the notion of Concept Modelling [Gärdenfors 2001], the self-developed *Concept Molecules* have been presented in section 4.3.8. The resulting data structure has been utilised within the Concept-based Context (CO2) modelling system (presented in the sections 4.1.1 and 5.2).

The result is a flexible, internal graph-based data structure that allows not only creating semantic relationships between concepts, it also enables the definition of pragmatic contexts; i.e. it is possible to freely define distinct meanings for concepts and to classify them for their application in distinct situations.

How can a modelling system contribute to improve knowledge acquisition in an adaptive system and which implications does it have on the adaptive behaviour of the overall system

Because of the flexibility provided by the Concept-based Context (CO2) modelling system (presented in section 5.2), distinct views on Web-based information spaces can be reached for distinct learner needs.

The results of an evaluation of the previous version of the CO2 system (the so-called Dynamic Background Library, DBL) showed that fully adapted hyperlinks were not the right solution. In essence, the overall goal of a DBL is to show a learner the 'best-suitable' hyperlink that leads to most relevant explanations of a certain topic. For that purpose (in the first version of the DBL), this hyperlink points at 'one' specific search engine.

The evaluation showed that several learners were disappointed because the hyperlink 'was not pointing to the right search engine': where some learners expected a textual explanation, others needed a translation, others needed pictures, and so on. Hence, what triggered the development of the CO2 system was this feedback of learners.

In summary, the evaluation on the usefulness of a DBL has proven that it enhances knowledge acquisition. Though, the most remarkable aspect within the context of this research goal is the fact that the implementation of the DBL works best if

stopping adaptation. Best results and acceptance are achieved if many distinct search engines are provided for each hyperlinked concept, i.e. the whole set of retrieval possibilities should be presented to learners thus that they are free to choose which 'lack of knowledge' should be filled by which type of explanation.

How can a modelling system support company members in their working tasks

The **"yo?"** system has been integrated experimentally into the Semantic Applications Unit of the Mistral system. The idea was to inspect if the modelling system would show difficulties when interacting with systems distinct from an adaptive engine for e-learning purposes. The integration could be conducted successfully because of the service-oriented nature of the solution.

Though, it is worth stating at this point that the same functions in the user modelling system were required and sufficient to meet the needs of both, an adaptive engine in the context of e-learning tasks and a workflow engine in the context of working tasks. The individual search data streams of users of Mistral's retrieval engine could be managed in the same way as the gaze-tracker data streams of the AdeLE system.

Hence, a user modelling system can be utilised at work-place (a) to enhance information retrieval needs by providing user information to personalise search results in terms of e.g. preferred media type or individual topics of interest, as well as (b) to provide additional personalised information according to workflow options, e.g. modelling individual task-paths, levels of expertise for problem-solving situations, or working contexts.

6.3 Related Work and Repositioning

This section presents some relevant aspects regarding the presented solution approach in comparison with existing related work. At this point it is highly relevant to emphasise that this comparison to related work is not an empirical evaluation of the overall practical functions of the **"yo?"** modelling system with the functions of the selected solution approaches and systems.

The aim is to identify most relevant similarities, and based on them, clarify those differences considered to be significant to the scope of this research work. It is hoped so to reach a clearer (re-)positioning of the **"yo?"** system regarding its innovative value compared with the contributions of other systems or approaches.

As the **DOPPELGÄNGER** system [Orwant 1996], the **"yo?"** system is also able to detect patterns from user actions, whereby the **"yo?"** system is open to adopt distinct machine learning methods for distinct purposes. As stated in [Orwant 1996], the DOPPELGÄNGER system includes 'cheap' reasoning techniques, i.e. simple and quick solutions for complex problems (see section 3.2.3). From this point of view, the simple but highly efficient Behaviour Tracking technique of the **"yo?"** system is an additional contribution to Orwant's cheap techniques.

Further, the notion of communities in the **"yo?"** system is also based on the idea of 'being member or not'. In DOPPELGÄNGER, this idea is implemented in a generic way and based on community 'voting' or degrees of membership. In **"yo?"** it follows the idea of 'fellow adoption and propagation', i.e. attributes in individual models or group models are given for granted and thus, depending on the task at hand, user information can be 'adopted' or 'propagated' (see section 4.2.1).

Personis is a user modelling server based on the idea of scrutiny interfaces that can be distributed to adaptive hypertext applications in order to provide distinct views on the stored user information (see e.g. [Kay 2002]). Because the **"yo?"** architecture and implementation includes a clear separation of usability and scrutabilty goals, its scrutable functions are similar to Personis. At this point it must be stated that scrutability is considered within the context of this book as a 'must-have feature' in any solution of user modelling systems.

Nonetheless, Judy Kay's solutions in the field of user modelling systems (e.g. **um** toolkit [Kay 1990]) comprise specialised libraries that can be deployed and combined depending on the task

at hand. Also Kay's newer theoretical user modelling framework **PLUS** can be seen as a library of specialised tools [Kay 2006]. Unfortunately, the available publications do not allow a deeper view on the granularity of these tools or on their technical details. As mentioned in section 4.3.3 of this book, the first version of the **"yo?"** modelling system contained a big number of highly specialised and reusable atomic functionalities; though, as stipulated in sub-chapter 4.4, the granularity of encapsulated functionalities (i.e. services) is a very critical design aspect.

The **"yo?"** system ensures a high availability, performance, flexibility and reusability of its parts through a balance of 'many' encapsulated specialised modules (for complex and rapid processing of internal data) and 'few' main service-oriented components (for simple access and composition at the level of data interchange).

As AdeLE within the context of service-oriented solutions for adaptive systems, the service-oriented system **APELS** [Conlan et al. 2003] also comprises an isolated user modelling component. Though, it (a) represents a generalised solution for adaptive educational hypermedia systems, (b) the system follows a strict metadata-driven approach, and therefore (c) a user model in APELS is abstracted and reduced to a repository of metadata-based user descriptions.

Further, **CUMULATE** and **I-HELP** are also learner modelling systems. In contrast to the **"yo?"** system, CUMULATE and I-HELP are agent-based solutions. Regarding the topic of agent-based solution approaches, as shown in section 4.2.3, the design of the **"yo?"** system also supports proactive collection of user data (and in turn, partitioning of models) through its Collector Factory module in the Tools component and the notion of system's siblings.

In sum, the three aforementioned systems are specialised solutions for the specific field of e-learning (i.e. application and domain dependent). And consequently, the most remarkable distinction of **"yo?"** is its multi-purposefulness.

Another system which focuses on the distribution of individual user models is given by [Heckmann 2005], **u2m.org**, which concentrates on ubiquitous computing and semantic technologies. In general, Heckmann's work focuses on the establishment of standardised sharing of distributed user information (UserML and GUMO).

The **"yo?"** system does not work explicitly with such standards at the level of message interchange. Nonetheless, **"yo?"** provides with its Formatter module in the Tools components a possibility to deploy functions for marshalling/unmarshalling of standardised messages.

As the **"yo?"** system, **LDAP-ums** is an application-independent user modelling system. In concrete, LDAP-ums is a user modelling server that focuses on and takes advantage of LDAP directories for an efficient storage and management of user data. On the top of this data management core, a set of specialised modules can be attached.

In contrast to this approach, the Data Persistence component of the **"yo?"** system utilises an abstracting database persisting layer. Further, the architecture of the **"yo?"** system is strictly service-based and does not permit to freely integrate any new service; among others, because it aims at minimising the difficulties of (too) high granularity of duties, e.g. too complex composition of services or loss of overall performance.

Finally, the multi-model and multi-purpose orientation of the **"yo?"** system needs more than a logical and hierarchical structuring of information. Thus, the **"yo?"** system comprises an overall solution which also decouples the persistence layer from the processing layer as well as allows extensibility of functionality through the deployment of new modules, but in addition, it stipulates how to deploy a new user modelling technique in accordance to the Human-Machine Transparency Dimensions.

This improves the reusability of the overall system and of the technique itself, because it 'calls upon' developers to consider its strict and clear separation of duties, i.e. it calls upon a decomposition of the technique into reasoning functions (to deploy into the Modeller component), work on data structures (Profiler), visualisation possibilities (View), support of a certain standard (Tools), etc. (see sections 4.2.3 and 4.3.6).

Naturally, many other frameworks, systems, tools or solution approaches could and should have been presented in this section. Though, the constraints of time have determined the presented selection.

So far, the selected systems have provided a general overview on the main issues of the **"yo?"** system in comparison with related work.

6.4 Open Issues and The Road Ahead

Undoubtedly, the specific research question concerning "How is it possible to support gaze-tracking in real-time in order to enhance the learning performance of users of the AdeLE system" (see section 6.2.2) represents the first and main open issue.

As the AdeLE research project finalises in after submitting this book, it could be assumed that there is enough time to evaluate its gaze-based Behaviour Tracking technique in order to empirically evaluate if it really enhances learner performance. Though, consider the following aspects. Firstly, the AdeLE team counts solely with one eye-tracking device, which costs (at present) many thousand Euros. Secondly, accurate results might only be gained during a long term observation of a big number of learners (as for any evaluation of behaviour-based learning). Due to these facts, just one feasible possibility remains, namely that some tests will be made with a restricted number of single test learners in order to experiment within a simple (low-cost) evaluation setup. It is hoped so to gain some insights into the *practicability of gaze-tracking methods under real-world conditions*.

Naturally, the road ahead will hopefully bring more opportunities to continue the investigation and experimentation with this fascinating device. Even more, I think it is absolutely conceivable that in the near future some eye-tracking techniques will be fully integrated into 'commonly used laptops' and therefore, utilising AdeLE's Behaviour Tracking techniques for learning could be as 'normal' as the usage of 'cookies' for current Web surfers. Though, coming back to the present, a critical (and open) issue arises within this context.

The acceptance of such a technique should be accurately evaluated. AdeLE's eye-tracking device looks in fact like a normal monitor and is therefore not as obtrusive as comparable devices. Nonetheless, in my role as lecturer of Web Design and Usability since many years, I have already conducted a large number of *usability evaluations* with video recorders and eye-tracking devices and can confirm that people do not behave 'as usually' when feeling observed. Here, the critical issue is not given by the intrusiveness of a recording device, rather it is the fact that the results of the recordings are used by third-parties.

Therefore, it is highly probable that learners using AdeLE's technology will not feel comfortable about demonstrating learning success, if they are aware of the fact that the collected data is streamed to 'some' remote adaptive system, which is just a machine that 'has the power' to decide if they have learned enough or not. To overcome these potential difficulties, some solutions are conceivable, e.g. displacing the whole analytical functionality to the client machine and giving learners the full control of the technique. Thus, this alternative could help learners to self-monitor their own performance, or even more, they can be use it to force themselves to 'stick on the chair and learn', i.e. it could be a possible solution for persons that need some 'pressure' to learn.

In fact, the **"yo?"** system shows that an optimised service-oriented solution approach increases the flexibility and reusability of distributed system components. Though, I admitted that 'three' application scenarios are not really enough proof to claim that it is really 'multi'-purpose. Therefore, an additional open issue is given by the fact that the road ahead should bring distinct application scenarios to verify the multi(!)-purposefulness of the system. For example, advantage should be taken of the highly flexible abstraction layers of the **"yo?"** system in order to foster the idea of separation of duties in adaptive systems. It makes sense to model only those things in an individual user model that are not common in other models; this avoids information overload and redundancy.

As stated in section 3.2.4, in order to efficiently support application-independence, a user modelling system might need to model e.g. the traits of the personal mobile phone of a user. Though, this device is just one among thousands of possible types of mobile phones. Thus, unmanageable information overload and redundancy might be the result if each personal phone is described in each individual user model. By utilising *distinct instances (siblings) of the "yo?" modelling system* (in this case e.g. one for modelling 'users' and one for modelling 'phones'), it is only required to administer the 'linkage' between an individual user model and its corresponding 'personal' phone model. This proposed solution sounds like just utilising the overlay technique, but the difference is that solely one system is needed from which two instances are used to model distinct things. Therefore, the notions of *multi-(agent-)modelling* and *multi-modelling* are also an interesting issue for future investigations.

To give an additional example of future work, let me mention at this point that (among others) IICM researchers have contributed to innovative and revolutionary solutions in the field of Knowledge Management (first and foremost, e.g. first versions and enhancements of the Hyperwave Information System[40]). Though, during the last years, IICM also recognises the shift from a knowledge-oriented into a competences-oriented society, and participates in research projects addressing the topic of Competence Management (such as the research project APOSDLE[41]). Thus, the field of *competence management* represents one possible future application area of the **"yo?"** system in order to further examine its multi-purpose capabilities (e.g. as a competence modeller).

Future experiments should also show if the provided functionality for the Format Factory in the Tools component of **"yo?"** shows relevant advantages for *distinct standards*. On the one hand, each new supported standard (e.g. IMS LIP, PAPI, GUMO, etc.) implies an additional coding effort, but on the other side, it makes the core functionality of the system independent of and scalable for distinct representation lan-

---

[40] See e.g. http://www.iicm.edu/hgbook or
http://www.hyperwave.com
[41] See e.g. http://www.aposdle.tugraz.at

guages. It is assumed that problems could arise concerning the maintenance of information for distinct external applications at the persistence layer, e.g. in terms of consistency and synchronicity of data as well as concurrent access to shared data.

In a last and broader context, the **"yo?"** modelling system has been primarily developed to support personalisation in Web-based adaptive systems. At present, the concept of personalisation is no longer supported solely by the notion of

"The Web is no longer the Web; It's your Web!" [Robertson 1997].

Rather, Web-based 'social software' applications as well as mash-ups, folksonomies and other technologies emerging from the present Web 2.0 era, are shifting the aforementioned notion towards the idea of what I coin as *"The Web is not only your Web; It's our Web"* to underline the current shift from my-Web to our-community-Web.

Against this background, the **"yo?"** system must still prove its community-based capabilities. The road ahead of the **"yo?"** system might include a new version of the Concept-based Context (CO2) modelling system, e.g. as a Community-oriented knowledge provider (perhaps 'CO3'?). For example, as stated in [Kobsa 1995], one of the frequently-found services of user modelling systems is the representation of assumptions about user's *misconceptions*. In [Lukas 1997], a method is presented to model misconceptions in an algebraic knowledge structure, which is based on the Knowledge Space Theory of [Doignon and Falmagne 1985] (see also [Albert and Hockemeyer 1997], [Doignon and Falmagne 1999]). This method is used in adaptive e-learning environments to detect contradictions while assessing the acquired knowledge of learners. *Contradictions and misconceptions* arise because a person takes some fact for correct (true) while the majority of 'colleagues' takes it for incorrect (false).

Based on this example, the road ahead of the CO2 system (which is an existing **"yo?"** sibling) could show if it is capable of modelling and managing individual vs. community-based mis-

conceptions, i.e. the intention is to extend the current CO2 principle of 'teacher-controlled annotating of concepts' to 'free tagging of learner communities' and to evaluate the implications on individual learning performance. The idea behind this future work is motivated by the assumption that misconceptions might propagate through an uncontrolled (i.e. unguided) usage of social software, and thus lead to the acquisition of 'wrong' knowledge.

Within the scope of e-learning, this dark side of social software was already coined e.g. in [Dron 2006]: *social software* presents the learner with recommended options, links and paths resulting from the behaviour of others learners, i.e. it assumes the teacher's role of providing control over the learning journey to the community members. Thus, a dynamic propagation of control is a property of the system and depends on the interactions and intentions of the learner community as a whole.

Therefore, community contributions in social software lead to a self-organising system, whereby learners may control their learning or delegate control to the community. As stated in [Dron 2006], these learning intentions in such a self-organising system should be somehow aligned (i.e. controlled):

"out of control, the wisdom of crowds can too easily become the stupidity of mobs". [Dron 2006]

The key message of this statement is that a guided control in such self-organising systems is required (e.g. by a tutor or editorial board), because the impact of the first contributions is so significant on the environment that it influences directly the ensuing contributions [Dron 2006]. Thus, autonomous learners (i.e. active community members that are used to contributing spontaneously with own opinions) may contribute unintentionally to an uncontrolled propagation of misconceptions.

These problems are also coined within the field of known theories, e.g. in *E-Learning Ecosystems*, as introduced in [Chang and Gütl 2007]. In an e-learning ecosystem, individuals may adopt or adapt their behaviours to contribute to or 'perturb the success of the learning environment'. This means that any alteration in the conditions of a self-organising ecosystem has direct impact on the behaviour and success of the system itself and of its components (thus, also on the learning performance of individuals) [Chang and Gütl 2007].

The CO2 system has shown that it can be successfully utilised for *exploratory learning* [diSessa et al. 1995]. Though, in contrast to diSessa's premise regarding 'learning does not have to be forced', the evaluation results on the usefulness of a dynamic background library (see e.g. section 5.2.2) have also shown that some test subjects just used the tool because it was prescribed for the evaluation.

On the one hand, some of them stated that the explorative character of the approach was very interesting and motivating, but on the other hand, some felt apprehension for a decreased contact to the teacher and to colleagues because of the combination of time pressure and distraction in hyperspace. Therefore, a future evaluation of the usefulness of the CO2 system and the **"yo?"** system should clarify which negative implications may have explorative and other constructivist or collaborative approaches on individual learning performance (here again, also considering the impact of social software). It should be evaluated to which extent self-adaptive systems and virtualisation of teaching tasks are better (or not) than a controlled intervention of teachers to *maintain a certain discipline* among the behaviours of learners, i.e. the authority degree of tutors as well as the impact of defining strict rules should be inspected.

Further, it should be also evaluated which are the implications of *lack of communication with teachers* in adaptive e-learning systems, more precisely the *lack of face-to-face contact* should be carefully observed and analysed. As indicated in [Frank et al. 2002], the degree of affinity between teacher and learners is the key issue in face-to-face interaction, i.e. teachers should be able to give learners 'human warmth' and convey an 'intimate attitude' in order to better help in solving personal problems.

Thus, a lack of face-to-face contact prevents teachers from proactively capture relevant learners' reactions, facial expressions or body language that may indicate a need of personal guidance. For example, according to [Carr-Chellman and Duchastel 2000], introverted learners have also problems of participation during a synchronised collaborative activity in a virtual environment, not only because of its real-time nature but also because of the potential presence of 'virtual' colleagues that are completely unknown and not 'visible', and thus, a greater social pressure for conformity in participation may exist.

In sum, the observations in the last paragraphs represent relevant personalisation issues regarding future evaluations of the multi-purposefulness of the **"yo?"** system in order to provide (a) useful individual information to adaptive engines, i.e. to other software components, (b) relevant, real-time observations of individual learner behaviours to teachers, e.g. through real-time sensors (such as eye-tracking devices, cameras, microphones, etc.), and (c) relevant personal information about the participating community to the community members themselves.

Hence, developers of user modelling systems should find a balance between '*knowing the user and creating assumptions about the user for machine-triggered adaptive purposes*' and '*carefully opening and sharing the content of user models for human-triggered pragmatic and social purposes*'. Among others, this second aspect implies the accurate and careful consideration of scrutability and privacy issues.

List of Figures

List of Tables

Bibliography

A

[Abowd et al. 2000] Abowd, G.D., Mynatt, E.D.: Charting Past, Present and Future Research in Ubiquitous Computing; in ACM Transactions on Computer-Human Interaction, Special issue on HCI in the new Millenium, Vol. 7 Nr. 1; pp. 29-58, 2000.

[ADL 2007] ADL: Advanced Distributed Learning: official Web site, URL: http://www.adlnet.gov; SCORM: Sharable Content Object Reference Model; 2007. URL http://www.adlnet.gov/scorm Last visit: 2008-02-21.

[Adomavicius and Tuzhilin 2001] Adomavicius, G., Tuzhilin, A.: Using Data Mining Methods to Build Customer Profiles; in IEEE COMPUTER, Vol. 34 Nr. 2 (0018-9162/01/); pp. 74-82, 2001.

[Akhras and Self 2000] Akhras, F.N., Self, J.: System Intelligence in Constructivist Learning; International Journal of Artificial Intelligence in Education, 11; pp. 344-376, 2000.

[Albert and Hockemeyer 1997] Albert, D., Hockemeyer, C.: Adaptive and dynamic hypertext tutoring systems based on knowledge space theory; in B. du Boulay & R. Mizoguchi (Ed.), Artificial Intelligence in Education: Knowledge and Media in Learning Systems; IOS Press, Amsterdam; pp. 553–555, 1997.

[Alepsis et al. 2006] Alepsis, E., Virvou, M., Kabassi, K.: Affective student modelling based on microphone and keyboard user actions; in Proceedings of the IEEE International Conference on Advanced Learning Technologies (ICALT 2006), in Kerkrade, The Netherlands; Kinshuk, R. Koper, P. Kommers, P. Kirschner, D. G. Sampson, W. Didderen (eds.), IEEE Computer Society Press; pp. 139-141, 2006.

[Alrifai et al. 2006] Alrifai, M., Dolog, P., Nejdl, W.: Learner profile management for collaborating adaptive eLearning applications; in Proceedings of the joint international workshop on Adaptivity, personalization and the semantic web (APS '06); ACM Press, New York, NY, USA; 31-34, 2006.

[Ambel et al. 2004] Ambel, M., Carletti, A., Casulli, L., Gineprini, M., Guastavigna, et al.: CMAPTOOLS - A Knowledge Modelling and Sharing Environment; in CMC 2004, in Pamplona, Spain; 2004. URL http://cmc.ihmc.us/papers/cmc2004-188.pdf Last visit: 2008-02-21.

[Amelung et al. 2006] Amelung, M., Piotrowski, M., Rösner, D.: EduComponents: experiences in e-assessment in computer science education; in Proceedings of the 11th annual SIGCSE conference on Innovation and technology in computer science education in Bologna-Italy (ITICSE '06); ACM Press, NY, USA; pp. 88-92, 2006.

[Andrews and Goodson 1995] Andrews, D.H., Goodson, L.A.: A Comparative Analysis of Models of Instructional Design; in Instructional Technology: Past, Present, and Future (2nd Edition); G-J. Anglin (ed.); Libraries Unlimited, Westport, USA; Chapter 13, pp. 161-182, 1995.

[Arabshahi et al. 1993] Arabshahi, P., Marks, R.J., Reed, R.: Adaptation of fuzzy inferencing: a survey; in Proceedings of IEEE Learning and Adaptive Systems, Nagoya University Workshop, Nagoya, Japan; 1993.

[Ardissono et al. 2002] Ardissono, L., Goy, A., Petrone, G., Segnan, M.: in Personalization in business-to-customer interaction; Communications of the ACM, Vol. 45, Nr. 5; ACM Press, New York, NY, USA; pp. 52-53, 2002.

[Assad et al. 2006] Assad, M., Carmichael, D.J., Kay, J., Kummerfeld, B.: Active Models for Context-Aware Services; TR Nr. 594 at School of Information Technologies, University of Sydney; 2006.

[Ashby 1956] Ashby, W.R.: An Introduction to Cybernetics; Chapman and Hall, London; 1956.

[Ashby 1960] Ashby, W.R.: Design for a Brain; Chapman and Hall, London, 2nd Edition; 1960.

[Ashby 1962] Ashby, W.R.: Principles of the self-organizing system; in Principles of Self-Organization, Pergamon Press, New York, N.Y.; pp. 255-278, 1962.

[Atif et al. 2003] Atif, Y., Benlamri, R., Berri, J.: Learning Objects Based Framework for Self-Adapting Learning; in Education and Information Technologies, Volume 8, Issue 4; Kluwer Academic Publishers, The Netherlands, pp. 345-368, 2003.

[Ausubel 1960] Ausubel, D. P.: The use of advance organizers in the learning and retention of meaningful verbal material; in Journal of Educational Psychology, Vol. 51; pp. 267-272, 1960.

[Azevedo-Tedesco 2003] Azevedo-Tedesco, P.: MArco: Building an Artificial Conflict Mediator to Support Group Planning Interactions; International Journal of Artificial Intelligence in Education, 13, IOS Press; pp. 117-155, 2003.

[Åström and Hägglund 1995] Åström, K.J., Hägglund, T.: The Control Handbook; in The Electrical Engineering Handbook Series; CRC Press, Inc., Boca Raton/FA; chapter Adaptive Control, page 824, 1995.

B

[Baeza and Ribeiro 1999] Baeza-Yates, R, Ribeiro-Neto, B.: Modern Information Retrieval; ACM Press, Addison-Wesley, USA; pp 262-267, 1999.

[Banathy 1992] Banathy, B. H.: A systems view of education; Englewood Cliffs, NJ: Educational Technology; 1992.

[Banathy and Jenlink 2003] Banathy, B.H., Jenlink, P.M.: SYSTEMS INQUIRY AND ITS APPLICATION IN EDUCATION; in Educational Technology Research and Development, 2003(25); pp. 37-57, 2003.

[Baumgartner 2003] Baumgartner, P.: E-Learning an Hochschulen: Didaktik, Modelle, Strategien und Perspektiven; In Proceedings of OCG e-Future, 2003 (in German language).

[Beasley 2002] Beasley, W.: From Context to Concept: The Implications for the Teaching of Chemistry, in Effective Teaching for Meaningful Learning; 2002. URL http://wwwcsi.unian.it/educa/teachmeth/wbeasley.html Last visit: 2008-01-25.

[Beaumont 1994] Beaumont I.: User modeling in the interactive anatomy tutoring system Anatom-Tutor; in User Modeling and User-Adapted Interaction (UMUAI); Vol. 4; pp. 121-145, 1994.

[Benitez et al. 2001] Benitez, A. B., Chang, S.F., Smith, J. R.: IMKA - A Multimedia Organization System Combining Perceptual and Semantic Knowledge; in Proceedings of ACM International Conference on Multimedia (ACM MM-2001), in Ottawa, Canada; 2001.

[Benjamins et al. 1998] Benjamins, V. R., Plaza, E., Motta, E., Fensel, D., Studer, R., Wielinga, B., et al.: IBROW3 - An intelligent brokering service for knowledge component reuse on the world-wide web; in Proceedings of 11th Banff Knowledge Acquisition for Knowledge-Based Systems Workshop (KAW98); 1998.

[Benyon 1993] Benyon, D. R.: Adaptive systems: a solution to usability problems; in User Modelling and User Adapted Interaction, Vol. 3 Nr. 1; pp. 65–87, 1993.

[Benyon and Murray 1993] Benyon, D. R., Murray, D. M.: Adaptive systems; from intelligent tutoring to autonomous agents. Knowledge-Based Systems, Vol. 6 Nr. 4; 197–219, 1993.

[Bianchi-Berthouze and Lissetti 2002] Bianchi-Berthouze, N., Lissetti, C.L.: Modeling Multimodal Expression of User's Affective Subjective Experience; in User Modeling and User-Adapted Interaction (UMUAI) Vol. 12; Kluwer Academic Publishers, The Netherlands; pp. 49-84, 2002.

[Bieber 2002] Bieber, G.: Openwings Blueprint, Version 1.1; General Dynamics Decision Systems Inc., Virginia, USA; 2002. URL http://openwings.org/download/other/Openwings_Blueprints.pdf Last visit: 2008-02-02.

[Bieber and Carpenter 2001] Bieber, G., Carpenter, J.: Openwings, A Service-Oriented Component Architecture for Self-Forming, Self-Healing, Network-Centric Systems (Rev 2.0); 2001. URL http://www.openwings.org Last visit: 2008-02-02.

[Bieber and Carpenter 2002] Bieber, G., Carpenter, J.: Introduction to Service-Oriented Programming, Rev 2.1; General Dynamics Decision Systems Inc., USA; 2002. URL http://www.openwings.org Last visit: 2008-02-02.

[Bieber and Carpenter 2003] Bieber, G., Carpenter, J.: Openwings Connector Service Specification Version 1.0 Final; 2003. URL http://www.openwings.org Last visit: 2008-02-02.

[Bieber and Crumpton 2003] Bieber, G., Crumpton, K.: Openwings Install Service Specification - Version 1.0 Final; 2003. URL http://www.openwings.org Last visit: 2008-02-02.

[Bieber and Thrash 2003] Bieber, G., Thrash, B.: Openwings Security Specification v1.0 Final, 2003. URL http://www.openwings.org Last visit: 2008-02-02.

[Bieber et al. 2003] Bieber, G., Nelson, M., Chang, L.: Openwings Interface Definition Specification Version 1.0 Final; 2003. URL http://www.openwings.org Last visit: 2008-02-02.

[Billsus and Pazzani 2000] Billsus, D., Pazzani, M.J.: User Modeling for Adaptive News Access; in User Modeling and User-Adapted Interaction (UMUAI) Vol. 10; Kluwer Academic (pub.), The Netherlands; pp. 147-180, 2000.

[Blank 1996] Blank, K.: Benutzermodellierung für Adaptive Interaktive Systeme: Architektur, Methoden, Werkzeuge und Anwendungen; PhD thesis at University of Stuttgart, Germany; 1996 (in German langauge).

[Bloedorn et al. 1996] Bloedorn, E., Mani, I., MacMillan, T.R.: Machine Learning of User Profiles: Representational Issues; in 13th National Conference on Artificial Intelligence, Portland, Oregon, USA; pp. 433-438, 1996.

[Bloom 1984] Bloom, B. S.: The 2 Sigma Problem: The Search for Methods of Group Instruction as Effective as One-to-One Tutoring; Educational Researcher, Vol. 13 Nr. 6; pp. 4-16, 1984.

[Bloom et al. 1956] Bloom, B.S., Engelhart, M.D., Furst, E.J., Hill, W.H., and Krathwohl, D.R.: Taxonomy of Education Objectives: The Classification of Educational Goals. David McKay, New York; 1956.

[Booth et al. 2007] Booth, D., Haas, H., McCabe, F., Newcomer I., Champion M., et al.: Web Services Architecture, W3C Working Group, 2007; URL http://www.w3.org/TR/ws-arch Last visit: 2008-02-19.

[Borst and Theunissen 1999] Borst, A., Theunissen, F.E.: Information theory and neural coding; in 'nature neuroscience' Vol. 2 Nr. 11 (Nov.1999); Nature America Inc. (pub.); pp. 974-957, 1999.

[Bødker 2000] Bødker, S.: Scenarios in User-Centered Design – Setting the Stage for Reflection and Action; in Interacting with Computers, Vol. 13; pp. 61-75, 2000.

[Brajnik and Tasso 1994] Brajnik, G., Tasso, C.: A shell for developing non-monotonic user modeling systems; in International Journal of Human-Computer Studies, Vol. 40 Nr. 1; pp. 31–62, 1994.

[Brandon 2002] Brandon, D.: Crud matrices for detailed object oriented design; in Journal of Computer Sciences in Colleges, Vol. 18, Nr. 2; Consortium for Computing Sciences in Colleges (pub.), USA; pp. 306-322, 2002.

[Brandwajn et al. 1979] Brandwajn, A., Hernandez, J.A., Joly, R., Kruchten, Ph.: Overview of the ARCADE system; in Proceedings of the 6th annual symposium on Computer Architecture; ACM Press; pp 42-49, 1979.

[Bransford et al. 2000] Bransford, J.D., Brown, A.L., Cocking, R.R. (eds.): How people learn: Brain, mind, experience and school; (expanded ed.); National Academy Press, Washington DC, USA; 2000.

[Britannica 2007] Encyclopaedia Britannica Online. URL http://www.britannica.com Last visit: 2008-01-25.

[Brown et al. 2006] Brown, E., Brailsford, T., Fisher, T., Moore, A., Ashman, H.: Reappraising cognitive styles in adaptive web applications; in Proceedings of the 15th international conference on World Wide Web (WWW '06); ACM Press, New York, NY, USA; pp. 327-335, 2006.

[Browne et al. 1990] Browne, D., Totterdell, P., Norman, M.: Adaptive User Interfaces. London: Academic Press; 1990.

[Brusilovsky 1994] Brusilovsky, P.L.: The Construction and Application of Student Models in Intelligent Tutoring Systems; in Journal of Computer and Systems Science International, Vol. 32 Nr. 1, pp. 70-89, 1994.

[Brusilovsky 1998] Brusilovsky, P.: Adaptive Educational Systems on the World-Wide-Web: A Review of Available Technologies; In Proceedings of Workshop on World Wide Web Based Tutoring, San Antonio; 1998.

[Brusilovsky 1996a] Brusilovsky, P.: Adaptive hypermedia, an attempt to analyze and generalize. In Multimedia, Hypermedia, and Virtual Reality (Lecture Notes in Computer Science, Vol. 1077); P. Brusilovsky, P. Kommers and N. Streitz (eds.), Springer-Verlag, Berlin; pp. 288-304, 1996.

[Brusilovsky 1996b] Brusilovsky, P. Methods and Techniques of Adaptive Hypermedia; in User Modelling and User Adapted Interaction (UMUAI), 6(2/3); pp. 87-129, 1996.

[Brusilovsky 1999] Brusilovsky, P.: Adaptive and Intelligent Technologies for Web-based Education; in Künstliche Intelligenz, Nr. 4; Special Issue on Intelligent Systems and Teleteaching; Rollinger, C., Peylo, C. (eds.); pp. 19-25, 1999.

[Brusilovsky 2000] Brusilovsky P. Adaptive hypermedia: From intelligent tutoring systems to Web-based education. Intelligent tutoring systems: Gauthier G, Frasson C, and van Lehn K (eds.); Springer Verlag; Berlin, 2000.

[Brusilovsky 2001] Brusilovsky, P.: Adaptive Educational Hypermedia; in Proceedings of 10th International PEG Conference, in Tampere, Finland; pp. 8-12, 2001.

[Brusilovsky 2003] Brusilovsky, P.: Developing adaptive educational hypermedia systems: From design models to authoring tools; In Authoring Tools for Advanced Technology Learning Environment; T. Murray, S. Blessing and S. Ainsworth (eds.) Kluwer Academic (pub.), Dordrecht; pp. 377-409, 2003.

[Brusilovsky 2004a] Brusilovsky, P.: Adaptive Educational Hypermedia: From generation to generation; in Proceedings of 4th Hellenic Conference on Information and Communication Technologies in Education, Athens, Greece; pp.19-33, 2004.

[Brusilovsky 2004b] Brusilovsky, P.: KnowledgeTree: A Distributed Architecture for Adaptive E-Learning; in Proceedings of the 13th International World Wide Web Conference (WWW 2004), Alternate track papers and posters, in New York, NY, USA; ACM 1-58113-912-8/04/0005; pp. 104-113, 2004.

[Brusilovsky and Miller 2001] Brusilovsky, P., Miller, P.: Course Delivery Systems for the Virtual University; In Access to Knowledge: New Information Technologies and the Emergence of the Virtual University; Tschang, T., Della Senta, T. (eds.), Elsevier Science, Amsterdam; pp. 167-206; 2001.

[Brusilovsky and Pesin 1994] Brusilovsky, P., Pesin, L.: ISIS-Tutor: An adaptive hypertext learning environment; in Proceedings of Japanese-CIS Symposium on knowledge-based software engineering (JCKBSE'94), in Tokyo; pp. 83-87, 1994.

[Brusilovsky and Peylo 2003] Brusilovsky, P., Peylo, C.: Adaptive and intelligent Web-based educational systems; In International Journal of Artificial Intelligence in Education 13 (2-4), Special Issue on Adaptive and Intelligent Web-based Educational Systems; P. Brusilovsky and C. Peylo (eds.); 159-172, 2003.

[Brusilovsky et al. 2005] Brusilovsky, P., Sosnovsky, S., Yudelson, M.: Ontology-based Framework for User Model Interoperability in Distributed Learning Environments; in Proceedings of World Conference on ELearning (E-Learn 2005), in Vancouver, Canada; pp. 2851-2855, 2005.

[Bull 1997] Bull, S.: See Yourself Write: A simple student model to make students think; in Proceedings of the 6th International Conference on User Modeling (UM97); Anthony Jameson, C´ecile Paris, and Carlo Tasso, (eds.), Springer Verlag, New York; 1997.

[Bull et al. 2003] Bull, S., Greer, J., McCalla, G.: The Caring Personal Agent; in International Journal of Artificial Intelligence in Education, Volume 13, Number 1, Special Issue: "Caring for the Learner" in honour of John Self; Paul Brna, Lewis Johnson, Helen Pain (eds.); pp. 21-34, 2003.

[Burke 2000] Burke, R.: Knowledge-based Recommender Systems; In Encyclopedia of Library and Information Systems, Vol. 69, Supplement 32; A. Kent (ed.), New York: Marcel Dekker; 2000.

[Burke 2002] Burke, R.: Hybrid Recommender Systems: Survey and Experiment; in User Modeling and User-Adapted Interaction (UMUAI) Vo. 12; Kluwer Academic Publishers, The Netherlands; pp.331-370, 2002.

[Burton 1992] Burton, I.: Adapt and Thrive; Downsview; Canadian Climate Centre, Ontario; 1992..

[Burton et al. 2003] Burton, J.K., Moore, D.M., Magliaro, S.G.: Behaviorism and Instructional Technology; Educational Technology Research and Development, 2003(25). p. 3-36, 2003.

[Bush 1945] Bush, V.: As We May Think; in The Atlantic Month, Volume 176, Number 1; pp. 101-108, 1945.

C

[Cannataro and Pugliese 2000] Cannataro, M., Pugliese, A.: An XML-Based Architecture for Adaptive Web Hypermedia Systems Using a Probabilistic User Model; In Proceedings of International Database Engineering and Applications Symposium (IDEAS'00), in Yokohama, Japan; pp. 257-265, 2000.

[Cannataro et al. 2001] Cannataro, M., Cuzzocrea, A., Pugliese, A.: A Probabilistic Adaptive Hypermedia System; in Proceedings of International Conference on Information Technology: Coding and Computing (ITCC '01); pp. 411-415, 2001.

[Carberry 2001] Carberry, S.: Techniques for Plan Recognition; in User Modeling and User-Adapted Interaction (UMUAI), Vol. 11; Kluwer Academic Publishers. Printed in the Netherlands; pp. 31-48, 2001.

[Carmichael et al. 2006] Carmichael, D ., Kay, J., Kummerfeld, B., Niu, W.: MyWorkPlace: Personalized information about a Ubiquitous Computing enabled building; in Proceedings of the 2nd International Workshop on Personalized Context Modeling and Management for UbiComp Applications (ubiPCMM 2006), in Orange County, California, USA; Jang, S., van Laerhoven, K., Lee, S.G., Mase, K. (eds); 2006.

[Carpenter and Grossberg 1988] Carpenter, G.A., Grossberg, S.: The ART of Adaptive Pattern Recognition by a Self-Organizing Neural Network; in IEEE - Computer: Innovative Technology for Computer Technologies; pp. 77-88, 1988.

[Carr-Chellman and Duchastel 2000] Carr-Chellman, A., Duchastel, P.: The Ideal Online Course; in British Journal of Educational Technology, Vol. 31 Nr. 3; pp. 229-241, 2000.

[Çetintemel et al. 2000] Çetintemel, U., Franklin, M.J., Giles, C.L.: Self-Adaptive User Profiles for Large-Scale Data Delivery; in Proceedings of the 16th International Conference on Data Engineering, in San Diego, California; pp. 622-633, 2000.

[Chang and Gütl 2007] Chang, V., Gütl, C.: E-Learning Ecosystem (ELES), A Holistic Approach for the Development of more Effective Learning Environment for Small-to-Medium Sized Enterprises (SMEs); Inaugural IEEE International Digital Ecosystems and Technologies Conference (IEEE DEST 2007), in Cairns, Australia; 2007.

[Chapin 1968] Chapin, J.R.: Problem areas in estimating costs of computer-assisted instruction; in Proceedings of the 1968 23rd ACM national conference; ACM Press; pp. 111-116, 1968.

[Chieu et al. 2004] Chieu, V.M., Milgrom, E., Frenay, M.: Constructivist Learning: Operational Criteria for Cognitive Flexibility; in Proceedings of IEEE International Conference on Advanced Learning Technologies (ICALT'04); pp. 221-225, 2004.

[Chin 1989] Chin, D.N.: KNOME: Modeling what the User Knows in UC; in User Models in Dialog Systems; Kobsa, A., Wahlster, W. (eds.), Springer (pub.), Berlin, Heidelberg; pp. 74-107, 1989.

[Chin 2001] Chin, D.: Empirical Evaluation of User Models and User-Adapted Systems; in User Modeling and User-Adapted Interaction (UMUAI), Vol. 11; Kluwer Academic Publishers, The Netherlands; pp. 181-194, 2001.

[Chiu and Webb 1998] Chiu, B.C., Webb, G.I.: Using Decision Trees for Agent Modeling: Improving Prediction Performance; in User Modeling and User-Adapted Interaction (UMUAI) Vol. 8; Kluwer Academic Publishers, The Netherlands; pp. 131-152, 1998.

[Chiu and Webb 1999] Chiu, B.C., Webb, G.: Dual-model: An Architecture for Utilizing Temporal Information in Student Modeling; in Proceedings of the Workshop on Machine Learning in User Modeling, Advanced Course on Artificial Intelligence (ACAI '99), in Chania, Greece; 1999.

[ChoiceStream 2004] ChoiceStream: "ChoiceStream Personalization Survey", Consumer Trends and Perceptions, 2004, URL http://www.choicestream.com Last visit: 2008-02-19.

[Cini and Valdeni 2002] Cini, A., Valdeni de Lima, J.: Adaptivity Conditions Evaluation for the User of Hypermedia Presentations Built with AHA!; in Proceedings of Adaptive Hypermedia and Adaptive Web-Based Systems: 2nd International Conference (AH 2002), in Malaga, Spain; p. 497, 2002.

[Claypool et al. 2001] Claypool, M., Brown, D., Le, P., Waseda, M.: Inferring User Interest; in IEEE Internet Computing: Engineering and Applying the Internet, Issue Nov-Dec 2001; pp. 32-39, 2001.

[Cohen et al. 1981] Cohen, P.R., Perrault, C.R., Allen, J. F.: Beyond question answering; in Strategies for Natural Language Processing; W. Lehnert and M. Ringle (eds.); pp. 245-274, 1981.

[Conlan et al. 2003] Conlan, O., Hockemeyer, C., Wade, V., Albert, D.: Metadata Driven Approaches to Facilitate Adaptivity in Personalized eLearning Systems"; in Journal of the Japanese Society for Information and Systems in Education (http://www.jsise.org); 2003.

[Cook and Kay 1994a] Cook, R., Kay, J.: The justified user model - a viewable, explained user model; in Proceedings of the 4th International Conference on User Modeling (UM94); Kobsa, A and D Litman (eds), MITRE, UM Inc, Hyannis, Massachusetts, USA; pp. 145-150, 1994.

[Cook and Kay 1994b] Cook, R., Kay, J.: The justified user model- a viewable, explained user model; Technical Report 483 (June 1994), Basser Department of Computer Science at University of Sydney; 1994.

[Cooper 1964] Cooper, W.S.: Fact Retrieval and Deductive Question-Answering Information Retrieval Systems; in Journal of ACM; ACM Press, Vol. 11, Nr. 2; pp 117-137, 1964.

[Cooper 1993] Cooper, P. A.: Paradigm shifts in designing instruction: From behaviorism to cognitivism to constructivism. Educational Technology, Vol. 33(5); pp. 12-19, 1993

[Cooperstein et al. 1999] Cooperstein, D., Delhagen, K., Aber, A., Levin, K.: Making Net Shoppers Loyal; in Forrester Research, Cambridge, MA, USA; 1999.

[Corbin and Frank 1966] Corbin, H.S., Frank, W.L.: Display oriented computer usage system; in Proceedings of the 1966 21st national conference, ACM Press; pp 515-526, 1966.

[Coulouris et al. 2005] Coulouris, G., Dollimore, J., Kindberg, T.: Distributed Systems: Concepts and Design; Pearson Education, 4. Edition; chapters 2,3,19; 2005.

[CRA 2002] Computing Research Association (2002) Grand Research Challenges in Computer Science and Engineering; 2002. URL http://www.cra.org/reports/gc.systems.pdf Last visit: 2008-01-19.

[Craik and Lockhart 1972] Craik, F. I. M., & Lockhart, R. S.: Levels of processing: A framework for memory research; in Journal of Verbal Learning and Verbal Behavior, 11; pp. 671-684, 1972.

[Cristea and DeBra 2002] Cristea, A., DeBra, P.: ODL Education Environments based on Adaptability and Adaptivity; in Proceedings of the AACE E-Learn'2002 conference, in Montreal, Canada; pp. 232-239, 2002.

[Cronbach 1957] Cronbach, L.J.: The two disciplines of scientific psychology; American Psychologist, Vol. 12; pp. 671-684, 1957.

[Crowell and Traegde 1967] Crowell, F.A., Traegde, S.C.: The role of computers in instructional systems: Past and future; in Proceedings of the 1967 22nd national conference; ACM Press; Washington, D.C., United States; pp. 417-425, 1967.

[Csinger et al. 1994] Csinger, A., Booth, K.S., Poole, D.: AI meets authoring: User models for intelligent multimedia; in Journal Artificial Intelligence Review, Computer Science Issue, Volume 8, Numbers 5-6; Springer Netherlands (pub.); pp. 447-468, 2004.

[Czarkowski and Kay 2003] Czarkowski, M., Kay, J.: Challenges of Scrutable Adaptivity; in Proceedings from the 11th International Conference in Artificial Intelligence in Education (AIED); Sydney, Australia; 2003.

D

[Daniel and Marquis 1979] Daniel, J., Marquis, C.: Interaction and independence: getting the mixture right; in Teaching at a Distance Vol. 15; pp. 25-44, 1979.

[Danine et al. 2006] Danine, A., Lefebvre, B., Mayers, A.: TIDES – Using Bayesian Networks for Student Modeling; in Proceedings of the IEEE International Conference on Advanced Learning Technologies (ICALT 2006), in Kerkrade, The Netherlands; Kinshuk and R. Koper and P. Kommers and P. Kirschner and D. G. Sampson and W. Didderen (eds.), IEEE Computer Society Press; pp. 1002-1006, 2006.

[deAssis et al. 2005] deAssis, A.S.F.R., Danchak, M.M., Polhemus, L.: Instructional design and interaction style for educational adaptive hypermedia; in Proceedings of the 2005 Latin American conference on Human-computer interaction (CLIHC '05); ACM Press, New York, NY, USA; pp. 289-294, 2005.

[deVrieze et al. 2006] de Vrieze, P., van der Weide, T.P., van Bommel, P.: GAM: A Generic Model for Adaptive Personalisation; Technical Report, Nr. ICIS--R06022, Radboud University Nijmegen, June 2006.

[deBra 2006] De Bra, P.: Web-based educational hypermedia; book chapter in Data Mining in E-Learning; C. Romero, S. Ventura (eds.),

Universidad de Cordoba, Spain, WIT Press, ISBN 1-84564-152-3; pp. 3-17, 2006.

[deBra and Calvi 1998] de Bra, P., Calvi, L.: AHA! An open Adaptive Hypermedia Architecture; The New Review of Hypermedia and Multimedia 4, 115–139, 1998.

[deBra et al. 1999] de Bra, P., Houben, G.J., Wu, H.: AHAM: A Dexter-based Reference Model for Adaptive Hypermedia; In Proceedings of the ACM Conference on Hypertext and Hypermedia. Darmstadt, Germany, pp. 147–156, 1999.

[deBra et al. 2003] de Bra, P., Aerts, A., Berden, B., de Lange, B., Rousseau, B., Santic, T., Smits, D., Stash, N.: AHA! The Adaptive Hypermedia Architecture; In Proceedings of the ACM Hypertext Conference. Nottingham, UK, pp. 81–84, 2003.

[deBra et al. 2004] De Bra, P., Aroyo, L., Chepegin, V.: The next big thing: Adaptive web-based systems; in Journal of Digital Information, Vol. 5 Issue 1, Article Nr. 247; 2004.

[Denaux et al. 2005a] Denaux, R., Aroyo, L., Dimitrova, V.: An approach for ontology-based elicitation of user models to enable personalization on the semantic web; in Proceedings of WWW '05: Special interest tracks and posters of the 14th international conference on World Wide Web, in Chiba, Japan; ACM Press (pub.), New York, NY, USA; pp. 1170-1171, 2005.

[Denaux et al. 2005b] Denaux, R., Dimitrova, V., Aroyo, L.: Integrating Open User Modeling and Learning Content Management for the Semantic Web; in User Modeling 2005, Book Series Lecture Notes in Computer Science, Vol. 3538/2005, Springer Berlin / Heidelberg (pub.); pp. 9-18 , 2005.

[Denenberg 1978] Denenberg, S.A.: 'A personal evaluation of the PLATO system'; In Journal SIGCUE Outlook, Vol. 12, Nr. 2, ACM Press, pp. 3-10, 1978

[deRosnay 1978] deRosnay, J.: History of Cybernetics and Systems Science; in Principia Cybernetica Web; 1978. URL http://pespmc1.vub.ac.be/CYBSHIST.html Last visit: 2008-02-12.

[Desmarais and Gagnon 2006] Desmarais, M., Gagnon, M.: Bayesian Student Models Based on Item to Item Knowledge Structures; in Innovative Approaches for Learning and Knowledge Sharing, 1st European Conference on Technology Enhanced Learning (Proceedings of EC-TEL 2006), in Crete, Greece; Nejdl W., Tochtermann K. (eds.), Lecture Notes in Computer Science Vol. 4227 (series), Springer (pub.); pp. 111-124, 2006.

[Dichev and Dicheva 2006] Dichev, C., Dicheva, D.: Using Contexts to Personalize Educational Topic Maps; in Proceedings of Adaptive Hypermedia and Adaptive Web-Based Systems, 4th International Conference (AH2006); Wade, A., Ashman, H., Smyth, B. (eds.); Springer Verlag, Berlin Heidelberg; pp. 269-173, 2006.

[Dietinger 2003] Dietinger, T.: Aspects of E-Learning Environments; PhD Thesis at Institute for Information Systems and Computer Media at Graz University of Technology, Austria; 2003.

[Dietinger and Maurer 1997] Dietinger, T., Maurer, H.: How Modern WWW Systems Support Teaching and Learning; in Proceedings of International Conference on Computers in Education (ICCE) 1997; Z. Halim, T. Ottmann, Z. Razak (eds.), Kuching, Sarawak Malaysia; pp. 37-51, 1997.

[Dietinger et al. 1998] Dietinger, T., Gütl, C., Maurer, H., Pivec, M., Schmaranz, K.: Intelligent Knowledge Gathering and Management as New Ways of an Improved Learning Process; in Proceedings of

WebNet 98, World Conference of the WWW; AACE, Internet and Intranet, Charlottesville, USA; pp. 244-249, 1998.

[diSessa et al. 1995] diSessa, A. A., Hoyles, C., Noss, R., & Edwards, L. D.: Computers and exploratory learning: Setting the scene; in A. A. diSessa, C. Hoyles, R. Noss, & L. D. Edwards (Eds.), Computers and exploratory learning, New York, Springer (pub.); pp. 1-12, 1995.

[Doane and Sohn 2000] Doane, S.M., Sohn, Y.W.: ADAPT: A Predictive Cognitive Model of User Visual Attention and Action Planning; in User Modeling and User-Adapted Interaction (UMUAI), Vol. 10; pp. 1-45, 2000.

[Doignon and Falmagne 1985] Doignon, J.-P., Falmagne, J.-Cl.: Spaces for the assessment of knowledge; in International Journal of Man-Machine Studies, Vol. 23; pp. 175–196., 1985.

[Doignon and Falmagne 1999] Doignon, J.-P., Falmagne, J.-Cl.: Knowledge Spaces; Springer–Verlag, Berlin; 1999.

[Douglas and Riding 1993] Douglas, G., Riding, R.J.: The effect of pupil cognitive style and position of prose passage title on recall; in Educational Psychology, Vol. 13; Goldstein, E.B. (eds.), Sensation and perception, Wadsworth Inc.; pp 385-393, 1993.

[Dreher und Maurer 2006] Dreher, H., Maurer, H.: The Worth of Anonymous Feedback; in Proceedings of 19th Bled eConference "eValues", in Bled, Slovenia; 2006.

[Dron 2006] Dron, J.: Social Software and the Emergence of Control; in Proceedings of the IEEE International Conference on Advanced Learning Technologies (ICALT 2006), in Kerkrade, The Netherlands; Kinshuk and R. Koper and P. Kommers and P. Kirschner and D. G. Sampson and W. Didderen (eds.), IEEE Computer Society Press; pp. 904-908, 2006.

[Duchowski 2000] Duchowski, A.: Eye-Based Interaction in Graphical Systems: Theory & Practice; in SIGGRAPH 2000, Course Notes; 2000.

[Duval 2001] Duval, E.: Metadata Standards: What, Who & Why; in Journal of Universal Computer Science (J.UCS – http://www.jucs.org), Volume 7, Number 7; pp. 591-601, 2001.

[Duval et al. 1998] Duval, E., Hendrikx, K., Olivié, H.: Building Hypermedia with Objects and Sets; in Journal of Universal Computer Science (J.UCS – http://www.jucs.org), Volume 4, Number 5; pp. 501-521, 1998.

[Durand 1998a] Durand, T.: Forms of incompetence; In Proceedings of the 4th International Conference on Competence-Based Management, Norwegian School of Management; Oslo, Norway; pp. 237-253, 1998.

[Durand 1998b] Durand, T.: The Alchemy of Competence; in Strategic Flexibility: Managing in a Turbulent Environment; G. Hamel, C. K. Prahalad, H. Thomas and D. O`Neal (eds.); pp. 325, 1998.

[Dwyer 1970] Dwyer, T.A.: Project Solo: a statement of position regarding CAI and creativity; in Journal SIGCUE Outlook, Vol. 4, Nr. 1, ACM Press; pp. 13-15, 1970.

E

[Ebner et al. 2006] Ebner, M., Scerbakov, N., Maurer, H.: New Features for eLearning in Higher Education for Civil Engineering; in J. of Universal Science and Technology of Learning (J.USTL), Vol. 0, No. 0; pp. 93-106, 2006.

[Edmonds 1981] Edmonds, E. A. (1981). Adaptive Man–Computer Interfaces. In Coombs, M.C. & Alty, J.L. Eds. Computing Skills and the User Interface. London: Academic Press, pp. 389–426.

[Edmonds 1987] Edmonds, E. A.: Adaptation, response and knowledge; Knowledge-Based Systems, Vol. 1 Nr. 1; Editorial, 1987.

[Eklund and Zeiliger 1996] Eklund, J., Zeiliger, R.: Navigating the Web: Possibilities and Practicalities for Adaptive Navigational Support; in Proceedings of Second Australian World Wide Web Conference, AusWeb96; 1996.

[Endrei et al. 2004] Endrei, M., Ang, J., Arsanjani, A., Chua, S., Comte, P., et al. : Patterns: Service-Oriented Architecture and Web Services; IBM Redbook; 2004. URL http://www.redbooks.ibm.com/redbooks/SG246303 Last visit: 2008-01-12.

F

[Far and Hashimoto 2000] Far, B.H., Hashimoto, A.H.: A Computational Model for Learner's Motivation States in Individualized Tutoring System; in Proceedings of the 8th International Conference on Computers in Education (ICCE 2000); pp. 21-24, 2000.

[Feeney and Hood 1977] Feeney, W.R., Hood, J.: Adaptive man/computer interfaces: information systems which take account of user style; SIGCPR Comput. Pers., Vol. 6, Nr. 3-4, ACM Press, New York, NY, USA; pp. 4-10, 1977.

[Felder and Silverman 1988] Felder, R. M., Silverman, L. K. Learning and Teaching Styles in Engineering Education; in J. Engineering Education, Vol. 78 Nr. 7; pp. 674-681, 1988.

[Fernandez et al. 2005] Fernandez, J., Fernandez, A., Pazos, J.: Optimizing Web services performance using caching; in Proceedings of the International Conference on Next Generation Web Services Practices (NWeSP'05); pp. 6-12, 2005.

[Ferry et al. 1997] Ferry, B., Hedberg, J., Harper, B.: How do Preservice Teachers use Concept Maps to Organize Their Curriculum Content Knowledge?; in Proceedings of ASCILITE 97, Perth, Western Australia, Australia; 1997. URL http://www.ascilite.org.au/conferences/perth97/papers/Ferry/Ferry.html Last visit: 2008-01-12.

[Fielding 2000] Fielding, R.T.: Architectural styles and the design of network-based software architectures; PhD Thesis at University of California, Irvine; 2000.

[Filippini-Fantoni et al. 2005] Filippini-Fantoni, S., Bowen, J.P., Numerico, T.: Personalization Issues for Science Museum Websites and E-learning; in E-learning and Virtual Science Centers; Subramaniam, R. (ed.), Idea Group (pub.), Hershey, USA; 2005.

[Finin and Drager 1986] Finin, T., Drager, D.: GUMS₁: A General User Modeling System; in Proceedings of Strategic Computing Natural Language Workshop of the Human Language Technology Conference 1986, Marina del Rey, CA, USA; pp. 224-230, 1986.

[Finin 1989] Finin, T.: GUMS - A General User Modeling Shell; in User Models in Dialog Systems; Kobsa, A., Wahlster, W. (eds.), Springer Verlag; pp. 411-430, 1989.

[Fink 2003] Fink, J.: User Modeling Servers - Requirements, Design, and Evaluation; PhD Thesis at University of Duisburg-Essen (Essen), Mathematics; Chapters 1, 2, 6; 2003.

[Fink and Kobsa 2000] Fink, J., Kobsa, A.: A Review and Analysis of Commercial User Modeling Servers for Personalization on the World Wide Web; in User Modeling and User-Adapted Interaction (UMUAI), Vol. 10; Kluwer Academic Publishers, The Netherlands; pp. 209-249, 2000.

[Fischer 1999] Fischer, G.: User Modeling: The Long and Winding Road; in Proceedings of User Modelling Conference (UM99), Banff, Canada; Judy Kay (ed.), Springer Verlag (pub.), Vienna, New York; pp. 349-355, 1999.

[Forehand 2005] Forehand, M.: Bloom's taxonomy: Original and revised; in Emerging perspectives on learning, teaching, and technology; M. Orey (ed.); 2005. URL http://www.coe.uga.edu/epltt/bloom.htm Last visit: 2008-02-19.

[Foshay 1998] Foshay, R.: Instructional philosophy and strategic direction of the PLATO system; PLATO, Inc., ERIC Document Reproduction Service; Edina, MN; pp. 464 603, 1998.

[Francois 1999] François, C.: Systemics and Cybernetics in a Historical Perspective; in Systems Research and Behavioral Science Syst. Res. 16; pp 203-219, 1999.

[Frank 1996] Frank, S. A.: The design of natural and artificial adaptive systems; in Adaptation; M. R. Rose and G. V. Lauder (eds.); Academic Press (pub.), New York; pp 451-505, 1996.

[Frank et al. 2002] Frank, M., Kurtz, G., Levin, N.: Implications of Presenting Pre-University Courses Using the Blended e-Learning Approach; in Educational Technology & Society, Vol. 5 Nr. 4; International Forum of Educational Technology & Society (IFETS), pubs.; pp. 137-147, 2002.

[Freyne and Smyth 2006] Freyne, J., Smyth, B.: Cooperating Search Communities; in Proceedings of the 4th International Conference on Adaptive Hypermedia and Adaptive Web-Based Systems (AH2006), in Dublin, Ireland; Wade, V., Ashman, H., Smyth, B. (eds.), LNCS 4018, Springer Verlag, Berlin Heidelberg; pp. 101-110, 2006.

[Friedman 2002] Friedman, R.: Caching web services in mobile ad-hoc networks: opportunities and challenges; in Proceedings of the 2nd ACM International Workshop on Principles of Mobile Computing (POMC '02), in Toulouse, France; ACM Press, New York, NY, USA; pp. 90-96, 2002.

[Fritz et al. 1992] Fritz, A., Galley, N., Groetzner, Ch.: Zum Zusammenhang von Leistung, Aktivierung und Motivation bei Kindern mit unterschiedlichen Hirnfunktionsstörungen; Zeitschrift für Neuropsychologie, 1, Heft 1; pp. 79-92, 1992. (in German language)

[Fröschl 2005] Fröschl, Ch.: User Modeling and User Profiling in Adaptive E-learning Systems: An approach for a service-based personalization solution for the research project AdeLE (Adaptive e-Learning with Eye-Tracking); Master's Thesis at Institute for Information Systems and Computer Media (IICM), Faculty of Computer Science, Graz University of Technology, Austria; 2005.

G

[Galley 2001] Galley, N.: Physiologische Grundlagen, Meßmethoden und Indikatorfunktion der okulomotorischen Aktivität; In Frank Rösler (ed.): Enzyklopädie der Psychologie, 4, Grundlagen und Methoden der Psychophysiologie; pp. 237-315, 2001. (in German language)

[García-Barrios 2002] García-Barrios, V.M.: Information Enhancing and Knowledge Organisation in Corporate Enterprises; M.Sc. Thesis at Institute for Information Processing and Computer Supported

New Media (IICM), Faculty of Computer Science, Graz University of Technology, Austria; chapters 2 to 5; 2002.

[García-Barrios 2006a] García-Barrios, V.M.: A Concept-based Enhancement of Didactical Goals and Learning Needs with a Dynamic Background Library: Semantics vs. Pragmatics; in Proceedings of the IEEE International Conference on Advanced Learning Technologies (ICALT 2006), in Kerkrade, The Netherlands; Kinshuk and R. Koper and P. Kommers and P. Kirschner and D. G. Sampson and W. Didderen (eds.), IEEE Computer Society Press; pp. 1-3, 2006.

[García-Barrios 2006b] García-Barrios, V.M.: Adaptive E-Learning Systems: Retrospection, Opportunities and Challenges; In ITI 2006 Proceedings of the 28th International Conference on Information Technology Interfaces, IEEE (Reg. 8), Cavtat/Dubrovnik, Croatia; V. Luzar and V. Hljuz Dobric (eds.), University Computing Centre SRCE, University of Zagreb (pub.), Zagreb, Croatia; pp. 53-58, 2006.

[García-Barrios 2006c] García-Barrios, V.M.: Finding the Missing Link: Enhancement of Semantic Representations through a Pragmatic Model, In Proceedings of the 6th International Conference on Knowledge Management (I-KNOW'06), Graz, Austria; K. Tochtermann and H. Maurer (eds.), Springer (pub.); pp. 296-303, 2006.

[García-Barrios 2006d] García-Barrios, V.M.: Personalised and Context-based Access to Corporate Knowledge: a Multi-modal and Multi-model Solution Approach for Learning Activities; In Proceedings of the 6th International Conference on Knowledge Management (I-KNOW'06), Graz, Austria; K. Tochtermann and H. Maurer (eds.), Springer (pub.); pp. 387-394, 2006.

[García-Barrios 2006e] García-Barrios, V.M.: Real-Time Learner Modeling: Using Gaze-Tracking in Distributed Adaptive E-Learning Environments; In Proceedings of the International Convention MIPRO 2006 (Opatija, Croatia), Volume IV - CE Computers in Education; M. Cicin-Sain and I. T. Prstacic and I. Sluganovic (eds.), MIPRO Croatian Society (pub.); pp. 185-190, 2006.

[García-Barrios and Gütl 2006] García-Barrios, V.M., Gütl, G.: Exploitation of MPEG-7 Descriptions on Multi-modal Meeting Data: First Results within MISTRAL Project; In Journal of Universal Knowledge Management (J.UKM – http://www.jukm.org), Volume 1, Issue 1; pp. 45-53, 2006.

[García-Barrios et al. 2002] García-Barrios, V.M., Gütl, C., Pivec, M.: Semantic Knowledge Factory: A New Way of Cognition Improvement for the Knowledge Management Process; in Proceedings of Society for Information Technology and Teacher Education; Nashville, USA. 2002.

[García-Barrios et al. 2004a] García-Barrios, V.M., Gütl, C., Mödritscher, F.: EHELP - Enhanced E-Learning Repository: The Use of a Dynamic Background Library for a Better Knowledge Transfer Process; In Proceedings of the International Conference on Interactive Computer Aided Learning (ICL 2004), Villach, Austria; M. Auer and U. Auer (eds.), Carinthia Tech Institute (pub); 2004.

[García-Barrios et al. 2004b] García-Barrios, V.M., Gütl, C., Preis, A., Andrews, K., Pivec, M., Mödritscher, F., Trummer, C.: AdELE: A Framework for Adaptive E-Learning through Eye Tracking; In Proceedings of the International Conference on Knowledge Management (I-KNOW'04), Graz, Austria; K. Tochtermann and H. Maurer (eds.), Springer (pub.); pp. 609-616, 2004.

[García-Barrios et al. 2005] García-Barrios, V.M., Mödritscher, F., Gütl, C.: Personalisation versus Adaptation? A User-centred Model Approach and its Application; In Proceedings of the International Conference on Knowledge Management (I-KNOW'05), Graz, Austria; K. Tochtermann and H. Maurer (eds.), Springer (pub.); pp. 120-127, 2005.

[Gärdenfors 2001] Gärdenfors, P.: Concept Learning: A Geometrical Model; in Proceedings of the Aristotelian Society, Vol. 101; pp. 163-183, 2001.

[Glaser 1984] Glaser, R.: Education and Thinking: The Role of Knowledge; in American Psychologist, Vol. 39; pp.93-104, 1984.

[Goldberg et al. 2002] Goldberg, J.H., Stimson, M.J., Lewenstein, M., Scott. N., Wichansky, A.M.: Eye Tracking in Web Search Tasks: Design Implications; in Proceedings of Symposium on ETRA2002, New Orleans, Louisiana, USA; pp. 51-58, 2002.

[Gonzalez et al. 2006] Gonzalez, C., Burguillo, J.C., Llamas, M.: A Qualitative Comparison of Techniques for Student Modeling in Intelligent Tutoring Systems; in Proceedings of 36th Annual Conference on Frontiers in Education (FIE2006), Session T1F: Computer and Web-based Software III, San Diego, CA, USA; pp. 13-18, 2006.

[Grcar et al. 2001] Grcar, M., Mladenic, D., Grobelnik, M.: User Profiling for the Web; in ComSIS Vol. 3 Nr. 2: pp.2-29, 2001.

[Greer et al. 1998] Greer, J., McCalla, G., Cooke, J., Collins, J., Kumar, V., Bishop, A., Vassileva, J.: The Intelligent HelpDesk: Supporting Peer Help in a University Course; in Proceedings of Intelligent Tutoring Systems (ITS'98), in San Antonio, TX, USA; LNCS No1452, Springer Verlag (pub.), Berlin; pp. 494-503, 1998.

[Greer et al. 2001] Greer, J., McCalla, G., Vassileva, J., Deters, R., Bull, S., Kettel, L.: Lessons Learned in Deploying a Multi-Agent Learning Support System: The I-Help Experience; in Proceedings of Artificial Intelligence in Education (AIED'2001), in San Antonio; IOS Press, Amsterdam; pp. 410-421, 2001.

[Green and Raphael 1968] Green, C.C., Raphael, B.: The use of theorem-proving techniques in question-answering systems; In Proceedings of the 1968 23rd ACM national conference; ACM Press; pp 169-181, 1968.

[Green et al. 1996] Green, T.R.G., Davies, S.P., Gilmore, D.J.: Delivering cognitive psychology to HCI - the problems of common language and of knowledge transfer, Vol. 8, No. 1; pp. 89-111, 1996.

[Gütl 2002] Gütl, C.: Ansätze zur modernen Wissensauffindung im Internet: Eine Annäherung an das Information Gathering and Organizing System xFIND (Extended Framework for INformation Discovery); Dissertation at Graz University of Technology, Austria; pp. 84-97, 2002. (In German language)

[Gütl 2007] Gütl, C.: e-Examiner - Towards a Fully-Automatic Knowledge Assessment Tool applicable in Adaptive E-Learning Systems; in Proceedings of the 2nd International Conference on Interactive Mobile and Computer Aided Learning (IMCL 2007), Jordan; 2007.

[Gütl and García-Barrios 2005a] Gütl, C., García-Barrios, V.M.: Towards an Advanced Modeling System Applying a Service-Based Approach; In Proceedings of the 5th IEEE International Conference on Advanced Learning Technologies (ICALT 2005), Kaohsiung, Taiwan; IEEE Computer Society Press (pub.); pp. 860-862, 2005.

[Gütl and García-Barrios 2005b] Gütl, C., García-Barrios, V.M.: The Application of Concepts for Learning and Teaching, In Proceedings of the International Conference on Interactive Computer Aided Learning (ICL 2005), Villach, Austria, M. Auer and U. Auer (eds.), Carinthia Tech Institute (pub.); 2005.

[Gütl and García-Barrios 2005c] Gütl, C., García-Barrios, V.M.: Semantic Meeting Information Application: A Contribution for Enhanced Knowledge Transfer and Learning in Companies; In Proceedings of 8th International Conference on Interactive Computer Aided Learning (ICL 2005), Villach, Austria, 2005.

[Gütl and García-Barrios 2005d] Gütl, C., García-Barrios, V.M.: MPEG-7 Requirements and Application for a Multi-Modal Meeting Information System: Insights and Prospects within the MISTRAL Research Project; In Proc. Of MPEG and Multimedia Metadata Community (2nd Workshop), 5th International Conference on Knowledge Management I-KNOW'05, Graz, Austria, 2005.

[Gütl and García-Barrios 2006] Gütl, C.; García-Barrios, V.M.: Smart Multimedia Meeting Information Retrieval for Teaching and Learning Activities; In Proceedings of the International Conference on Society for Information Technology and Teacher Education (SITE 2006); C. M. Crawford and R. Carlsen and K. McFerrin and J. Price and R. Weber and D. A. Willis (eds.), AACE (pub.); 2006.

[Gütl and Pivec 2003] Gütl, C., Pivec, M.: A Multimedia Knowledge Module Virtual Tutor Fosters Interactive Learning; in Journal of Interactive Learning Research (JILR), Vol. 14 Nr. 2; pp. 231-258, 2003.

[Gütl and Safran 2006] Gütl, C., Safran, C.: Personalized Access to Meeting Recordings for Knowledge Transfer and Learning Purposes in Companies; in Proceedings of the 4th International Conference on Multimedia and Information and Communication Technologies in Education (m-ICTE 2006), Seville, Spain; 2006.

[Gütl et al. 2002] Gütl, C., Pivec, M., García-Barrios, V.M.,: Quality Metadata Scheme xQMS for an Improved Information Discovery Process for Scholar Work within the xFIND Environment; in Proceedings of SITE 2002, AACE; Nashville, USA. 2002.

[Gütl et al. 2004] Gütl, C., García-Barrios, V.M., Mödritscher, F.: Adaptation in E-Learning Environments through the Service-Based Framework and its Application for AdeLE; In Proceedings of the World Conference on E-Learning in Corporate, Government, Healthcare, and Higher Education (E-Learn 2004); J. Nall and R. Robson (eds.), AACE (pub); pp. 1891-1898, 2004.

[Gütl et al. 2005a] Gütl, C., Dreher, H., Williams, R.: E-TESTER: a Computer-based Tool for Auto-generated Question and Answer Assessment, In Proceedings of the World Conference on E-Learning in Corporate, Government, Healthcare, and Higher Education (E-Learn 2005); G. Richards (ed.), AACE (pub.); pp. 2929-2936, 2005.

[Gütl et al. 2005b] Gütl, C., Pivec, M., Trummer, C., García-Barrios, V.M., Mödritscher, F., Pripfl, J., Umgeher, M.: AdeLE (Adaptive e-Learning with Eye-Tracking): Theoretical Background, System Architecture and Application Scenarios; In European Journal of Open, Distance and E-Learning (EURODL); 2005.

H

[Hafri et al. 2003a] Hafri, Y., Djeraba, C., Stanchev, P., Bachimont, B.: A Markovian Approach for Web User Profiling and Clustering; in Proceedings of PAKDD 2003; K.Y. Whang, J. Jeon, K. Shim, J. Srivatava (eds.), Springer-Verlag, Berlin Heidelberg; pp. 191-202, 2003.

[Hafri et al. 2003b] Hafri, Y., Djeraba, C., Stanchev, P., Bachimont, B.: A Web User Profiling Approach; in Proceedings of APWeb 2003; X. Zhou, Y. Zhang, and M.E. Orlowska (eds.), Springer-Verlag; pp. 227-238, 2003.

[Halasz 1988] Halasz, F.G.: Reflections on NoteCards: seven issues for the next generation of hypermedia systems; In Journal Commun. ACM, Vol. 31, Nr. 7; ACM Press; pp. 836-852, 1988.

[Halasz and Schwartz 1990] Halasz, F., Schwartz, M.: The Dexter hypertext reference model; In Proceedings of the NIST Hypertext

Standardization Workshop. Gaithersburg, MD, USA, pp. 95–133, 1990.

[Halasz and Schwartz 1994] Halasz, F., Schwartz, M.: The Dexter hypertext reference model: Hypermedia; Communications of the ACM 37(2); pp. 30–39, 1994.

[Hasebrook and Maurer 2004] Hasebrook, J.P., Maurer, H.: Learning Support Systems for Organizational Learning; World Scientific Publishing Co. Pte. Ltd.; Jain. L. C. (ed.), Series of Innovative Intelligence, Vol. 8; 2004.

[Haubelt et al. 2004] Haubelt, A., Bullinger, E., Sauter, T., Allgöwer, F., Gilles, E.D.: Systems Biology - A Glossary from two Perspectives: What Biology Says and Systems Theory Understands - and Vice Versa; Institute for Systems Theory in Engineering, & Institute for Systems Dynamics and Control Engineering, University of Stuttgart, 2004; URL http://www.sysbio.de/projects/glossary/glossar.shtml Last visit: 2008-02-21.

[Heckmann 2005] Heckmann, D.: Ubiquitous User Modeling; Dissertation Thesis at Saarland University; Saarbrücken, Germany; 2005.

[Helic 2006] Helic, D.: Technology-Supported Management of Collaborative Learning Processes; in International Journal of Learning and Change, Vol. 1, Issue 3; pp. 285-298, 2006.

[Helic et al. 2005] Denis Helic, Janez Hrastnik, Hermann Maurer: An Analysis of Application of Business Process Management Technology in E-Learning Systems, In Proceedings of E-Learn 2005; AACE, Charlottesville, USA; pp. 2937-2942, 2005.

[Helic et al. 2005b] Helic, D., Maurer, H., Scerbakov, N.: 'A Didactics Aware Approach to Knowledge Transfer in Web-based Education', In Claude Ghaoui, Mitu Jain, Vivek Bannore (Editors), Studies in Fuzziness and Soft Computing, Volume 178/2005, Chapter 9; Publisher: Springer-Verlag GmbH; pp. 233-260, 2005.

[Henze and Nejdl 2004] Henze, N., Nejdl, W.: A Logical Characterization of Adaptive Educational Hypermedia; in New Review of Hypermedia and Multimedia (NRHM), Vol. 10, Iss. 1; pp. 77-113, 2004.

[Henze et al. 2004] Henze, N., Dolog, P., Nejdl, W.: Reasoning and Ontologies for Personalized E-Learning in the Semantic Web; in Educational Technology & Society, Vol. 7 Nr. 4; pp. 82-97, 2004.

[Heylighen 2004] Heylighen, F.: Web Dictionary of Cybernetics and Systems; Principia Cybernetica Web; 2004. URL http://pespmc1.vub.ac.be Last visit: 2008-02-12.

[Hicks and Tochtermann 2001] Hicks, D., Tochtermann, K.: Personal Digital Libraries and Knowledge Management; in Journal of Universal Computer Science, Vol. 7 Nr. 7; Springer Pub.Co.; pp. 550-565, 2001.

[Hill et al. 2003] Hill, J.R., Wiley, D., Miller, L., Han, S.: Exploring Research on Internet-based Learning: From Infrastructure to Interactions; Educational Technology Research and Development, 2003(16). pp. 433-460, 2003.

[Hoeppner et al. 1983] Hoeppner, W., Christaller, T. Marburger, H., Morik, K., Nebel, B., O'Leary, M., Wahlster, W.: Beyond Domain Independence: Experience with the Development of a German Language Access System to Highly Diverse Background Systems; In 8th International Conference on Artificial Intelligence; pp. 588-594, 1983.

[Hof et al. 1998] Hof, R., Green, H., Himmelstein, L.: Now it's YOUR WEB; in Business Week; pp. 68-75, 1998.

[Hoffman 1978] Hoffman, L.L.: Test gen: A tool for emancipation from traditional instruction; in Proceedings of the 1978 annual conference, ACM Press; pp 845-848, 1978.

[Holland 1962] Holland, J.H.: Outline for a Logical Theory of Adaptive Systems; in J. ACM, Vol. 9, Nr. 3; ACM Press; pp. 297-314, 1962.

[Holland 1992] Holland, J.: Adaptation in Natural and Artificial Systems: An Introductory Analysis with Applications to Biology, Control, and Artificial Intelligence; First MIT (Massachusetts institute of Technology) Press Edition; USA, Chapter 1-3 and 9, 1992.

[Holmén 2000] Hollmén, J.: User profiling and classification for fraud detection in mobile communications networks; PhD dissertation at Department of Computer Science and Engineering, Helsinki University of Technology, (Espoo, Finland); 2000.

[Holmberg 1989] Holmberg, B.: Theory and practice of distance education; London: Routledge, 1989.

[Honey 1986] Honey, P.: The Manual of Learning Styles; Peter Honey, Maidenhead, Berks, 1986.

[Hothi and Hall 1998] Hothi, J., Hall, W.: An Evaluation of Adapted Hypermedia Techniques Using Static User Modelling; in Proceedings of the 2nd Workshop on Adaptive Hypertext and Hypermedia of the Hypertext'98 conference, in Pittsburg, USA; 1998.

[Huang et al. 2002] Huang, P., Lenders, V., Minnig, P., Widmer, M.: Mini: A Minimal Platform Comparable to Jini for Ubiquitous Computing; International Symposium on Distributed Objects and Applications (DOA); 2002.

[Hudlicka and McNeese 2002] Hudlicka, E., McNeese, M.D.: Assessment of User Affective and Belief States for Interface Adaptation: Application to an Air Force Pilot Task; User Modeling and User-Adapted Interaction (UMUAI) Vol. 12; Kluwer Academic Publishers, The Netherlands; pp.1-47, 2002.

I

[IDC 2003] IDC: Analyse the Future; U.S. Corporate and Government eLearning Forecast 2002-2007, IDC is a subsidiary of IDG (International Data Group), USA; 2003, URL http://www.idc.com - Last visit: 2008-01-12.

[IDS 2002] IDS - Instruction at FSU: A Guide to Teaching and Learning Practices; Instructional Development Services, Florida State University; 2002.

[Isermann et al. 1992] Isermann, R., Lachmann, K.-H., Matko, D.: Adaptive Control Systems; Prentice International Series in Systems and Control Engineering, M. J. Grimble (ed.); United Kingdom; pp 1-14, 1992.

[Ivory and Hearst 2001] Ivory, M.Y., Hearst, M.A.: The State of the Art in Automating Usability Evaluation of User Interfaces; ACM Computing Surveys (CSUR) Vol. 33 Nr. 4; ACM Press, USA; pp. 470-516, 2001.

J

[Jacob 1995] Jacob, R. J. K.: Eye tracking in advanced interface design; in Advanced Interface Design and Virtual Environments; Oxford University Press, Oxford; pp. 258-288, 1995.

[Jain et al. 2002] Jain, L. C., Howlett, R. J., Ischalkaranje, N. S., Tonfoni, G.: "Virtual Environments for Teaching & Learning"; World Scientific Publishing; Jain. L.C. (ed), Series of Innovative Intelligence, Vol. 1; Preface, 2002.

[Jameson 1996] Jameson, A.: Numerical Uncertainty Management in User and Student Modeling: An Overview of Systems and Issues; in User Modeling and User Adapted Interaction (UMUAI), Vol. 5, Special issue on Numerical Uncertainty Management in User and Student Modeling; 1996.

[Jameson 2001] Jameson, A.: Systems That Adapt to Their Users: An Integrative Perspective; Saarbrücken: Saarland University; 2001.

[Jameson 2003] Jameson, A.: Adaptive Interfaces and Agents; in Human-Computer Interaction Handbook, J. A. Jacko & A. Sears (Eds.), Erlbaum (Pub.), Mahwah, NJ; pp. 305-330, 2003.

[John and Mooney 2001] John, R.I,. Mooney, G.J.: Fuzzy User Modeling for Information Retrieval on the World Wide Web; in Knowledge and Information Systems Vol. 3; Springer-Verlag, London Ltd.; pp. 81-95, 2001.

[Johnson and Johnson 2003] Johnson, D.W., Johnson, R.T.: COOPERATION AND THE USE OF TECHNOLOGY; in Adaptive Instructional Systems; Educational Technology Research and Development (30); pp. 785-811, 2003.

[JXTA 2007] JXTA Project: Technology Documents; 2007; URL: http://spec.jxta.org Last visit: 2008-01-12.

K

[Kaenampornpan and O'Neill 2004] Kaenampornpan, M., O'Neill, E.: An Integrated Context Model: Bringing Activity to Context; in Proceedings of UbiComp 2004, Japan; 2004.

[Kahneman 1966] Kahneman, D., Beatty, J.: Pupil diameter and load on memory; Science, Vol. 154; pp. 1583-1585, 1966.

[Kalmey and Niccolai 1981] Kalmey, D.L., Niccolai, M.J.: A model for a CAI learning system; in SIGCSE '81: Proceedings of the twelfth SIGCSE technical symposium on Computer science education, St. Louis, Missouri, United States; ACM Press, NY, USA; pp. 74-77, 1981.

[Karampiperis and Sampson 2005] Karampiperis, P., Sampson, D.: Adaptive Learning Resources Sequencing in Educational Hypermedia Systems; in Educational Technology & Society Journal, Vol. 8 Nr. 4; pp. 128-147, 2005.

[Karypidis and Lalis 2006] Karypidis, A., Lalis, S.: Automated context aggregation and file annotation for PAN-based computing; In Journal ‚Personal Ubiquitous Computing', volume 11, number 1; Springer-Verlag (pub.), London, UK; pp. 33- 44, 2006.

[Kass and Finin 1988] Kass, R., Finin, T.: Modeling the user in natural language systems"; in Computational Linguistics Vol. 4 Nr. 3; pp. 5-22, 1988.

[Kay 1990] Kay, J.: um - a user modelling toolkit; in 2nd International User Modelling Workshop, Hawaii; USA; pp. 11, 1990.

[Kay 1994] Kay, J.: Lies, damned lies, and stereotypes: pragmatic approximations of users; in Proceedings of the 4th International Conference on User Modeling (UM1994); A. Kobsa & D. Litman (eds.); pp. 73-78, 1994.

[Kay 1995] Kay, J.: The um toolkit for cooperative user modelling; in User Modeling and User-Adapted Interaction (UMUAI), Vol. 4, Nr. 3; 149-196, 1995.

[Kay 2000] Kay, J.: User modeling for adaptation; in User Interfaces for All; Stephanidis, C. (ed.), Salvendy, G. (General ed.); Human Factors Series, Lawrence Erlbaum Associates; pp. 271-294, 2000.

[Kay 2000a] Kay,J.: Stereotypes, Student Models and Scrutability; in Proceedings of the 5th International Conference on Intelligent Tutoring Systems (ITS 2000), pp. 19–30, 2000.

[Kay and Crawford 1993] Kay, J., Crawford, K.: Metacognitive processes and learning with intelligent educational systems; In Cognitive Science Down Under; P. Slezak (ed.), Ablex; pp. 63-77, 1993.

[Kay and Lum 2005] Kay, J., Lum, A.: Ontology-based User Modelling for the Semantic Web; in Proceedings of 10th International Conference on User Modelling (UM2005), Workshop on Pesonalisation on the Semantic Web (PerSWeb2005); pp. 15-23, 2005.

[Kay et al. 2002] Kay, J., Kummerfeld, B., Lauder, P.: Personis: A Server for User Models; in Proceedings of Adaptive Hypermedia and Adaptive Web-Based Systems (AH 2002); pp. 203-212, 2002.

[Kay et al. 2005] Kay, J., Kummerfeld, B., Carmichael, D., Quigley, A.: The Scrutable Personalised Pervasive Computing Environment; in Proceedings of Intelligent User Interfaces (IUI 2005), Workshop for Multi-User and Ubiquitous User Interfaces (MU3I); San Diego, USA; 2005.

[Kay 2006] Kay, J.: Scrutable adaptation: because we can and must; in Proccedings of 4th International Conference for Adaptive Hypermedia and Adaptive Web-Based Systems (AH 2006), in Dublin, Ireland; Wade, V, H Ashman and B Smyth, (eds.), Springer (pub.); pp. 11-19, 2006.

[Keegan 1980] Keegan, D.: On defining distance education; in Distance Education, Vol. 1 Nr. 1; 13–36, 1980.

[Keegan 1986] Keegan, D.: The foundations of distance education (second ed.), London: Routledge; 1986.

[Kelly and Tangney 2006] Kelly, D., Tangney, B.: Adapting to intelligence profile in an adaptive educational system; in Journal Interacting with Computers, Volume 18, Number 3; Elsevier Science Inc. (pub.), New York, NY, USA; pp. 385-409, 2006.

[Khoi et al. 2006] Khoi, A.P., Tari, Z., Bertok, P.: Optimizing Web Services Performance by Using Similarity-Based Multicast Protocol; in Proccedings of 4th European Conference on Web Services (ECOWS '06); pp. 119-128, 2006.

[Kinshuk and Goh 2003] Kinshuk, Goh, T. T.: Mobile Adaptation with Multiple Representation Approach as Educational Pedagogy; in Wirtschaftsinformatik 2003 - Medien - Märkte – Mobilität, Uhr W., Esswein W., Schoop E. (eds.), Physica-Verlag (pub.), Heidelberg, Germany; pp. 747-763, 2003.

[Kinshuk and Lin 2003] Kinshuk, Lin, T.: User Exploration Based Adaptation in Adaptive Learning Systems; in International Journal of Information Systems in Education, Vol. 1 Nr. 1; ISSN 1348-236X; pp. 22-31, 2003.

[Kinshuk et al. 2001] Kinshuk, Han B, Hong H, Patel A: Student Adaptivity in TILE: A Client-Server Approach; in Proceedings of IEEE ICALT 2001. Madison, Wisconsin; pp. 297, 2001.

[Klann et al. 2003] Klann, M., Eisenhauer, M., Oppermann, R., Wulf, V.: Shared initiative: Cross-fertilisation between system adaptivity and adaptability; in Universal Access in HCI; Constantine Stephanidis (ed.), Lawrence Erlbaum Associates Mahwah; pp. 562-566, 2003.

[Kobsa 1990a] Kobsa, A.: Modeling The User's Conceptual Knowledge in BGP-MS, a User Modeling Shell System; Computational Intelligence (6), pp. 193-208, 1990.

[Kobsa 1990b] Kobsa, A.: User Modeling in Dialog Systems: Potentials and Hazards; in AI and Society, The Journal of Human and Machine Intelligence, Vol. 4; pp. 214-231, 1990.

[Kobsa 1993] Kobsa, A.: User Modeling: Recent Work, Prospects and Hazards; in Adaptive User Interfaces: Principles and Practise; Schneider-Hufschmidt, M., Kühme, T. Malinowski, U. (eds.), North-Holland, Amsterdam, The Netherlands; 1993.

[Kobsa 1995] Kobsa, A., Pohl, W.: The User Modeling Shell System BGP-MS; in User Modeling and User-Adapted Interaction, Vol. 4, Nr. 2; pp. 59-106, 1995.

[Kobsa 2001] Kobsa, A.: Generic User Modeling Systems; in User Modeling and User-Adapted Interaction, Volume 11: Kluwer Academic Publishers, The Netherlands; pp. 49-63, 2001.

[Kobsa and Fink 2003] Kobsa, A.; Fink, J.: Performance Evaluation of User Modeling Servers under Real-World Workload Conditions; In Proceedigns of User Modelling (UM) 2003, in Johnstown, USA; 2003.

[Kobsa and Fink 2006] Kobsa, A., Fink, J.: An LDAP-Based User Modeling Server and its Evaluation; in User Modeling and User-Adapted Interaction: The Journal of Personalization Research, Vol. 16, Nr. 2; pp.129-169, 2006.

[Kobsa and Pohl 1995] Kobsa, A., Pohl,. W.: The User Modeling Shell System BGP-MS; in User Modeling and User-Adapted Interaction, Vol. 4, Nr. 2; pp. 59-106, 1995.

[Kobsa and Schreck 2003] Kobsa, A., Schreck, J.: Privacy through Pseudonymity in User-Adaptive Systems; in ACM Transactions on Internet Technology, Vol. 3 Nr. 2; ACM Press; pp. 149-183, 2003.

[Kobsa and Teltzrow 2006] Kobsa, A., Teltzrow, M.: Convincing Users to Disclose Personal Data; in Proceedings of the CHI2006 Workshop on Privacy-Enhanced Personalization; Alfred Kobsa, Ramnath K. Chellappa, and Sarah Spiekermann (eds.); pp. 39-41, 206.

[Kobsa et al. 2001] Kobsa, A., Koenemann, J., Pohl,. W.: Personalized Hypermedia Presentation Techniques for Improving Online Customer Relationships; in The Knowledge Engineering Review, volume 16 number 2; Cambridge University Press, United Kingdom; pp. 111-155.

[Koch 2000] Koch, N.P.: Software Engineering for Adaptive Hypermedia Systems: Reference Model, Modeling Techniques and Development Process; PhD Thesis at Ludwig-Maximilians-University of Munich, Germany; 2000.

[Koch and Wirsing 2002] Koch, N., Wirsing, M.: The Munich Reference Model for Adaptive Hypermedia Applications; in Adaptive Hypermedia and Adaptive Web-Based Systems: Second International Conference, AH 2002, Vol. 2347 of Lecture Notes in Computer Science. Malaga, Spain, p. 213, 2002.

[Kokkinaki 1997] Kokkinaki, A.I.: On Atypical Database Transactions: Identification of Probable Frauds using Machine Learning for User Profiling; in Proceedings of IEEE Knowledge and Data Engineering Exchange Workshop (KDEX '97); pp. 107-113, 1997.

[Kolbitsch and Maurer 2005] Kolbitsch, J., Maurer, H.: Dynamic Adaptation of Content and Structure in Electronic Encyclopaedias; 2005; URL http://www.kolbitsch.org/research/papers/2005-JoDI-Content_Adaptation_in_Encyclopaedias.pdf Last visit: 2008-02-25.

[Koyama et al. 2001] Koyama, A., Barolli, L., Tsuda, A., Cheng, Z.: An Agent-based Personalized Distance Learning System; In Proceedings of the 15th International Conference on Information Networking (ICOIN'01); Beppu City, Oita, Japan; pp. 895-899, 2001.

[Kravcik and Gasevic 2006] Kravcik, M., Gasevic, D.: Adaptive hypermedia for the semantic web; in Proceedings of the joint international workshop on Adaptivity, personalization and the semantic web, APS '06; Odense, Denmark; ACM Press (pub.), New York, NY, USA; pp. 3-10, 2006.

[Krottmaier 2004] Krottmaier, H.: The Need for Sharing User-Profiles in Digital Libraries; in Proceedings of the 8th ICCC/IFIP International Conference on Electronic Publishing (ELPUB 2004); pp. 323-330, 2004.

[Kurhila et al. 2001] Kurhila, J., Miettinen, M., Niemivirta, M., Nokelainen, P., Silander, T., Tirri, H.: Bayesian Modeling in an Adaptive On-Line Questionnaire for Education and Educational Research; in Proceedings of the 10th International PEG Conference, pp. 194-201, 2001.

L

[Langley 1999] Langley, P.: User Modeling in Adaptive Interfaces; in Proceedings of the 7th International Conference on User Modeling, Banff, Alberta, Canada; Springer Verlag; 357-370, 1999.

[Lankhorst et al. 2002] Lankhorst, M.M., van Kranenburg, H., Salden, A., Peddemors, A.J.H.: Enabling Technology for Personalizing Mobile Services; in Proceedings of 35th Annual Hawaii International Conference on System Sciences (HICSS'02), Volume 3; Big Island, Hawaii; pp. 87-94, 2002.

[Lascio et al. 1999] Di Lascio, L., Fischetti, E., Gisolfi, A.: A Fuzzy-Based Approach to Stereotype Selection in Hypermedia; in User Modeling and User-Adapted Interaction (UMUAI), Vol. 9; Kluwer Academic Publishers, The Netherlands; pp. 285-320, 1999.

[Lendaris 1964] Lendaris, G.G.: On the Definition of Self-Organizing Systems; in Proceedings of IEEE, 52, 1964.

[Lenders et al. 2001] Lenders, V., Huang, P., Muheim, M.: Hybrid Jini for Limited Devices; in Proceedings of IEEE International Conference on Wireless LANs and Home Networks (ICWLHN 2001), Singapore; 2001.

[Lennon and Maurer 1994] Lennon, J., Maurer, H.: Applications and Impact of Hypermedia Systems: An Overview; In Journal of Universal Computer Science (J.UCS – http://www.jucs.org), Volume 0, Number 0; pp 54-107, 1994.

[Liaw et al. 2003] Liaw, S.S., Ting, I.H., Tsai, Y. C.: Developing a conceptual model for designing a Web assisted information retrieval system; in Proceedings of the 2003 International Conference on Computer-Assisted Instruction (ICCAI2003); 2003.

[Li and Ji 2005] Li, X., Ji, Q.: Active Affective State Detection and Assistance with Dynamic Bayesian Networks; in IEEE Transactions on Systems, Man, and Cybernetics, Special Issue on Ambient Intelligence, Vol. 35, Nr. 1; pp. 93-105, 2005.

[Li et al. 2001] Li, Q., Yang, J., Zhuang, Y.: Web-based Multimedia Retrieval: Balancing out between Common Knowledge and Personalized Views; in Proceedings of 2nd International Conference on Web Information System Engineering, Kyoto, Japan; pp. 92-101, 2001.

[Littlejohn et al. 2003] Littlejohn, A., Buckingham-Shum, S.: Reusing Online Resources: A Sustainable Approach to eLearning; In Journal of Interactive Media in Education, Special Issue on Reusing Online Resources; ISSN: 1365-893X, 2003.

[Liu et al. 2006] Liu, J., Zheng, Q., Chan. F.: A Method for User Behavior Modeling Based on Web Page Metadata; in Proceeding of 10th International Conference on Computer Supported Cooperative Work in Design; pp. 1-6, 2006.

[Lock and Kudenko 2006] Lock, Z., Kudenko, D.: Interactions between Stereotypes; in Proceedings of the 4th International Conference on Adaptive Hypermedia and Adaptive Web-Based Systems (AH2006), in Dublin, Ireland; Wade, V., Ashman, H., Smyth, B. (eds.), LNCS 4018, Springer Verlag, Berlin Heidelberg; pp. 172-181, 2006.

[Lockee et al. 2003] Lockee, B., Moore, D.M., Burton, J.: Foundations of Programmed Instruction; Educational Technology Research and Development 2003(25); pp. 545-569, 2003.

[Lukas 1997] Lukas, J.: Modellierung von Fehlkonzepten in einer algebraischen Wissensstruktur; in Journal Kognitionswissenschaft, Vol. 6, Nr. 4; Springer Berlin / Heidelberg (pub.); pp. 196-204; 1997. (In German)

M

[Mackinnon and Wilson 1996] Mackinnon, L., Wilson, M.: User Modelling for Information Retrieval from Multidatabases; in Proceedings of the 2nd ERCIM (European Research Consortium for Informatics and Mathematics) Workshop on User Interfaces for All, in Prague, Czech Republic; Stephanidis, C, (ed.); pp. 7-8, 1996.

[Mader 2007] Mader , H.: Visualizing Multidimensional Metadata; Master's Thesis at Institute for Information Systems and Computer Media (IICM) at Graz University of Technology, Austria; 2007.

[Martens 2003] Martens, A.: Centralize the Tutoring Process in Intelligent Tutoring Systems; in Proceedings of the 5th International Conference on New Educational Environments, ICNEE Lucerne, Switzerland; 2003.

[Maurer 1985] Maurer, H.A.: Authoring systems for computer assisted instruction; in ACM '85 Proceedings of the 1985 ACM annual conference on The range of computing: mid-80's perspective, Denver, Colorado, USA; ACM Press, NY, USA; pp. 551-561, 1985.

[McLoughlin 1999] McLoughlin, C.: The implications of the research literature on learning styles for the design of instructional material; in Australian Journal of Educational Technology, Vol. 15, Nr. 3; pp. 222-241, 1999.

[Melnikov 1978] Melnikov, G. P.: Systemology and Linguistic Aspects of Cybernetics; Translated from the Russian by J.A. Cooper, Brunel University, U. K., Gordon And Breach (New York, London, Paris, Montreux, Tokyo, Melbourne); 1988, URL http://www.philol.msu.ru/~lex/melnikov/meln_en/title_e.htm Last visit: 2008-02-21; Original in Russian language: http://www.philol.msu.ru/~lex/melnikov/meln_r/titl.htm Last visit: 2008-02-21.

[Merrian-Webster 2007] Merrian-Webster Online Search. URL http://www.m-w.com Last visit: 2008-01-21.

[Merten and Conati 2006] Merten, C., Conati, C.: Eye-tracking to model and adapt to user meta-cognition in intelligent learning environments; in Proceedings of the 11th international conference on Intelligent user interfaces; ACM Press, New York, NY, USA; pp. 39-46, 2006.

[Middleton et al. 2004] Middleton, S.E., Shadbolt, N.R., De Roure, D.C.: Ontological user profiling in recommender systems; in ACM Journal Transactions of Information Systems, Vol. 22, Nr. 1; ACM Press, New York, NY, USA; pp. 54-88, 2004.

[Millan and Perez 2002] Millán, E., Perez, J. L.: A Bayesian Diagnostic Algorithm for Student Modeling and its Evaluation; in User Modeling and User-Adapted Interaction (UMUAI), Vol. 12; Kluwer Academic Publishers, The Netherlands; pp. 281-330, 2002.

[Mobasher et al. 2002] Mobasher, B., Dai, H., Luo, T., Nakagawa, M.: Using Sequential and Non-Sequential Patterns in Predictive Web Usage Mining Tasks; in Proceedings of the IEEE International Conference on Data Mining (ICDM'02), in Maebashi City, Japan; pp. 669-672, 2002.

[Mödritscher et al. 2004a] Mödritscher, F., García-Barrios, V.M., Gütl, C.: Enhancement of SCORM to support adaptive E-Learning within the Scope of the Research Project AdeLE; In Proceedings of the World Conference on E-Learning in Corporate, Government, Healthcare, and Higher Education (E-Learn 2004), Washington, DC, USA; J. Nall and R. Robson (eds.), AACE (pub.); pp. 2499-2505, 2004.

[Mödritscher et al. 2004b] Mödritscher, F., Gütl, C., García-Barrios, V.M., Maurer, H.: Why do Standards in the Field of E-Learning not fully support Learner-centred Aspects of Adaptivity?; In Proceedings of the World Conference on Educational Multimedia, Hypermedia and Telecommunications (ED-MEDIA 2004); L. Cantoni and C. McLoughlin (eds.), AACE (pub.); pp. 2034—2039, 2004

[Mödritscher et al. 2004c] Mödritscher, F., García-Barrios, V.M., Gütl, C.: The Past, the Present and the Future of adaptive E-Learning: An Approach within the Scope of the Research Project AdeLE, In Proceedings of the International Conference on Interactive Computer Aided Learning (ICL 2004), Villach, Austria; M. Auer and U. Auer (eds.), Carinthia Tech Institute (pub); 2004.

[Mödritscher et al. 2005] Mödritscher F, García-Barrios V-M, Maurer, H.: The Use of a Dynamic Background Library within the Scope of adaptive e-Learning; In Proceedings of the World Conference on E-Learning in Corporate, Government, Healthcare, and Higher Education (E-Learn 2005); G. Richards (ed.), AACE (pub.); pp. 3045-3052, 2005.

[Mödritscher et al. 2006a] Mödritscher, F., García-Barrios, V.M., Gütl, C., Helic, D.: The first AdeLE Prototype at a Glance. In Proceedings of the World Conference on Educational Multimedia, Hypermedia and Telecommunications (ED-MEDIA 2006), Orlando, USA; E. Pearson and P. Bohman (eds.), AACE (pub); pp. 791-798, 2006.

[Mödritscher et al. 2006b] Mödritscher, F., Spiel, S., García-Barrios, V.M.: Assessment in E-Learning Environments: A Comparison of three Methods; In Proceedings of the International Conference on Society for Information Technology and Teacher Education (SITE 2006), Orlando, FLA, USA; AACE (pub.); pp. 108-113, 2006.

[Moore 1973] Moore, M.G.: Toward a theory of independent learning and teaching; in Journal of Higher Education, Vol. 44; pp. 66-69, 1973.

[Morik and Rollinger 1985] Morik, K., Rollinger, C-R.: The Real-Estate Agent - Modeling Users by Uncertain Reasoning; in AI Magazine, Vol. 6; pp. 44-52, 1985.

[Muehlen et al. 2005] Muehlen, M., Nickerson, J.V., Swenson, K.D.: Developing Web Services Choreography Standards - The Case of REST vs. SOAP; in Decision Support Systems, Vol. 40 Nr 1; pp. 9-29, 2005.

[Müller and Wiesinger 2006] Müller, H., Wiesinger, F.: Modelling "user understanding" in simple communication tasks; in Proceedings of the international workshop in conjunction with AVI 2006 on Context in advanced interfaces (CAI '06); ACM Press, New York, NY, USA; pp. 39-43, 2006.

[Mullen 2003] Mullen, K.: A mathematical theory of adaptation: Senior project submitted to the Division of Natural Sciences and Mathematics of Bard College, 2003. URL http://www.nat.vu.nl/~kate/DOC.pdf Last visit: 2008-02-21.

[Murray 1987] Murray, D.M.: Embedded User Models; In Proceedings of the 2nd IFIP International Conference on Human-Computer Interaction (INTERACT' 87); pp. 229-235, 1987.

N

[Nakakoji et al. 2003] Nakakoji, K., Yamada, K., Yamamoto, Y., Morita, M.: A Conceptual Framework for Learning Experience Design; in Proceedings of the 1st Conference on Creating, Connecting and Collaborating through Computing (C5'03); pp. 76-83, 2003.

[Nelson 1970] Nelson, T.H.: No More Teacher's Dirty Looks; Article in Computer Decisions, 9 (8), pp. 16-23, 1970.

[Nelson 1974] Nelson, T.H.: Computer Lib/Dream Mashines, 1974. Microsoft Press; 1974.

[Ng 2006] Ng, A.: Optimising Web services performance with table driven XML; In Proceedings of the 2006 Australian Software Engineering Conference (ASWEC'06); pp. 10-20; 2006.

[Nickull 2005] Nickull, D.: Service Oriented Architecture; Whitepaper, Adobe Systems Inc.; 2005; URL http://www.adobe.com/enterprise/pdfs/Services_Oriented_Architec ture_from_Adobe.pdf Last visit: 2008-02-21.

[Nirmalani and Stock 2003] Nirmalani Gunawardena, C., Stock McIsaac, M.: Distance Education; Educational Technology Research and Development, Vol. 25; pp. 355-395, 2003.

[NISO 2006] NISO Web Services and Practices Working Group: Best Practices for Designing Web Services in the Library Context - A Recommended Practice of the National Information Standards Organization; NISO Press, National Information Standards Organization (pub.), Bethesda, Maryland, USA; pp.5-7, 2006.

[Niu et al. 2003a] Niu, X., McCalla, G., Vassileva, J.: Purpose-based user modeling in Multi-agent Portfolio Management System; in Proceedings of the 9th International Conference on User Modeling (UM2003); P. Brusilovsky et al. (eds.), Springer-Verlag (pub.); pp. 398-402, 2003.

[Niu et al. 2003b] Niu, X., McCalla, G., Vassileva, J.: Purpose-based Expert Finding in a Portfolio Management System; in Proceedings of the Workshop on Business Agents and the Semantic Web (BaseWeb03) at the 16th Canadian Conference on Artificial Intelligence, in Halifax, Canada; pp. 11-14, 2003.

[NLII 2004] NLII: Learning Object Repositories; National Learning Infrastructure Initiative, USA; 2004. URL http://elearning.utsa.edu/guides/LO-repositories.htm Last visit: 2008-01-12.

[Novak 2003] Novak, J.D.: The Promise of New Ideas and New Technology for Improving Teaching and Learning; in Journal of Cell Biology Education, Vol. 2; pp. 122-132, 2003.

O

[Ohene-Djan et al. 2003] Ohene-Djan, J., Gorle, M., Bailey, C. P., Wills, G. B., Davis, H. C.: Understanding Adaptive Hypermedia: An Architecture for Personalisation and Adaptivity. In Proceedings of Workshop on Adaptive Hypermedia and Adaptive Web-Based Systems AH2003, Nottingham, UK, 2003.

[OMG 2006] Object Management Group: Common Object Request Broker Architecture; 2006. URL http://www.omg.org Last visit: 2008-01-12.

[Oppermann 1994] Oppermann, R.: Adaptively supported adaptability; in International Journal of Human Computer Studies, Vol. 40 Nr. 3; pp. 455–472, 1994.

[Orwant 1991] Orwant, J.L.: The Doppelganger User Modelling System; in Proceedings of IJCAI Workshop W4: Agent Modelling for Intelligent Interaction, in Sydney, Australia; pp. 164-168, 1991.

[Orwant 1993] Orwant, J.: Doppelganger goes to school: machine learning for user modelling; M.Sc. Thesis at MIT Media Laboratory; 1993.

[Orwant 1995] Orwant, J.: Heterogeneous Learning in the Doppelgänger User Modeling System; in User Modeling and User-Adapted Interaction, 4/2; pp. 107–130, 1995.

[Orwant 1996] Orwant, J.: For want of a bit the user was lost: cheap user modeling; in IBM Systems Journal, Vol. 35; IBM Corp. (pub.), Riverton, NJ, USA; pp. 398-416, 1996.

[OWL 2007] W3C: OWL Web Ontology Language Guide; W3C Recommendation, February the 10th of 2004. URL http://www.w3.org/TR/owl-guide; Last visit: 2008-01-12.

[Oxford 2007] Oxford English Dictionary; 2007. URL http://www.oed.com Last visit: 2008-01-12.

P

[Paiva et al. 1995] Paiva, A., Self, J., Hartley, R.: Externalising learner models; in Proceedings of the World Conference on Artificial Intelligence in Education, in Washington DC, USA; J. Greer (ed.), AACE (pub.); 1995.

[Papanikolaou et al. 2002] Papanikolaou, K.A., Grigoriadou, M., Magoulas, G.D., Kornilakis, H.: Towards new forms of knowledge communication: the adaptive dimension of a web-based learning environment; in Computers & Education, Nr. 39;: Elsevier Science Ltd. (pub.); pp. 333-360, 2002.

[Papanikolaou et al. 2003] Papanikolaou K.A., Grigoriadou M., Kornilakis H., Magoulas G.D.: Personalising the Interaction in a Web-based Educational Hypermedia System: the case of INSPIRE; in User-Modeling and User-Adapted Interaction (UMUAI), Vol. 13 Nr. 3; pp. 213-267, 2003.

[Papanikolaou et al. 2004] Papanikolaou K.A., Grigoriadou M.: Designing Adaptive Instruction in the context of Adaptive Educational Hypermedia Systems; in Proceedings of the 4th Panhellenic conference on New technologies in education Conference (ETPE'04); Athens, Greece; 2004. URL http://hermes.di.uoa.gr/lab/CVs/papers%5Cpapanikolaou%5CPapa nikolaou-G-ETPE2004.pdf Last visit: 2008-01-12.

[Papazoglou 2003] Papazoglou, M.P.: Service -Oriented Computing: Concepts, Characteristics and Directions; in Proceedings of the 4th International Conference on Web Information Systems Engineering (WISE'03); 2003; URL

http://csdl.computer.org/dl/proceedings/wise/2003/1999/00/1999000 3.pdf Last visit: 2008-02-24.

[Paramythis and Loidl-Reisinger 2003] Paramythis, A., Loidl-Reisinger, S.: Adaptive Learning Environments and e-Learning Standards; in Electronic Journal of e-Learning Vol. 2 Nr. 11; pp. 181–194, 2003.

[Park and Lee 2003] Park O, Lee J.: Adaptive Instructional Systems; Educational Technology Research and Development, Vol. 25; pp. 651-684, 2003.

[Parvez and Blank 2007] Parvez, S.M., Blank, G.D.: A pedagogical framework to integrate learning style into intelligent tutoring systems; in J. Comput. Small Coll., Volume 22, Number 3; Consortium for Computing Sciences in Colleges (pub.), USA; pp. 183-189, 2007.

[Pask 1976] Pask, G.: Styles and strategies of learning; in British Journal of Educational Psychology, Vol. 46: pp. 128-148, 1976.

[Pavlov 1927] Pavlov, I. P.: Conditioned reflexes. London: Clarendon Press; 1927.

[Pazzani and Billsus 1997] Pazzani, M., Billus, D.: Learning and Revising User Profiles: The Identification of Interesting Web Sites; in Machine Learning Vol. 27; Kluwer Academic Publishers, The Netherlands; pp. 313-331, 1997.

[Perugini and Ramakrishnan 2003] Perugini, S., Ramakrishnan, N.: Personalizing Web Sites with Mixed-Initiative Interaction; in (IEEE) IT PROFESSIONAL Vol. 5 Nr. 2; IEEE Published by the IEEE Computer Society; pp. 9-15, 2003.

[Peters 1971] Peters, O.: Theoretical aspects of correspondence instruction; In The Changing World of Correspondence Study University Park, PA: Pennsylvania State University; O. Mackenzie & E. L. Christensen (eds.); 1971.

[Peterson et al. 2003] Peterson, E.R., Deary, I.J., Austin, E.J.: On the assessment of cognitive style: Four red herrings; in Personality and Individual Differences, Vol. 34; pp. 899-904, 2003.

[Picard 1997] Picard, R.W.: Affective Computing; MIT Press, Cambridge, MA, USA; pp. 292, 1997.

[Pivec 2000] Pivec, M.: Knowledge Transfer in On-line Learning Environments; PhD work at Graz University of Technology, Austria; 2000.

[Pivec et al. 2004] Pivec, M., Preis, A., García-Barrios, V.M., Gütl, C., Müller, H., Trummer, C., Mödritscher, F.: Adaptive Knowledge Transfer in E-Learning Settings on the Basis of Eye Tracking and Dynamic Background Library; In Proceedings of EDEN 2004, Budapest, Hungary; A. Szücs and I. Bo (eds.), University of Technology and Economics (pub.); 2004.

[Pivec et al. 2005] Pivec, M., Pripfl, J., Gütl, C., García-Barrios, V.M., Mödritscher, F., Trummer, C.: AdeLE first prototype: experiences made; In Proceedings of the International Conference on Knowledge Management (I-KNOW'05), Graz, Austria; K. Tochtermann and H. Maurer (eds.), Springer (pub.); pp. 673-681, 2005.

[Prensky 2001] Prensky, M.: Digital Game-Based Learning; McGraw-Hill Companies, 1st edition; p. 414, 2001.

[Pretschner and Gauch 1999] Pretschner, A., Gauch, S.: Ontology Based Personalized Search; in Proceedings of 11th IEEE International Conference on Tools with Artificial Intelligence, Chicago, Illinois; pp. 391-398, 1999.

[Pripfl 2006] Pripfl, J.: User Behaviour Detection by Means of Eye-Tracking; 28th International Conference on Information Technology Interfaces, IEEE (Reg. 8), Cavtat/Dubrovnik, Croatia; University Computing Centre SRCE, University of Zagreb, Croatia; 2006.

Q, R

[Rasmussen 1998] Rasmussen, K. L.: Hypermedia and learning styles: Can performance be influenced?; in Journal of Multimedia and Hypermedia, Vol. 7, Nr. 4, pp. 291-308, 1998.

[Rasmussen and Davidson-Shivers 1998] Rasmussen, K. L., Davidson-Shivers, G.V.: Hypermedia and learning styles: Can performance be influenced?, in Journal of Educational Multimedia and Hypermedia, Vol. 7, Nr. 4, AACE (pub.), USA; pp. 291-308, 1998.

[Rayner and Riding 1998] Rayner, S., Riding, R. J.: Towards a categorisation of cognitive styles and learning styles; in Educational Psychology, Vol. 17; pp 5-28, 1998.

[Reeve and Sherman 1993] Reeve, H.K., Sherman, P.W.: Adaptation and the goals of evolutionary research; in Quarterly Review of Biology 68; pp 1-32, 1993. URL http://www.bio.unc.edu/courses/2003fall/biol133/Reeve%28QRB%29.pdf Last visit: 2008-01-12.

[Reigeluth 1996] Reigeluth, C. M.: A new paradigm of ISD?; in Educational Technology & Society, Vol. 36, Nr. 3; pp. 13-20, 1996.

[Reiser 1987] Reiser R-A. Instructional technology: A history. Instructional technology: Foundations; Lawrence Erlbaum Associates, New Jersey, 1987.

[Resnick and Varian 1997] Resnick, P., Varian, H. R.: Recommender systems; in Journal (Commun. ACM, Vol. 40, Nr. 3; ACM Press (pub.), New York, USA; pp. 56-58, 1997.

[Rich 1979] Rich, E.. User Modeling via Stereotypes; in Cognitive Science, vol. 3; pp. 329-354, 1979.

[Richter 1995] Richter, G.: Adaptive Systeme: Computer passen sich an; Der GMD Spiegel 2'95, Adaptive Systeme in offenen Welten, Fraunhofer-Gesellschaft and GMD-Forschungszentrum Informationstechnik GmbH, Gemany; 1995. (In German language)

[Riding 1991] Riding, R.J.: Cognitive Styles Analysis Users' Manual, Birmingham, Learning & Training Technology; 1991.

[Riding and Cheema 1991] Riding, R.J., Cheema, I.: Cognitive styles - An overview and integration; in Educational Psychology Vol. 11, Nrs. 3 and 4; pp. 193–215, 1991.

[Riding and Douglay 1993] Riding, R., Douglas, G.: The Effect of Cognitive Style and Mode of Presentation on Learning Performance; in British Journal of Educational Psychology Vol. 63; pp. 297-307, 1993.

[Riding and Rayner 1998] Riding, R., Rayner, S.: Cognitive styles and learning strategies; London, David Fulton Publishers; 1998.

[Riding and Watts 1997] Riding, R.J., Watts, M.: The effect of cognitive style on the preferred format of instructional material; in Educational Psychology Vol. 17, Nr. 1 and 2; pp. 179-183, 1997.

[Robertson 1997] Robertson, N.: A Personalized Web; in Internet World, April 1997; pp. 32-34, April 1997.

[Rollett et al. 2001] Rollett, H., Ley. T., Tochtermann, K.: Supporting Knowledge Creation: Towards a Tool for Explicating and Sharing Mental Models; in Proceedings of the 2nd European Conference on Knowledge Management, in Bled, Slovenia; pp. 569-582, 2001.

[Rosenberg 2001] Rosenberg, M.J.: e-Learning, Strategies for Delivering Knowledge in the Digital Age; McGraw-Hill., New York, USA, 2001.

[Rouse 1981] Rouse, W.B.: Human-Computer Interaction in the Control of Dynamic Systems; In Journal ACM Comput. Surv., Vol. 13, Nr. 1; ACM; pp. 71-99, 1981.

S

[Sadler-Smith and Riding 1999] Sadler-Smith, E., Riding, R.: Cognitive style and instructional preferences; in Instructional Science, Volume 27, Issue 5; Kluwer Academic Publishers, The Netherlands; pp. 355-371, 1999.

[Sadiq et al. 2002] Sadiq, S., Sadiq, W., Orlowska, M.: Workflow Driven e-Learning: Beyond Collaborative Environments; in Proceedings of the World Congress on Networked Learning in a global environment, Challenges and Solutions for Virtual Education, Technical University of Berlin, Germany; 2002.

[Safran 2006] Safran, C.: A Concept-Based Information Retrieval Approach for User-oriented Knowledge Transfer; Master's Thesis at Institute for Information Systems and Computer Media (IICM), Graz University of Technology, Austria; 2006.

[Safran et al. 2006] Safran, C., García-Barrios, V.M., Gütl, C.: A Concept-based Context Modelling System for the Support of Teaching and Learning Activities; In Proceedings of the International Conference on Society for Information Technology and Teacher Education (SITE 2006); C. M. Crawford and R. Carlsen and K. McFerrin and J. Price and R. Weber and D. A. Willis (eds.), AACE (pub.); pp. 2395-2402, 2006.

[Salvucci and Goldberg 2000] Salvucci, D.D., Goldberg, J.H.: Identifying Fixations and Saccades in Eye-tracking Protocols; in Proceedings of the Eye Tracking Research and Applications Symposium, ACM Press, New York, USA; pp. 71-78, 2000.

[Scharl 2000] Scharl, A.: A Classification of Web Adaptivity: Tailoring Content and Navigational Systems of Advanced Web Applications; 22nd International Conference on Software Engineering (Second ICSE Workshop on Web Engineering); S. Murugesan and Y. Deshpande (eds.); Limerick: University of Limerick; pp. 18-27, 2000.

[Schilit et al. 1994] Schilit B., N. Adams and R. Want: Context - Aware Computing Applications; in Proceedings of Workshop on Mobile Computing Systems and Applications, IEEE; 1994.

[Schmaranz 2002] Schmaranz, K.: Dinopolis - A Massively Distributable Componentware System; Habilitation thesis at Institute for Information Systems and Computer Media (IICM), Faculty of Computer Science at Graz University of Technology, Austria; 2002.

[Schmeck 1988] Schmeck, R.R.: Learning Strategies and Learning Styles; New York, Plenum Press, 1988.

[Schmidt 2005] Schmidt A.: Bridging the Gap Between Knowledge Management and E-Learning with Context-Aware Corporate Learning Solutions; in Proceedings of LOKMOL 2005; Kaiserslautern; pp. 170-176, 2005.

[Schreck 2003] Schreck, J.: Security and Privacy in User Modeling; Human-Computer Interaction Series Nr. 2; Karat, J. and Venderdonckt, J. (eds.), Kluwer Academic Publishers, Dordrecht / Boston / London; 2003.

[Scouller 1998] Scouller, K.: The influence of assessment method on students' learning approaches: Multiple choice question examination versus assignment essay; in Journal Higher Education, Vol. 35; pp. 453-472, 1998.

[Self 1974] Self, J.A.: Student Models in Computer Aided Instruction. In International Journal of Man-Machine Studies, 6, pp. 261-276, 1974.

[Self 1977a] Self, J.A.: Artificial Intelligence Techniques in Computer Assisted Instruction. The Australian Computer Journal, 9 (3), pp. 118-127, 1977.

[Self 1977b] Self, J.A.: Concept Teaching. Artificial intelligence, 9, pp.197-221, 1977.

[Self 1988] Self, J.A.: Bypassing the Intractable Problem of Student Modelling. In Frasson C. and Gauthier, G. (Eds.) Intelligent Tutoring Systems : At the Crossroads of Artificial Intelligence and Education, Norwood, NJ : Ablex, pp. 107-123, 1988.

[Self 1994] Self, J.: Formal Approaches to Student Modelling; book chapter in Student Modelling: the key to individualize knowledge based instruction; McCalla, G.I. and Greer, J. (eds.), Springer-Verlag Berlin; pp. 295-352. 1994.

[Sevarac 2006] Sevarac, Z.: Neuro Fuzzy Reasoner for Student Modeling; in Proceedings of 6th IEEE International Conference on Advanced Learning Technologies (ICALT 2006), in Kerkrade, The Netherlands; IEEE Computer Society; Kinshuk, Koper R., Kommers P., Kirschner P., Sampson D.G., Didderen W. (eds.), Los Alamitos, CA, USA; pp. 740-744, 2006.

[Sewart 1987] Sewart, D.: Staff development needs in distance education and campus-based education: Are they so different?; Sewart, D. (Ed.); London: Croom Helm; 1987.

[Shapiro and Niederhauser 2003] Shapiro A, Niederhauser D.: Learning from Hypertext: Research Issues and Findings; in Educational Technology Research and Development, Vol. 25; pp. 605-62, 2003.

[Shepherd et al. 2002] Shepherd, M., Watters, C., Marth, A.T.: Adaptive User Modeling for Filtering Electronic New; in Proceedings of 35th Annual Hawaii International Conference on System Sciences (HICSS'02) Volume 4; pp. 102-111, 2002.

[Shute and Towle 2003] Shute, V., Towle, B.: Adaptive E-Learning; in EDUCATIONAL PSYCHOLOGIST, Vol. 38 Nr. 2; Lawrence Erlbaum Associates, Inc.; pp. 105-114, 2003.

[Sison and Shimura 1998] Sison, R., Shimura, M.: Student Modeling and Machine Learning; in International Journal of Artificial Intelligence in Education, Vol. 9; p.p. 128–158, 1998.

[Skinner 1974] Skinner, B. F.: About behaviourism; New York: Knopf; 1974.

[Slagle 1965] Slagle, J.R.: Experiments with a deductive question-answering program; in Journal Commun. ACM, ACM Press, Vol. 8, Nr. 12; pp 792-798, 1965.

[Sleeman 1985] Sleeman, D.: UMFE: a user modelling front-end subsystem; in International Journal Man-Machine Studies, Vol. 23 Nr. 1; Academic Press Ltd. (pub.), London, UK; pp. 71-88, 1985.

[Smit et al. 1999] B. Smit, I. Burton, R.J.T. Klein, R. Street: The Science of Adaptation: A Framework for Assessment; in Mitigation and Adaptation Strategies for Global Change, Volume 4, Issue 3-4; pp 199-213, 1999.

[Smit et al. 2001] Smit, B., Pilifosova, O., Burton, I., Challenger, B. Huq, S., Klein, R.J.T., Yohe, G., Adger, N., Downing, T., Harvey, E., Kane, S., Parry, M., Skinner, M., Smith, J. Wandel, J.: "Adaptation to Climate Change in the Context of Sustainable Development and Equity"; in "Climate Change 2001: Impacts, Adaptation and Vulnerability", Chapter 18, Intergovernmental Panel of Climatic Change (UNEP & WMO), 2001.

[Smith 1970] Smith, L.B.: A Survey of Interactive Graphical Systems for Mathematics; In Journal ACM Comput. Surv., ACM Press, Vol. 2, Nr. 4; pp. 261-301, 1970.

[Smith 1970a] Smith, L. B.: The use of interactive graphics to solve numerical problems; In Journal Commun. ACM, ACM Press, volume 13, number 10; pp 625-634, 1970.

[Smith and Zeng 2004] Smith, T.R., Zeng, M.L.: Building Semantic Tools for Concept-based Learning Spaces: Knowledge Bases of Strongly-Structured Models for Scientific Concepts in Advanced Digital Libraries; in Journal of Digital Information, Vol. 4/4, Art. Nr. 263; 2004.

[Smith et al. 1996] Smith, J.B., Ragland, S.E., Pitts, G.J.: A process for evaluating anticipatory adaptation measures for climate change; Water, Air, and Soil Pollution, 92; pp. 229-238, 1996.

[Smithers and Smit 1997] Smithers, J., Smit, B.: Human adaptation to climatic variability and change; Global Environmental Change, Vol. 7, Nr. 2; pp. 129-146, 1997.

[Simmons 1965] Simmons, R.F.: Answering English questions by computer: a survey; in Journal Commun. ACM, ACM Press, Vol. 8, Nr. 1; pp 53-70, 1965.

[Sollazzo 2001] Sollazzo, T.: Ontology-based Services for the Semantic Web - Services in the Area of Software Tools and Supply Chain Management; Master's thesis at University of Karlsruhe, Germany; 2001

[Sparck-Jones 1984] Sparck Jones, K.: User Models and Expert Systems; in Technical Report 61, Computer Laboratory, University of Cambridge, England; 1984.

[Specht and Burgos 2006] Specht, M., Burgos, D.: Implementing Adaptive Educational Methods with IMS Learning Design; in Proceedings of Workshops held at the Fourth International Conference on Adaptive Hypermedia and Adaptive Web-Based Systems (AH2006), ADALE Workshop on Adaptive Learning and Learning Design; Stephan Weibelzahl and Alexandra Cristea (eds.), ISSN 1649-8623, National College of Ireland, Dublin, Ireland; pp. 241-251, 2006.

[Specht and Kahabka 2000] Specht, G., Kahabka, T.: Information Filtering and Personalisation in Databases Using Gaussian Curves; in Proceedings of International Database Engineering and Applications Symposium (IDEAS'00), Yokohama, Japan; pp. 16-24, 2000.

[Specht and Oppermann 1999] Specht, M., Oppermann, R.: User Modeling and Adaptivity in Nomadic Information Systems; In Proceedings of the 7th GI-Workshop 'Adaptivität und Benutzermodellierung in Interaktiven Softwaresystemen'; Universität Magdeburg, Germany; pp. 325-328, 1999.

[Spector and Ohrazda 2003] Spector, J.M., Ohrazda, C.: Automating instructional design: Approaches and limitations; in Educational Technology Research and Development, Vol. 26; pp. 685-700, 2003.

[Stach et al. 2004] Stach, N., Cristea, A., De Bra, P.: Authoring of Learning Styles in Adaptive Hypermedia; in Proceedings of the 13th

International World Wide Web Conference (WWW'04), NY-USA; IEEE; pp. 114-123, 2004.

[Stakhiv 1993] Stakhiv, E.: Evaluation of IPCC Adaptation Strategies; Fort Belvoir, Institute for Water Resources, U.S. Army Corps of Engineers, draft report; 1993.

[Steinhardt 2005] Steinhardt, S.: High Performance Learning Spaces; Key Note Speech at International Conference for Interactive Computer Aided Learning (ICL 2005), in Villach, Austria; 2005; URL http://www.cti.ac.at/online-lab/ICL_Archive/2005/presentations Last visit: 2008-02-19.

[Stelmaszewska et al. 2005] Stelmaszewska, H., Blandorf, A., Buchanan, G.: Designing to Change User's Information Seeking Behaviour: A Case Study; book chapter in Adaptable and Adaptive Hypermedia Systems; Chen, S.Y, Magoulas, G.D. (eds.), IRM Press (pub.), USA; pp. 1-18, 2005.

[Stephanidis et al. 1997] Stephanidis, C., Paramythis, A., Savidis, A., Sfyrakis, M., Stergiou, A., Leventis, A., Maou, N., Paparoulis, G., Karagiannidis, C.: Developing web browsers accessible to all: Supporting user-adapted interaction; in Proceedings of 4th European Conference for the Advancement of Assistive Technology (AAATE `97), Porto Carras, Greece; pp. 233-237, 1997.

[Stephanidis et al. 1998] Stephanidis, C., Paramythis, A., Sfyrakis, M., Stergiou, A., Maou, N., Leventis, A., Paparoulis, G., Karagiannidis, C.: Adaptable and Adaptive User Interfaces for Disabled Users in the AVANTI Project; in Proceedings of the 5th International Conference on Intelligence in Services and Networks (IS&N '98), Intelligence in Services and Networks: Technology for Ubiquitous Telecommunications Services, Antwerp, Belgium; S. Trigila, A. Mullery, M. Campolargo, H. Vanderstraeten & M. Mampaey (eds.); Springer, Lecture Notes in Computer Science, 1430, Berlin; pp. 153-166, 1998.

[Stephanidis 2001] Stephanidis, C.: Adaptive Techniques for Universal Access; in User Modeling and User-Adapted Interaction (UMUAI) Vol. 11; Kluwer Academic Publishers, The Netherlands; pp. 159-179, 2001.

[Stewart et al. 2006] Stewart, C., Cristea, A., Celik, I., Ashman, H.: Interoperability between AEH user models; in Proceedings of the joint international workshop on Adaptivity, personalization and the semantic web, APS '06 in Odense, Denmark; ACM Press, New York, NY, USA; pp 21-30, 2006.

[Sue et al. 2004] Sue, P.C., Weng, J.F., Su, J.M., Tseng S.S.: A New Approach for Constructing the Concept Map; in Proceedings of 4th IEEE International Conference on Advanced Learning Technologies (ICALT'04); pp. 76-80, 2004.

[SUN 2006] SUN: Jini Technology; 2006. URL http://www.jini.org Last visit: 2008-02-12 and http://wwws.sun.com/software/jini Last visit: 2008-02-12.

[Suzumura et al. 2005] Suzumura, T., Takase, T., Tatsubori, M.: Optimizing Web services performance by differential deserialization; in Proceedings. 2005 IEEE International Conference on Web Services (ICWS 2005), Vol. 1; pp. 185 - 192, 2005.

T

[Takase and Tatsubori 2004] Takase, T.; Tatsubori, M.: Efficient Web services response caching by selecting optimal data representation; in Proceedings of 24th International Conference on Distributed Computing Systems 2004; pp. 188-197, 2004.

[Tanenbaum and vanSteen 2007] Tanenbaum, A. S., van Steen, M.: Distributed Systems: Principles and Paradigms; Pearson Prentice Hall, 2nd Edition, USA; chapters 1,2, 10-13; 2007.

[Terrase et al. 2006] Terrasse, M.N., Savonnet, M., Leclercq, E., Grison, T., Becker, G.: Do we need metamodels AND ontologies for engineering platforms?; in Proceedings of the 2006 international workshop on Global integrated model management (GaMMa '06); ACM Press, New York, NY, USA; pp. 21-28, 2006.

[Terry and Ramasubramanian 2003] Terry, D.B., Ramasubramanian, V.: Caching XML Web Services for Mobility; ACM Queue Vol. 1, Nr. 1; Microsoft Research; 2003, UR L http://www.acmqueue.com/modules.php?name=Content&pa=show page&pid=38 Last visit: 2008-02-21.

[Thevenin and Coutaz 1999] Thevenin, D., Coutaz, J.: Plasticity of User Interfaces: Framework and Research Agenda; in Proceedings of Human–Computer Interaction INTERACT'99; Angela Sasse and Chris Johnson (eds.), IOS Press (pub.); pp. 1-8, 1999.

[Thimbleby 1990] Thimbleby, A.: User Interface Design; Workingham: Addison Wesley; 1990.

[Thorndike 1913] Thorndike, E. L.: Educational psychology: The psychology of learning. New York: Teachers College Press; 1913.

[Tochtermann et al. 2005] Tochtermann, K., Granitzer, M., Sabol, V.: MISTRAL - Cross-Media Techniques for Extracting Semantics from Multimedia and their Application; in Proceedings of SEMANTICS 2005 Conference, in Vienna, Austria; 2005.

[Tonge 1966] Tonge, F.M.: A view of artificial intelligence; in Proceedings of the 1966 21st national conference, ACM Press; pp. 379-382, 1966.

[Trastour and Bartolini 2001] Trastour, D., Bartolini, C.: Approach to Service Description for Matchmaking and Negotiation of Services; in Proceedings of SWWS'01, Stanford, USA; 2001.

[Travis 1962] Travis, L.E. : In defense of artificial intelligence research; in Journal Communications of the ACM; ACM Press, Volume 5, Number 1; pp. 6-7, 1962.

[Trevellyan and Browne 1986] Trevellyan, R., Browne, D.P.: A self-regulating adaptive system; in Proceedings of the SIGCHI/GI conference on Human factors in computing systems and graphics interface; ACM Press; pp. 103-107, created 1986, published 1987.

[Tsiriga and Virvou 2003] Tsiriga, V., Virvou, M.: Initializing Student Models in Web-Based ITSs: A Generic Approach; in Proceedings of the 3rd IEEE International Conference on Advanced Learning Technologies (ICALT 2003); pp. 42–46, 2003.

U

[Uther 2001] Uther, J.B.: On the Visualisation of Large User Models in Web Based Systems; PhD dissertation at The University of Sydney; 2001.

V

[Vassileva et al. 2003] Vassileva, J., McCalla, G., Greer, J.: Multi-Agent Multi-User Modeling in I-Help; in User Modeling and User-Adapted Interaction (UMUAI) Vol. 13; Kluwer Academic Publishers, The Netherlands; pp. 179-210, 2003.

[Vergara 1994] Vergara, V.: PROTUM - A Prolog Based Tool for User Modeling; Technical report at Department of Information Science of University of Konstanz; 1994.

[Voß et al. 1999] Voß, A., Nakata, K., Juhnke, M.: Concept Indexing; GROUP 1999, Phoenix, Arizona, USA; pp. 1-10, 1999.

[Vygotsky 1978] Vygotsky, L.: Mind in society. Cambridge, MA: Harvard University Press, 1978.

W

[Waern and Rudstrom 2001] Waern, A., Rudström, A.: Can Readers Understand their Profiles? A Study of Human Involvement in Reader Profiling; in Proceedings of 34th Annual Hawaii International Conference on System Sciences (HICSS-34) Volume 4, in Maui, Hawaii; pp. 4012-4021, 2001.

[Wahlster and Kobsa 1986] Wahlster, W., Kobsa, A.: Dialog-based User Models; in Journal Proceedings of the IEEE Special Issue on Natural Language Processing, July 1986; pp. 948 -960, 1986.

[Wahlster and Kobsa 1989] Wahlster, W., Kobsa, A.: User Models in Dialog Systems; in User Models in Dialog Systems; Kobsa, A., Wahlster, W. (eds.), Springer Verlag, Berlin; pp. 4-34, 1989.

[Wang and Fung 2004] Wang, G., Fung, C.K.: Architecture paradigms and their influences and impacts on component-based software systems; in Proceedings of the 37th Annual Hawaii International Conference on System Sciences; 2004. URL http://csdl.computer.org/comp/proceedings/hicss/2004/2056/09/2056 90272a.pdf Last visit: 2008-01-12.

[Wang and Zhang 2007] Wang, X., Zhang, S.: Performance Evaluation & Optimization about Lookup Service in Jini Architecture; Report at The University of Wisconsin, Madison; Computer Sciences; 2007. URL http://www.cs.wisc.edu/~wxd/report/cs736.pdf Last visit: 2008-02-21.

[Wardrip-Fruin 2004] Wardrip-Fruin, N.: What hypertext is; in Proceedings of the 15th ACM conference on Hypertext and hypermedia (HYPERTEXT '04) in Santa Cruz, CA, USA; ACM Press (pub.), New York, NY, USA; pp. 126-127, 2004.

[Watson et al. 1996] Watson, R.T., Zinyowera, M.C., Moss, R.H.: Climate Change 1995 - Impacts, Adaptations and Mitigation of Climate Change - Scientific-Technical Analyses; Contribution of Working Group II to the Second Assessment Report of the Intergovernmental Panel on Climate Change, Cambridge: Cambridge University Press; 1996.

[Webb et al. 2001] Webb, G.I., Pazzani, M.J., Billsus, D.: Machine Learning for User Modeling; in User Modeling and User-Adapted Interaction (UMUAI) Vol. 11; Kluwer Academic Publishers, The Netherlands; pp. 19-29, 2001.

[Wedemeyer 1977] Wedemeyer, C.A.: Independent study; in The International Encyclopedia of Higher Education Boston: Northeastern University; A.S. Knowles (ed.); 1977.

[Weibelzahl 2002] Weibelzahl, S.: Evaluation of Adaptive Systems; Dissertation work at Faculty I of the University of Trier; pp. 20-24, 2002.

[Weiser 1991] Weiser, M.: The computer for the twenty-first century; in Scientific American; 1991.

[Weißenberg et al. 2004] Weißenberg, N., Voisard, A., Gartmann, R.: Using ontologies in personalized mobile applications; in Proceedings of the 12th annual ACM international workshop on Geographic information systems (GIS 2004), Washington DC, USA; ACM Press, New York, NY, USA; pp. 2-11, 2004.

[Westin 1970] Westin, A.F.: Privacy and Freedom; Atheneum, Bodley Head; 1970.

[Wilson 1997] Wilson, B. G.: Reflections on constructivism and instructional design; in Instructional development paradigms; C. R. Dills & A. J. Romiszowski (Eds.); Englewood Cliffs, NJ: Educational Technology Publications; pp. 63-80, 1997.

[Williams and Dreher 2004] Williams, R., Dreher, H.: Automatically Grading Essays with Markit© Proceedings of Informing Science 2004 Conference, Rockhampton, Queensland, Australia, June 25-28, 2004.

[Woods 1993] Woods, D. D.: The price of flexibility; in Proceedings of the 1993 International Workshop on Intelligent User Interfaces; Orlando; pp. 19–25, 1993.

[WordNet 2007] WordNet, a lexical database for the English language; 2007. URL http://wordnet.princeton.edu Last visit: 2008-01-11.

[Wu 2002] Wu, H.: A reference Architecture for Adaptive Hypermedia Applications; Ph.D. thesis at Technical University of Eindhoven (ISBN: 90-386-0572-2), 2002.

[Wu and De Bra 2001] Wu, H., De Bra, P.: Sufficient Conditions for Well-behaved Adaptive Hypermedia Systems; in Proceedings of the 1st Asia-Pacific Conference on Web Intelligence: Research and Development; Lecture Notes in Artificial Intelligence, Vol. 2198, Springer; pp. 148-152, 2001.

[Wu et al. 2001] Wu, H., De Kort, E., De Bra, P.: Design Issues for General-Purpose Adaptive Hypermedia Systems; in Proceedings of the ACM Conference on Hypertext and Hypermedia, in Aarhus, Denmark; pp. 141-150, 2001.

X

[xFIND 2007] xFIND: Extended Framework for Information Discovery; official Web site; 2007. URL http://xfind.iicm.edu Last visit: 2008-01-12.

Y

[Yoder et al. 1998] Yoder, J.W., Johnson, R.E., Wilson, Q.D.: Connecting Business Objects to Relational Databases; in Proceedings of the 5th Conference on the Pattern Languages of Programs (PloP'98), Monticello, Italy; pp. 1-41, 1998.

[Yudelson et al. 2004] Yudelson, M., Brusilovsky, P., Sosnovsky, S.: Accessing Interactive Examples with Adaptive Navigation Support; in Proceedings of 4th IEEE International Conference on Advanced Learning Technologies (ICALT'04); pp. 842-843, 2004.

Z

[Zakaria et al. 2002] Zakaria, M.R., Moore, A., Ashman, H., Stewart, C., Brailsford, T.J.: The Hybrid Model for Adaptive Educational Hypermedia; in Proceedings of Adaptive Hypermedia and Adaptive Web-Based Systems: 2nd International Conference (AH 2002), in Malaga, Spain; P. De Bra, P. Brusilovsky, and R. Conejo (eds.), Springer-Verlag Berlin Heidelberg; pp. 580-585, 2002.

[Zakaria et al. 2003] Zakaria, M.R., Moore, A., Stewart, C., Brailsford, T.J.: "Pluggable" user models for adaptive hypermedia in education; in Proceedings of the 14th ACM conference on Hypertext and hypermedia (HYPERTEXT '03), in Nottingham, UK; ACM Press, New York, NY, USA; pp. 170-171, 2003.

[Zimmermann et a. 2003] Zimmermann, A., Lorenz, A., Specht, M.: User Modeling in Adaptive Audio-Augmented Museum Environments; in Proceedings of User Modeling (UM 2003); P. Brusilovsky et al. (eds.), LNAI 2702, Springer-Verlag Berlin Heidelberg; pp. 403-407, 2003.

[Zimmermann et al. 2005] Zimmermann, A., Specht, M., Lorenz, A.: Personalization and Context Management; in User Modeling and User-Adapted Interaction (UMUAI) Vol. 15; Sringer Verlag; pp. 275–302, 2005.

[Zukerman and Albrecht 2001] Zukerman, I., Albrecht, D.W.: Predictive Statistical Models for User Modeling; in User Modeling and User-Adapted Interaction (UMUAI), Vol. 11; Kluwer Academic Publishers, The Netherlands; pp. 5-18, 2001.

www.ingramcontent.com/pod-product-compliance
Lightning Source LLC
LaVergne TN
LVHW022311060326
832902LV00020B/3391